Nanomaterials for the Environmental Remediation of Water and Soil

Nanomaterials for the Environmental Remediation of Water and Soil

Guest Editors

Roberta G. Toro
Daniela Caschera
Abeer M. Adel

MDPI

Basel • Beijing • Wuhan • Barcelona • Belgrade • Novi Sad • Cluj • Manchester

Guest Editors

Roberta G. Toro
National Council of Research
Institute for the Study of
Nanostructured Materials
Montelibretti
Italy

Daniela Caschera
National Research Council
Institute for the Study of
Nanostructured Materials
Montelibretti
Italy

Abeer M. Adel
National Research Center
Department of Paper
and Cellulose
Dokki, Giza
Egypt

Editorial Office
MDPI AG
Grosspeteranlage 5
4052 Basel, Switzerland

This is a reprint of the Special Issue, published open access by the journal *Materials* (ISSN 1996-1944), freely accessible at: www.mdpi.com/journal/materials/special_issues/nanomaterials_environmental_remediation_water_soil.

For citation purposes, cite each article independently as indicated on the article page online and using the guide below:

Lastname, A.A.; Lastname, B.B. Article Title. *Journal Name* **Year**, *Volume Number*, Page Range.

ISBN 978-3-7258-3852-3 (Hbk)
ISBN 978-3-7258-3851-6 (PDF)
https://doi.org/10.3390/books978-3-7258-3851-6

Contents

About the Editors

Roberta G. Toro

Roberta G. Toro is a permanent researcher at the CNR Institute for the Study of Nanostructured Materials in Montelibretti (Rome), where she has been working since 2009. She graduated in Industrial Chemistry (cum laude) from the University of Catania in 2001 and earned her Ph.D. in Chemistry in 2005. She is a licensed Chemist.

Dr. Toro's research spans inorganic chemistry and materials science, with a particular focus on the synthesis and characterization of innovative nanomaterials. Her work includes the development of semiconducting metal oxide nanoparticles with photocatalytic properties for environmental applications, such as water remediation. She is also investigating organic/inorganic hybrid systems and the doping of metal oxide nanoparticles with rare earth ions, exploring their photoluminescent behavior. Additionally, she works on the fabrication of advanced surfaces with superhydrophobic and superoleophilic properties.

She specializes in the design, synthesis, and characterization of oxide-based nanostructured thin films using Metal–Organic Chemical Vapor Deposition (MOCVD). Her analytical expertise includes X-ray Diffraction (XRD), Scanning Electron Microscopy (SEM) with Energy-Dispersive X-ray Spectroscopy (EDX), Atomic Force Microscopy (AFM), Fourier-Transform Infrared Spectroscopy (FTIR), UV-Vis Spectroscopy, Fluorescence, and Thermal Analysis.

She is author of about 80 articles in international journals with more than 1750 citations and an H-index 26 and of numerous communications in national and international conferences.

Daniela Caschera

D. Caschera graduated in Chemistry in 2001 at Università La Sapienza (Rome) and was awarded a Ph.D. in Industrial and Mechanical Engineering by Università of Roma Tre (Rome) in 2008. Currently, she works as senior researcher at the Institute for the Study of Nanostructured Materials (ISMN) of the Italian National Research Council (Rome). Since 2002, her research activity has focused on surface modification by plasma treatments, CVD-PVD deposition of carbon-based thin films and synthesis, and the structural, morphological, and functional characterizations of nanostructured materials. Recently, she has been involved in the fabrication of hybrid nanocomposites, from the green treatment of agri-waste, for environmental remediation and sustainability. She has been involved, even as PI, in several national and international projects, in the field of the fabrication and characterization of nanostructured materials, for advanced technological applications. She is responsible for ISMN-CNR Surface Lab Infrastructure, which includes facilities for thin film and nanosystem preparation and functional characterizations. She is the author of about 60 articles in international journals with more than 1900 citations and an H-index 24 and of numerous communications in national and international conferences.

Abeer M. Adel

Prof. A.M. Adel took her degree in 1994 in Cairo University with a thesis entitled "Non Conventional Bleaching for some Agriculture Wastes" and her PhD in 2001 on "Recycling of Old Corrugated Container". From 1991, she has worked for NRC in Cairo and is professor in the Cellulose and Paper Department of the National Research Center. She has great experience in the field of materials science and nanomaterials. In particular, in recent years, she has been involved in the study of the treatment of natural polymers and the development of sustainable functional polymers, nanocomposites, and membranes, even starting from the recycling of agri-waste, for applications in the pulp and paper industry, for packaging and environmental remediation. She has been involved in several national and international projects on the use of agricultural residues for the preparation of biodegradable nanocellulose composites, most as principal or deputy investigator. She has been supervisor of several PhD students. In 2000, she won the ASRT prize and the Prof. Dr. Yehia Abd El-Latif award in cellulose, paper, and wood. She is co-author of about 45 scientific works (h-index=22), published in peer-reviewed international journals, and she attended many national and international conferences for poster presentations or oral communications.

Preface

"Extract, produce, consume, and reproduce": this is the new economic model to pursue. It replaces the traditional concept of waste with that of resources, aiming to reduce the consumption of raw materials while increasing efficiency in material use, with a focus on maximizing reuse and recycling. Treating waste with waste is an efficient method of environmental protection, as it allows the simultaneous treatment of multiple type of wastes, resulting in a positive environmental impact.

In recent decades, with the advancement of industrial, agricultural, and urban activities, the levels of groundwater and soil pollution have increased enormously. In this regard, the application of nanotechnology for the remediation of contaminants has shown promising results. Nanomaterials used as catalysts, chemical oxidants, and adsorbents guarantee the rapid detection and consequent detoxification of various contaminants such as pharmaceuticals, pesticides, drugs, aromatic heterocycles, volatile organic compounds, heavy metals, and inorganic ions from water, air, and contaminated land sites.

This Special Issue focuses on the prospects and challenges of nanoremediation and environmental clean-ups, presenting several interesting contributions from researchers exploring the use of waste-derived precursors for the preparation of nanomaterials aimed at the remediation of contaminated water and soil. In particular, the presented research explores green technologies and natural resources in the fabrication of nanoparticles and hybrid nanocomposites, combining them with natural derivates such as clay (e.g., zeolite, sepiolite) or cellulose (from agri-waste recovery, textile support) to be used in water remediation, pollutant adsorption, and oil/water separation.

Several contributions also discuss methodologies for evaluating metal contamination in soil, and their accumulation in aquatic plants, starting from the analysis of trace elements in aerial parts and plant roots. Additionally, the use of natural derivates such as mucilage-based flocculants or bacterial metabolite products is presented as a sustainable alternative to synthetic materials, due to their non-toxic and biodegradable nature, offering many technological and environmental benefits.

Special attention is also devoted to the equilibrium and kinetic modeling of adsorption mechanisms, as well as to the mechanical properties of the membrane and composite materials proposed for remediation procedures.

The Guest Editors would like to thank all the colleagues who have contributed to the success of this Special Issue. Their work on safe-by-design nanomaterials and the enhancement in their functional properties offers innovative and eco-friendly solutions for environmental decontamination. We hope their findings and research will significantly contribute to the preservation and restoration of the environment for future generations.

Roberta G. Toro, Daniela Caschera, and Abeer M. Adel
Guest Editors

materials

MDPI

Article

Removal of Cesium and Strontium Ions from Aqueous Solutions by Thermally Treated Natural Zeolite

Marin Șenilă *, Emilia Neag *, Claudiu Tănăselia and Lacrimioara Șenilă

INCDO-INOE 2000, Research Institute for Analytical Instrumentation, 67 Donath Street, 400293 Cluj-Napoca, Romania

* Correspondence: marin.senila@icia.ro (M.Ș.); emilia.neag@icia.ro (E.N.); Tel.: +40-264-420-590 (E.N.)

Abstract: The radionuclides of cesium (Cs) and strontium (Sr) are dangerous products of nuclear fission that can be accidentally released into wastewater. In the present work, the capacity of thermally treated natural zeolite (NZ) from Macicasu (Romania) to remove Cs^+ and Sr^{2+} ions from aqueous solutions in batch mode was investigated by contacting different zeolite quantities (0.5, 1, and 2 g) of 0.5–1.25 mm (NZ1) and 0.1–0.5 mm (NZ2) particle size fractions with 50 mL working solutions of Cs^+ and Sr^{2+} (10, 50, and 100 mg L^{-1} initial concentrations) for 180 min. The concentration of Cs in the aqueous solutions was determined by inductively coupled plasma mass spectrometry (ICP-MS), whereas the Sr concentration was determined by inductively coupled plasma optical emission spectrometry (ICP-OES). The removal efficiency of Cs^+ varied between 62.8 and 99.3%, whereas Sr^{2+} ranged between 51.3 and 94.5%, depending on the initial concentrations, the contact time, the amount, and particle size of the adsorbent material. The sorption of Cs^+ and Sr^{2+} was analyzed using the nonlinear form of Langmuir and Freundlich isotherm models and pseudo-first-order (PFO) and pseudo-second-order (PSO) kinetic models. The results indicated that the sorption kinetics of Cs^+ and Sr^{2+} on thermally treated natural zeolite was described by the PSO kinetic model. Chemisorption dominates the retention of both Cs^+ and Sr^{2+} by strong coordinate bonds with an aluminosilicate zeolite skeleton.

Keywords: clinoptilolite; radionuclides; volcanic tuff; porous materials; water treatment; chemisorption; ion-exchange

Citation: Șenilă, M.; Neag, E.; Tănăselia, C.; Șenilă, L. Removal of Cesium and Strontium Ions from Aqueous Solutions by Thermally Treated Natural Zeolite. *Materials* **2023**, *16*, 2965. https://doi.org/10.3390/ma16082965

Academic Editors: Roberta G. Toro, Daniela Caschera and Abeer M. Adel

Received: 13 March 2023
Revised: 3 April 2023
Accepted: 6 April 2023
Published: 7 April 2023

1. Introduction

Considering the current trend of diminishing power plants based on fossil fuels because of high carbon dioxide emissions and the increasing energy demand, nuclear power plants are considered one of the best alternatives to produce cheap and clean energy [1–3]. Security issues represent the primary concern that is still limiting the development of this sector because of the dangerous radioactive wastes that nuclear power plants produce. Thus, the production of radioactive wastes could lead to significant ecological problems. Among the wastes generated in nuclear power plants, about 85% are liquid radioactive wastes [4]. The radionuclides of cesium (Cs) and strontium (Sr) are the most critical products of nuclear fission [5].

The Cs and Sr radionuclides have high chemical and biological radiotoxicity, damaging the human body [6], and thus, proper management of wastes from nuclear power plants is mandatory. The common approaches for radioactive Cs^+ and Sr^{2+} removal from aqueous solutions include ion exchange, adsorption, precipitation, electrolysis, and redox methods [7]. Zeolites have particular characteristics that make these materials suitable candidates for the treatment of wastewater containing Cs^+ and Sr^{2+}. They possess a high cation-exchange capability, good adsorption capacity, and thermal and chemical stability [8–13], and exhibit ion-exchange and chemisorption properties [14]. Zeolites are complex, crystalline inorganic polymers, and their structure consists of three-dimensional frameworks of SiO_4 and AlO_4 tetrahedral. The aluminum ion fills the location in the center of the tetrahedron of four oxygen atoms, and the isomorphous replacement of Si^{4+} by Al^{3+} induces a negative charge in the lattice [15–17]. The exchangeable alkali or alkaline earth metal cations present within the zeolite structure, such as sodium, potassium, magnesium, and calcium, balance this net negative charge. These cations may be exchanged by other cations present in the contact solution [18–20]. Compared to synthetic zeolites and other adsorbing materials, clinoptilolite is widespread, cost-effective, and an appropriate adsorbent for large-scale wastewater applications [21].

Several studies were carried out for the removal of radioactive Cs and Sr [1–6,22]. At the same time, other sorption experiments were performed for non-radioactive Cs^+ and Sr^{2+} removal to calculate the mass concentration that can be retained by the adsorbents [23,24]. The reported studies on the Cs^+ and Sr^{2+} removal from liquid media by various adsorbent materials disclosed different results, probably due to their specific chemical compositions, structure, pore volumes, or cation-exchange capacity. Also, the water matrices from where cations are removed play an essential role in the removal efficiency. Kubota et al. [22] studied the removal of cesium, strontium, and iodine from water. The results revealed a successful reduction of ^{134}Cs and ^{85}Sr from river water by zeolite and bentonite, whereas ^{85}Sr was scarcely removed from seawater. To the best of our knowledge, research aiming to solve the pollution problems of radioactive Cs^+ and Sr^{2+} using zeolite-derived adsorbents is very scarce compared with studies that address the removal of other metal cations. Considering that a natural and cost-effective material is used, with high applicability, this represents a significant gap in this topic. Thus, the contribution to eliminating radio pollution is an original and essential research topic. This study aims to assess the sorption capacity of natural zeolite towards radionuclides after a minimum pretreatment, representing a novelty and following the zero-pollution and zero-carbon action plan [25].

The physicochemical properties and structure of zeolites influence their performance. Thus, new information related to the sorption behaviors of natural zeolites from different quarries is necessary. Therefore, this study aimed to (1) investigate the influence of process parameters, such as zeolite quantity, zeolite particle size, contact time, initial concentration, and competing ions over the removal of Cs^+ and Sr^{2+} from contaminated aqueous solutions by thermally treated zeolites; (2) assess the contribution of ion-exchange and chemisorption processes to the total retention of Cs^+ and Sr^{2+} by natural zeolites; (3) analyze the sorption kinetics and equilibrium data of Cs^+ and Sr^{2+} removal onto two different grain sizes of natural zeolites.

2. Materials and Methods

2.1. Chemicals

Ultrapure water (18 MΩ cm^{-1}) and analytical reagent-grade chemicals were used to prepare the solutions. The zeolite samples were digested using Emsure® ACS premium grade acids HNO_3 65%, HCl 37%, and HF 40%, supplied from Merck (Darmstadt, Germany). The chemical analyses were performed according to the methodology described in previous work [12]. A single component Cs standard solution of 1000 mg L^{-1} purchased from Merck (Darmstadt, Germany) was used for external calibration of the inductively coupled plasma mass spectrometer, ICP-MS, ELAN DRCII (Perkin-Elmer, Woodbridge, ON, Canada). CRM BCS-CRM 375/1 soda feldspar (Bureau of Analyzed Samples, Middlesbrough, UK) was analyzed for quality control in total metal and oxides concentration with satisfactory accuracy.

2.2. Zeolite Collection, Preparation, and Characterization

The zeolitic tuff was obtained from a quarry located in Macicasu, Cluj County, Romania. Particle size fractions of 0.5–1.25 mm (NZ1) and 0.1–0.5 mm (NZ2) were obtained by crushing and granulometric separation on sieves and then activated at 200 °C for 3 h. The aliquot parts from each granulometric fraction were ground in the laboratory using a micronization system (PilotMill-2 FPS1015, Como, Italy) to obtain a particle size of <0.05 mm for the analytical characterization (determination of major oxides, metals, exchangeable cations content). Total surface area, pore radius, and total pore volumes were measured on fractions of 0.5–1.25 mm and 0.1–0.5 mm particle sizes.

The content of major elements (Al, Ca, Mg, K, Na, Fe, Mn) in the zeolite samples were measured after microwave acid digestion of 1 g of zeolite sample with 14 mL mixture of HNO_3 65%: HCl 37%: HF 40% (3:9:2, *v:v:v*) using an Xpert system (Berghof, Eningen, Germany). A volume of 20 mL of saturated H_3BO_3 was added to neutralize the excess of HF. The resulting slurry was filtered through a 0.45 μm membrane filter and diluted with ultrapure water to a final volume of 100 mL. The resulting solutions were analyzed for metals using a dual viewing ICP-OES Optima 5300DV (Perkin Elmer, Norwalk, CT, USA), in radial viewing, after external calibration over the range 0–20 mg L^{-1} for each analyzed element. The concentrations of major oxides were calculated from the total elements concentration [12]. The SiO_2 and loss of ignition (LOI) were measured by the gravimetric method [26], in triplicate. The cation-exchange capacity (CEC) [27] was estimated from the sum of major cations (K^+, Na^+, Ca^{2+}, and Mg^{2+}) concentrations extracted in 1 M ammonium acetate solution and measured by ICP-OES in radial viewing mode. Zeolitic tuff morphology (total surface area, total pore volume, and average pore radius) was evaluated according to the methodologies described in our previous work [12].

2.3. Batch Sorption Experiments

Stock solutions of Cs^+ (1000 mg L^{-1}) and Sr^{2+} (1000 mg L^{-1}) were prepared by dissolving 1.266 g CsCl and 1.809 g $SrCl_2$, respectively, in 1000 mL of ultrapure water. The salts were of analytical purity and purchased from Merck, Darmstadt, Germany. The working solutions with initial Cs^+ and Sr^{2+} concentrations of 10, 50, and 100 mg L^{-1} were prepared by appropriate dilution of the concentrated stock solutions with ultrapure water. The experiments were performed in batch mode, contacting different quantities of NZ1 and NZ2 (0.5, 1, and 2 g) with 50 mL working solutions of Cs^+ and Sr^{2+} at a stirring rate of 200 rpm at room temperature (22 ± 2 °C). Samples were taken at different contact times (1, 3, 5, 7, 10, 30, 60, 90, 120, and 180 min) and further filtered using 0.45 μm polyethersulfone (PES) syringe filters (Whatman, Clifton, NJ, USA). The concentration of Sr was determined using ICP-OES, whereas Cs concentration, due to its low analytical signal in ICP-OES, was determined using ICP-MS. All the experiments were performed in triplicate, and the average value was used.

The Cs^+ and Sr^{2+} amount in the adsorbent phase, q_e (mg/g) was calculated using Equation (1), whereas removal efficiency, E (%) was calculated using Equation (2):

$$q_e = \frac{(C_o - C_e)}{m} \cdot \frac{V}{1000} \quad (1)$$

$$E(\%) = \frac{(C_o - C_e)}{C_o} \cdot 100 \quad (2)$$

where q_e is the Cs^+ and Sr^{2+} amount adsorbed per gram of adsorbent at equilibrium (mg g^{-1}), V is the volume of solution (mL), m is the weight of natural zeolite (g), C_e is the equilibrium Cs^+ and Sr^{2+} concentrations (mg L^{-1}), and C_0 is the initial Cs^+ and Sr^{2+} concentrations (mg L^{-1}) [28].

In order to appraise the possible release of Cs^+ and Sr^{2+} from natural zeolite before its use in the batch experiments, and to account for these concentrations in the calculation of the removal efficiency, 0.5, 1, and 2 g, of the NZ1 and NZ2 were mixed with 50 mL of ultrapure water at a stirring rate of 200 rpm for 2 h at room temperature, then the resulting solutions were filtered by 0.45 μm PES syringe filter (Whatman, Clifton, NJ, USA). To evaluate if the retention of Cs^+ and Sr^{2+} on zeolites is based on ion exchange or chemisorption, the zeolite that previously retained Cs^+ and Sr^{2+} when contacted with 50 mL of solution with a concentration of 100 mg L^{-1} for 180 min was subjected to desorption. For this, zeolite was mixed with ammonium acetate solution 1 M at a ratio solid: liquid of 1:50, at a stirring rate of 200 rpm for 3 h at room temperature, and finally the resulting solutions were filtered by a 0.45 μm PES filter. The quantities of Cs and Sr desorbed from the solid phase were determined considering the initial amount of zeolite and the final volume of extraction solutions [12,29].

In order to evaluate the influence of the zeolite structure on the sorption mechanisms, the batch experimental data for Cs^+ and Sr^{2+} removal were correlated with the XRF and FTIR results performed on zeolite tuff before and after the sorption of cations.

The zeolite samples containing the sorbed cations were separated from the aqueous solution by filtration and then dried for 2 h at 105 °C. The infrared spectra of the zeolite samples were recorded on 1% KBr pellets in the 4000–400 cm^{-1} range, with a spectral resolution of 2 cm^{-1}, using a BX II Fourier transform infrared spectrophotometer (Perkin Elmer, Waltham, MA, USA). XRF analyses were performed using a Bruker Tracer 5i portable X-ray fluorescence with a 5 kV and 4-watt X-ray source, 8 μm Be window, and 8 mm spot collimator.

2.4. Equilibrium and Kinetics Studies

The sorption experiments and kinetics studies were carried out in batch mode by adding 0.5 g of NZ1 and NZ2 to 50 mL of various Cs^+ and Sr^{2+} solutions of different initial concentrations for 180 min.

The isotherm parameters for Cs^+ and Sr^{2+} sorption by thermally treated natural zeolite were estimated using nonlinear forms of Langmuir and Freundlich isotherm models. The equations of the used models were presented in our previous paper [12]. The nonlinear forms of pseudo-first (PFO) and pseudo-second-order (PSO) kinetic models were used to investigate the sorption kinetics of Cs^+ and Sr^{2+} on NZ1 and NZ2. The nonlinear regression analysis of Cs^+ and Sr^{2+} sorption on thermally treated natural zeolite was performed using the OriginPro software, version 2020b, OriginLab Corporation, Northampton, MA, USA.

3. Results and Discussion

3.1. Characteristics of Thermally Treated Volcanic Tuffs

The major oxides content in the zeolites is presented in Table 1. The Si/Al ratio > 4 and Na + K > Ca in zeolitic samples indicate the presence of clinoptilolite as a major constituent [30].

Table 1. Chemical composition of zeolitic tuff from Macicasu quarry (average value ± stdev. (n = 3 parallel determinations).

Parameter	NZ1	NZ2
pH	8.31 ± 0.25	8.40 ± 0.28
SiO_2 (%)	65.8 ± 2.2	66.1 ± 2.3
Al_2O_3 (%)	12.3 ± 0.8	12.5 ± 0.7
CaO (%)	3.77 ± 0.07	3.94 ± 0.08
MgO (%)	0.63 ± 0.05	0.60 ± 0.04
K_2O (%)	1.44 ± 0.13	1.42 ± 0.11
Na_2O (%)	1.35 ± 0.10	1.21 ± 0.12
Fe_2O_3 (%)	2.07 ± 0.12	2.05 ± 0.10
MnO (%)	0.03 ± 0.004	0.03 ± 0.004
LOI (%)	2.33 ± 0.65	2.21 ± 0.83
Others (%)	1.97	1.54

Generally, the concentration of major oxides (Table 1) was not affected by the grain sizes of the zeolite samples. Moreover, their concentration was not significantly different than that measured in the zeolitic tuff from the same quarry studied in previous work [31], except for K_2O, which was 30% higher in the current work. Also, X-ray diffraction (XRD) analysis showed clinoptilolite as the main zeolite mineral and quartz, muscovite, and albite as minor phases in the zeolitic tuff from this quarry [31]. Moreover, SEM–EDX studies revealed that the zeolite sample contains mainly clinoptilolite [31].

The crystalline structure of clinoptilolite is stable at high temperatures. This was confirmed by thermogravimetric analysis (TGA) conducted at a temperature of up to 1000 °C. According to the TGA analysis, the samples showed high thermal stability [30], which is characteristic of clinoptilolite [32].

The total surface area is 46.1 $m^2\ g^{-1}$ in NZ1 and 43.8 $m^2\ g^{-1}$ in NZ2, whereas the total pore volume is 0.093 $cm^3\ g^{-1}$ in NZ1, and 0.082 $cm^3\ g^{-1}$ in NZ2, respectively. The average pore radius is 36 Å in NZ1 and 36 Å in NZ2, less than 50 nm which classifies the samples as mesoporous materials [32]. No important differences between larger (0.5–1.25 mm) and smaller grain size (0.1–0.5 mm) samples were observed since the main contribution to the total surface area is given by the internal specific surface area of zeolite which is not meaningly modified by grinding. The surface area and total pore volume are important physical parameters in the sorption process. Concentrations of exchangeable cations are presented in Table 2.

Table 2. Concentrations of exchangeable cations (CEC) of the NZ1 and NZ2 thermally treated samples (n = 3 parallel determinations).

Zeolite	Na^+	K^+	Ca^{2+}	Mg^{2+}	CEC
			mEq 100 g^{-1}		
NZ1	3.2	24.5	61.7	3.3	92.6
NZ2	2.6	22.7	64.8	3.6	93.8

The main contribution of the exchangeable major cations to the total CEC decreased in the order $Ca^{2+} > K^+ >> Na^+ \cong Mg^{2+}$. The theoretic CEC value calculated as the total content of K, Na, Ca, and Mg in NZ1 is 239.7 meq/100 g, which consists of 30.6 meq K^+/100 g, 43.5 meq Na^+/100 g, 134.6 meq Ca^{2+}/100 g, and 31.0 meq Mg^{2+}/100 g, whereas in NZ2 the theoretical CEC value is 239.4 meq/100 g, containing 30.1 meq K^+/100 g, 39.0 meq Na^+/100 g, 140.7 meq Ca^{2+}/100 g, and 29.5 meq Mg^{2+}/100 g. These results show that K^+ is the most mobile cation. On average, 78% of K, 46% of Ca, 7% of Na, and 11% of Mg cations from the studied zeolitic materials are in an exchangeable form and can be involved in ion-exchange processes.

3.2. Release of Cs^+ and Sr^{2+} from Natural Zeolites to the Aqueous Solution

In order to evaluate the possible release of Cs^+ and Sr^{2+} from natural zeolite into the aqueous solution during the batch experiments, different NZ1 and NZ2 quantities (0.5, 1, and 2 g) were mixed with 50 mL of ultrapure water for 180 min. The concentration of released Cs^+ was in all cases below LOQ (0.5 μg L^{-1}). In the case of Sr^{2+}, the measured released concentrations varied between 3.1 μg L^{-1} (<LOQ) and 1.02 μg L^{-1} (maximum concentration released from zeolite NZ2).

The concentrations of Sr^{2+} measured in these experiments were subtracted from the equilibrium concentrations of Sr^{2+} measured in the batch experiments at the corresponding contact time and zeolite grain size.

3.3. Removal Efficiency (E%) from Contaminated Solutions

The experiments were carried out using different NZ1 and NZ2 dosages (0.5, 1, 2 g) which were contacted with 50 mL of Cs^+ and Sr^{2+} solution (concentrations of 10, 50, 100 mg L^{-1}) for 180 min. The removal of Cs^+ and Sr^{2+} by different NZ1 quantities is presented in Figures 1 and 2.

The removal of Cs^+ and Sr^{2+} by different NZ2 quantities is presented in Figures 3 and 4.

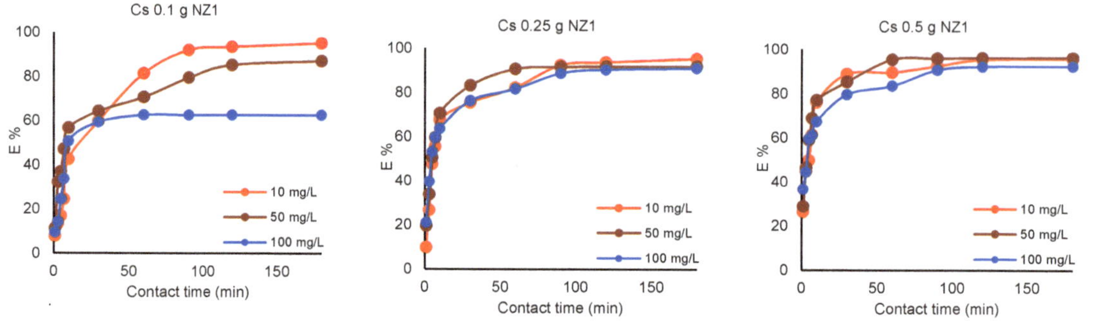

(a) (b) (c)

Figure 1. Removal efficiency (%) of Cs^{+} by different NZ1 quantities; (**a**) 0.1 g, (**b**) 0.25 g, (**c**) 0.5 g; V = 50 mL; 200 rpm; 22 ± 2 °C, 180 min.

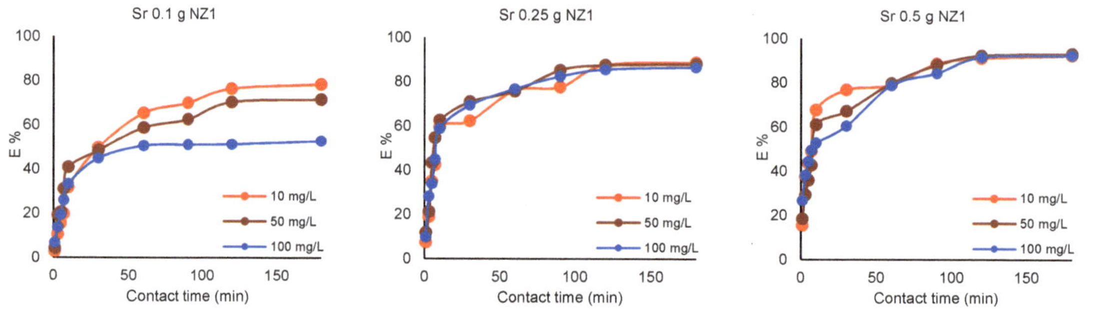

(a) (b) (c)

Figure 2. Removal efficiency (%) of Sr^{2+} by different NZ1 quantities; (**a**) 0.1 g, (**b**) 0.25 g, (**c**) 0.5 g; V = 50 mL; 200 rpm; 22 ± 2 °C, 180 min.

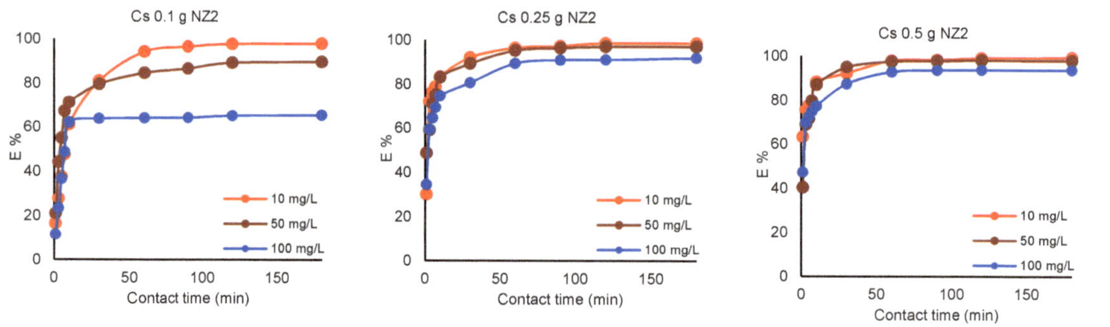

(a) (b) (c)

Figure 3. Removal efficiency (%) of Cs^{+} by different NZ2 quantities; (**a**) 0.1 g, (**b**) 0.25 g, (**c**) 0.5 g; V = 50 mL; 200 rpm; 22 ± 2 °C, 180 min.

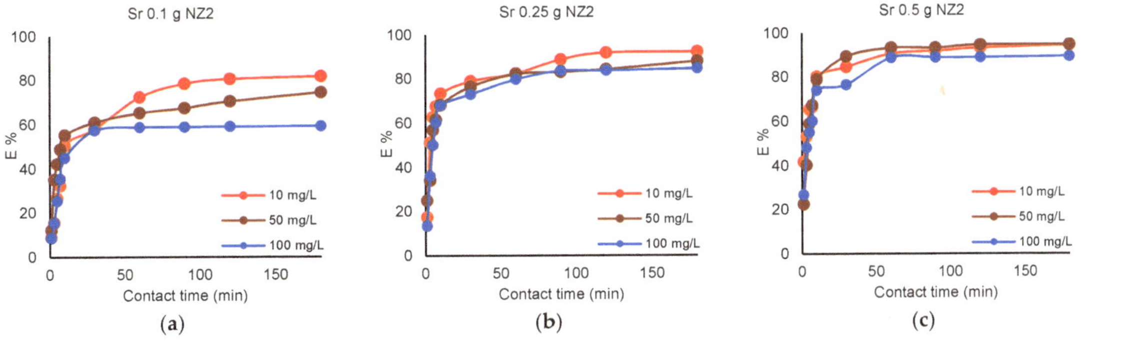

Figure 4. Removal efficiency (%) of Sr^{2+} by different NZ2 quantities; (**a**) 0.1 g, (**b**) 0.25 g, (**c**) 0.5 g; V = 50 mL; 200 rpm; 22 ± 2 °C, 180 min.

3.3.1. Influence of Zeolite Quantities on Removal Efficiency

The influence of the zeolite dosage on the Cs^+ and Sr^{2+} removal efficiency is mainly observed when a smaller amount (0.1 g) of adsorbent was used. Thus, when 0.1 g NZ1 is used as an adsorbent for Cs^+, the maximum removal efficiency (E%) is 63.6% at an initial concentration of 100 mg L^{-1}, 87.2% at an initial concentration of 50 mg L^{-1}, and reaches 95.3% when the initial concentration is only 10 mg L^{-1}. Increasing NZ1 dosage to 0.25 g leads to a maximum E% for Cs^+ of 92%, regardless of the initial concentrations of the solution. A higher dosage of 0.5 g NZ1 increased the maximum E% for Cs^+ to 96% when the initial concentrations of the solution were 10 mg L^{-1} and 50 mg L^{-1} and 93% when the initial concentration was 100 mg L^{-1}. In the case of Sr^{2+}, when 0.1 g NZ1 is used, the maximum (E%) is 51.3% when the initial concentration of the solution is 100 mg L^{-1}, 71.6% when the initial concentration of the solution is 50 mg L^{-1} and 79.5% when the initial concentration is 10 mg L^{-1}. The 0.25 g NZ1 dosage results in a maximum E% of 88%, whereas the dosage of 0.5 g of NZ1 conducts an increasing removal efficiency of 93%.

As presented in Figures 3 and 4, if 0.1 g of NZ2 is used for Cs^+ removal, a maximum E% of 65.6% is obtained at a concentration of 100 mg L^{-1}, 89.8% maximum E% if the initial concentration of the solution is 50 mg L^{-1}, and 98.1% maximum E% at the initial concentration of 10 mg L^{-1}. If 0.25 g NZ2 is used, E% for Cs^+ reach about 92% at the concentration of 100 mg L^{-1}, and about 97–98% at lower concentrations of 50 mg L^{-1} and 10 mg L^{-1}, respectively. The dosage of 0.5 g NZ2 removed over 99% Cs^+ from the solution with an initial concentration of 10 mg L^{-1}. In the case of Sr^{2+}, when 0.1 g NZ2 is used, the maximum E% is 59.1% at the initial concentration of 100 mg L^{-1}, 74.3% if the initial concentration of the solution is 50 mg L^{-1}, and 81.7% when the initial concentration is 10 mg L^{-1}. The 0.25 g NZ2 dosage results in a maximum removal efficiency of about 92%, whereas the dosage of 0.5 g NZ1 increases removal efficiency to over 94%. Results demonstrated that adsorbed Cs^+ and Sr^{2+} ions gradually increased with the adsorbent dosage, both for NZ1 and NZ2.

As the concentration of Cs^+ and Sr^{2+} increases in the solution, the driving force increases to overcome the mass transfer resistance of strontium from the aqueous solution into zeolite, which increases the sorption capacity of the zeolite.

3.3.2. Influence of Contact Time

The removal efficiency of Cs^+ increased within 60 min of removal when the solutions were contacted with 0.1 g of zeolites reaching the maximum at this contact time. By increasing the zeolite amount, the removal efficiency of Cs^+ rapidly increased and reached equilibrium after 30 min. The removal efficiency of Sr^{2+} slowly increased as compared with Cs^+ removal. When the solutions were contacted with 0.1 g of zeolites, the maximum removal efficiency was achieved after 120 min. Furthermore, the removal efficiency increased with the increase of the zeolite quantity. Li et al. [6] reported a maximum removal efficiency for Sr^{2+} at 30 min.

3.4. Influence of Zeolite Grain Sizes on Removal Efficiency (E%)

The effect of particle size of NZ1 and NZ2 is presented in Figures 1–4. Cs^+ and Sr^{2+} removal efficiencies are strongly influenced by zeolite particle size. So, the samples with smaller particle sizes were more efficient for cations sorption than the samples with larger particle sizes due to their larger external surface area available for the sorption. The particle size reduction increases the external surface area available for the interaction with ions in the solution and leads to a shorter diffusion path length. The diffusion of cations from the external to interparticle sites in zeolites is slow because of their interaction with the surface functional group [33]. This finding was also reported in previous studies [12,34], which showed that smaller particle sizes have an increased number of active sites.

3.5. Effect of Interfering Ions

Sorption properties and zeolite selectivity are strongly influenced by exchangeable cation properties, solution concentration, presence of other cations, and zeolite characteristics. Zeolites with a high Si/Al ratio have a higher affinity for cations with lower charge densities, such as Cs, while zeolites with a lower Si/Al ratio favor the retention of cations with higher charge densities [24]. Usually, the zeolite used in our experiments with a high Si/Al (>4), should prefer Cs^+ and Sr^{2+} to cations such as Na^+, K^+, Ca^{2+}, and Mg^{2+}. To evaluate the preference of natural zeolite used in this study for the Cs^+ and Sr^{2+} sorption in the presence of competitor cations, a 50 mL multicomponent solution containing 100 mg L^{-1} of Na^+, K^+, Ca^{2+}, Mg^{2+}, Cs^+, and Sr^{2+} was contacted for 60 min with 0.1 g of NZ1 in batch mode at a stirring rate of 200 rpm at room temperature (22 $\pm$ 2 °C). After filtration, Cs^+ and Sr^{2+} concentrations in the solution were measured. Compared with the removal from single component solutions, a decrease to 73.5% from the maximum sorption capacity for Cs^+ and 64.7% from the maximum sorption capacity of Sr^{2+} was recorded.

3.6. Retention Mechanisms

Ion exchange and chemisorption are the two main mechanisms that contribute to the retaining of cations by zeolites [15]. Ion exchange involves the replacement of the exchangeable cations (Na^+, K^+, Ca^{2+}, and Mg^{2+}) from the zeolite crystalline framework with other cations present in the contact solution. The cations retained in zeolite by ion exchange are weakly bounded and can be replaced by $NH_4{}^+$. In contrast, if the cations are retained by stronger chemical bonds (chemisorption) cannot be replaced by $NH_4{}^+$ [15,29]. The zeolites previously maintained in a 50 mL solution of 100 mg L^{-1} Cs^+ and Sr^{2+} were contacted with a 1 M ammonium acetate solution (ratio 1:50) for desorption to assess the Cs^+ and Sr^{2+} retained due to the ion exchange process. The results revealed that 44.4% of Cs^+ and 38.2% of Sr^{2+} were desorbed from zeolite being replaced by $NH_4{}^+$, which means that these cations are retained mainly by chemisorption in this type of zeolite. In addition, the data were analyzed using the Dubinin–Radushkevich (D–R) isotherm model to investigate whether the sorption process of Cs^+ and Sr^{2+} was physical or chemical. The mean free energy (E) values calculated from the D-R isotherm model for Cs^+ sorption on NZ1 and NZ2 were 9.2 kJ mol^{-1} and 12.1 kJ mol^{-1}, respectively. Regarding Sr^{2+} sorption on NZ1 and NZ2, the E values were 8.0 kJ mol^{-1} and 9.0 kJ mol^{-1}, respectively. Thus, the process of Cs^+ and Sr^{2+} sorption on both zeolites occurred by chemisorption.

The ICP results have been linked with IR spectroscopic studies to assess the retention mechanisms. FTIR spectra of the zeolitic tuff sample before and after Cs^+ and Sr^{2+} sorption were recorded in the range of 400–4000 cm^{-1} to identify the presence of functional groups responsible for the sorption process (Figure 5). There are several absorption peaks, indicating the complex nature of the zeolite. The peaks around 1630, and 3622 cm^{-1} were attributed to H–O–H bending O–H bonding (hydroxyl groups) vibrations and water absorption bands of the clinoptilolite, respectively [35,36]. The main characteristic bands of clinoptilolite were found at the wavelengths of 464 cm^{-1} resulting from the stretching vibrations of the Al–O bonds, at 604 cm^{-1} which is assigned to the vibration of the external linkage of the tetrahedral, at 790 and 1063 cm^{-1} (the most intense band characteristic to silicates) which are ascribed to Si–O–Si and Al–O–Si bonds due to asymmetric stretching vibrations [15,36].

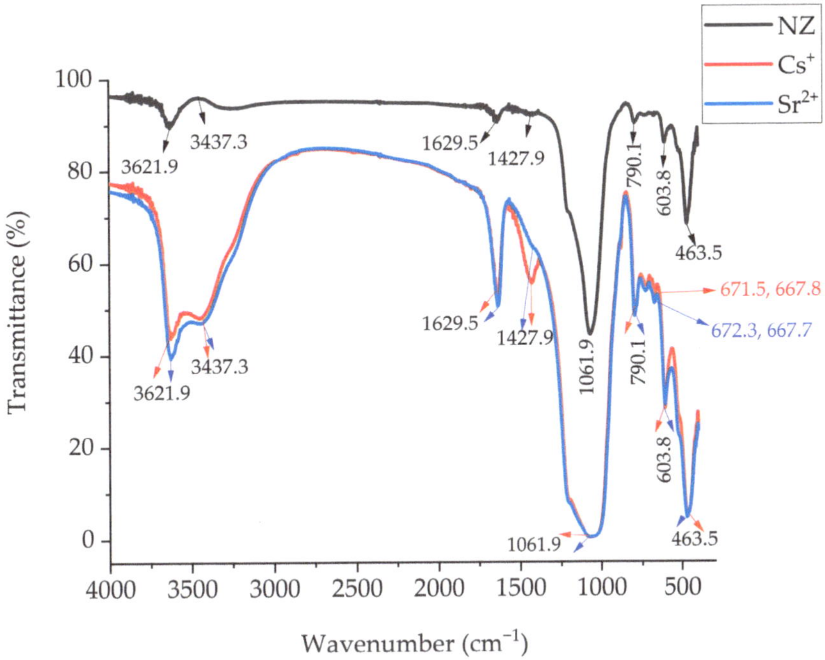

Figure 5. FTIR spectra of the zeolitic tuff sample before and after Cs^+ and Sr^{2+} sorption.

The most important changes in IR spectra following the Cs^+ and Sr^{2+} sorption are observed in the range of the pseudo-lattice vibrations, that is, 700–600 cm^{-1}, as it has been previously reported for other heavy metals sorption [29]. Thus, the bands in the range of 700–655 cm^{-1} attributed to the vibrations of rings in the zeolite structure have been investigated in more detail. In this range, the band at 672 cm^{-1} is observed when Cs^+ is sorbed, whereas the band at 668 cm^{-1} appears when Sr^{2+} is sorbed. Consequently, these two bands are characteristic of sorption through symmetrical stretching vibrations of the Si–O bond existing in the rings of the zeolite.

The change in the intensity of these bands can give information related to the amount of Cs^+ and Sr^{2+} sorbed in the zeolite structure. Even if an increase in the intensity of the two bands was observed with an increase of Cs^+ and Sr^{2+} initial concentration, a more precise quantification of the amount of Cs^+ and Sr^{2+} sorbed on zeolite can be obtained after ICP measurement of the two cations remained in the solution.

As estimated from the adsorption and desorption of the two cations with 1 M ammonium acetate solution, chemisorption dominates the retention for the monovalent Cs^+ and bivalent cation of Sr^{2+}, and this can be due to a lesser extent by the zeolite structure.

A possible explanation of the higher contribution of chemisorption to the retention of Cs^+ and Sr^{2+} is that the stable inner-sphere complexes are formed at the zeolite surface [37]. The higher ionic radius of Cs^+ and Sr^{2+} compared with the major cations from zeolite may prevent the diffusion of these ions in the zeolite framework to participate in the ion exchange process. Thus, the two cations form mainly strong coordinate bonds with an aluminosilicate sorbent skeleton.

The spectra of zeolite before and after the sorption of Cs^{+} and Sr^{2+} were registered by XRF (Figure 6). There was no Cs^{+} in the zeolite before sorption, but its signal was detected after sorption. In the case of Sr^{2+}, a signal was observed even before extraction, due to a concentration of about 200 mg kg^{-1} in the natural zeolite, which significantly increased after the extraction.

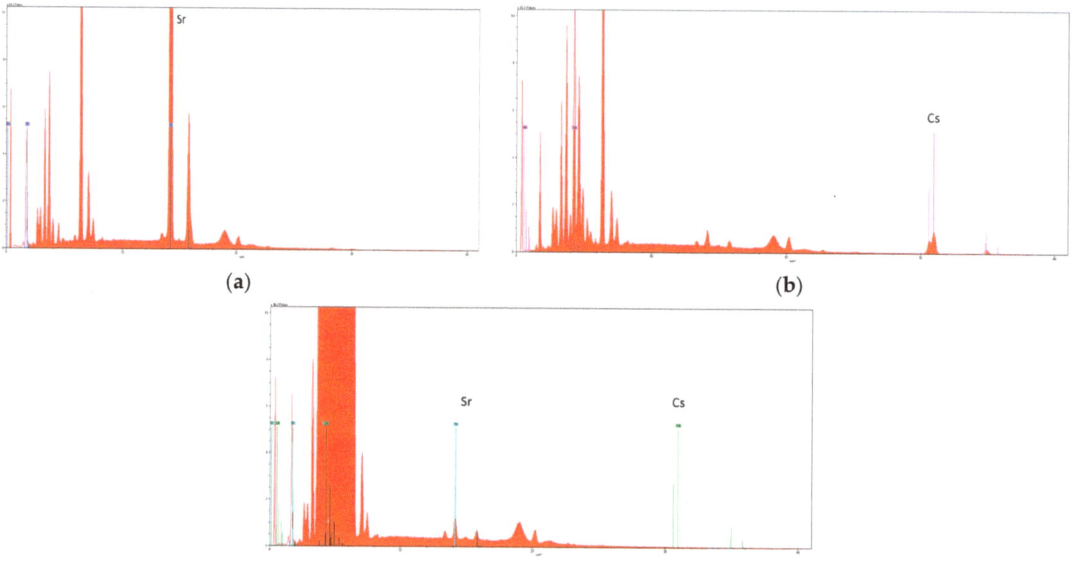

Figure 6. XRF spectra of zeolite after the sorption of Sr^{2+} (**a**), after the sorption of Cs^{+} (**b**), and before the sorption experiments (**c**).

3.7. Adsorption Isotherms and Kinetics

The nonlinear fitting of Langmuir and Freundlich isotherm models for the Cs^{+} and Sr^{2+} sorption on NZ1 and NZ2 at different initial concentrations is presented in Figure 7.

The Langmuir isotherm parameters, q_{max} (the maximum sorption capacity, mg g^{-1}) and K_L (Langmuir constant, L mg^{-1}), and the Freundlich isotherm parameters, K_F (related to sorption capacity, $mg^{1-1/n}$ $L^{1/n}$ g^{-1}) and $1/n$ (the sorption intensity) [12] are listed in Table 3. Table 3 shows that the experimental data for Cs^{+} and Sr^{2+} sorption on NZ1 and NZ2 were slightly better fitted using the Langmuir isotherm model than with the Freundlich isotherm model. For instance, the values of R^2 determined using the Langmuir isotherm model for Cs^{+} sorption on NZ1 and NZ2 were 0.9945 and 0.9932, respectively. These values were slightly higher than that determined using the Freundlich isotherm model (R^2 = 0.9722 and R^2 = 0.9916, respectively). A similar trend was observed for Sr^{2+} sorption by both NZ1 and NZ2. This indicates that the zeolite surface is homogeneous, and monolayer sorption of Cs^{+} and Sr^{2+} on a thermally treated zeolite surface can take place [5]. The obtained results are in accordance with previous studies [38,39]. Mimura and Akiba [8] investigated the removal of Cs^{+} and Sr^{2+} on synthetic zeolite P and found that the sorption of the studied ions obeyed the Langmuir isotherm model. Also, Lee et al. [39] found that the Langmuir isotherm model fitted better than the Freundlich and Dubinin–Radushkevich isotherms models, the sorption of Cs^{+} and Sr^{2+} on zeolitic materials synthesized from Jeju volcanic rocks.

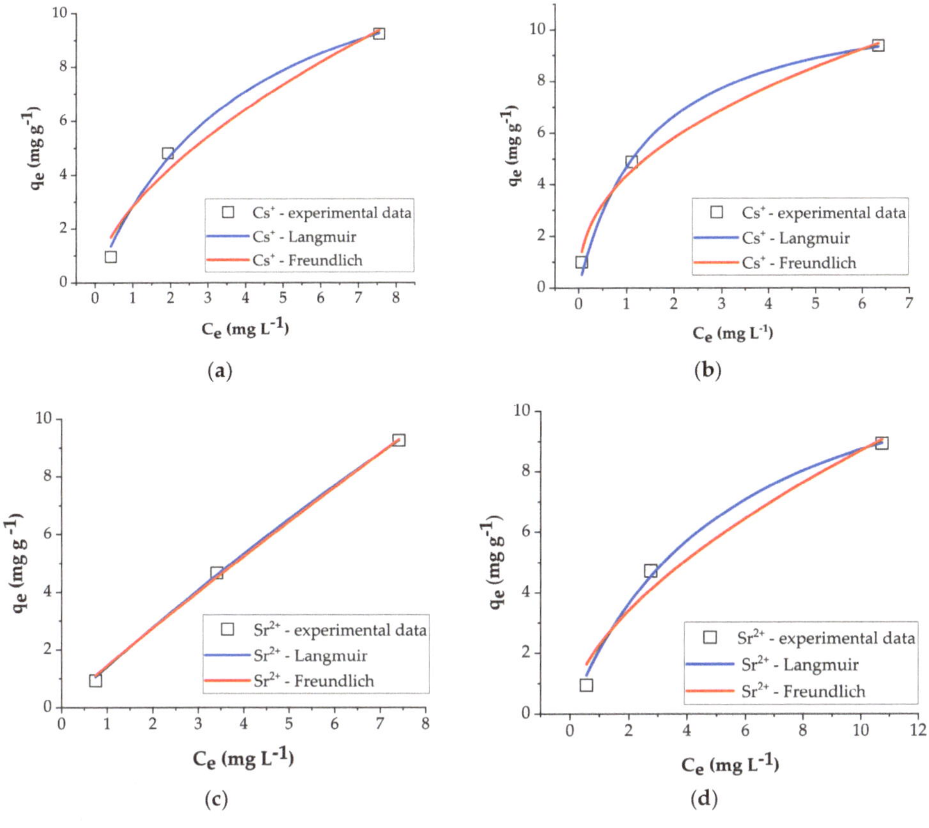

Figure 7. Nonlinear fitting of Langmuir and Freundlich isotherm models for the Cs^+ sorption on NZ1 (**a**), Cs^+ sorption on NZ2 (**b**), Sr^{2+} sorption on NZ1 (**c**), and Sr^{2+} sorption on NZ2 (**d**).

Table 3. Langmuir and Freundlich parameters for Cs^+ and Sr^{2+} sorption on NZ1 and NZ2 at different initial concentrations.

Cation	Zeolite	Langmuir Isotherm			Freundlich Isotherm		
		q_{max}	K_L	R^2	K_F	n	R^2
		($mg\ g^{-1}$)	($L\ mg^{-1}$)		($mg^{1-1/n}\ L^{1/n}\ g^{-1}$)		
Cs^+	NZ1	14.22	0.25	0.9945	2.83	0.59	0.9722
Cs^+	NZ2	11.46	0.70	0.9932	4.36	0.42	0.9916
Sr^{2+}	NZ1	68.88	0.02	0.9993	1.45	0.92	0.9985
Sr^{2+}	NZ2	13.46	0.19	0.9956	2.29	0.58	0.9726

According to the data presented in Table 3, the maximum sorption capacity for Cs^+ was 14.22 mg g^{-1} on NZ1 and 11.46 mg g^{-1} on NZ2, whereas the maximum sorption capacity for Sr^{2+} was 68.88 mg g^{-1} on NZ1 and 13.46 mg g^{-1} on NZ2.

In a study on the removal of Sr^{2+} from an aqueous solution using synthetic zeolite, Arraisi et al. [23] obtained a maximum adsorption capacity of 2.36 mmol g^{-1}, which decreased to 1.60 mmol g^{-1} when a binary solution with Ba^{2+} was used. Bakusi et al. [40] evaluated the adsorption of Sr^{2+} from an aqueous media by TiO2-modified zeolite and reported an adsorption capacity of 9.632 mg g^{-1}. Belviso et al. [41] used zeolites synthesized from volcanic ash for Cs removal and reported maximum adsorption capacities in the range of 31.75–63.69 mg g^{-1}. Borai et al. [42] used zeolites impregnated with calix [4]arene for Cs removal and obtained maximum adsorption capacities ranging between 1.201 and 1.865 mmol g^{-1}. Lee et al. [43] prepared geopolymers containing nano-crystalline zeolites and reported a maximum adsorption capacity for Cs^+ of 15.24 mg g^{-1}. Abdel Moamen et al. [44] reported maximum adsorption capacities of 15.32–19.29 mEq g^{-1} for Sr^{2+} and 5.74–6.72 mEq g^{-1} for Cs^+ on synthetic nano-sized zeolite using the nonlinearized form of Langmuir isotherm model. Synthesized zeolite Y had a maximum adsorption capacity of 2.401 mmol g^{-1} for Cs+ and 1.527 mmol g^{-1} for Sr^{2+} [45]. Adsorption studies on a modified zeolite with nano iron particles for strontium removal found adsorption capacities in the range of 9.08–32.83 mg g^{-1} from solution with initial concentrations of Sr^{2+} ranging between 50 and 200 mg L^{-1} [46]. A maximum adsorption capacity of 84.12 mg g^{-1} for Sr^{2+} removal on modified zeolite with nano iron particles was obtained [46]. Lee et al. [39] reported maximum adsorption capacities of 154.84 mg/g for Sr^{2+} and 144.01 mg g^{-1} for Cs^+ removal by zeolitic materials. Higher adsorption capacities ranging between 54.5 and 45.2 mg g^{-1} for Cs^+ and between 30 and 21 mg g^{-1} for Sr^{2+} by clinoptilolite-rich zeolites (with smaller particle sizes) were reported by Sterba et al. [47]. Our study demonstrated that the natural zeolite from the studied quarry is highly effective for Cs^+ and Sr^{2+} removal after a minimum pretreatment. Both particle fractions, 0.5–1.25 mm (NZ1) and 0.1–0.5 mm (NZ2), can be easily prepared by crushing and granulometric separation on sieves in a factory and then activated at a temperature of only 200 °C to eliminate water from zeolite pores. The cost of the zeolite used in this study was 0.2 USD kg^{-1}, much less than that of synthetic zeolites, (approximately 50 US dollars kg^{-1}).

The kinetic parameters, q_t (the amount adsorbed at time t, mg g^{-1}), k_1 (the first-order rate constant, min^{-1}), and k_2 (the second-order rate constant, g mg·min^{-1}) obtained by applying the nonlinearized form of the PFO and PSO for Cs^+ and Sr^{2+} sorption on NZ1 and NZ2 are presented in Table 4.

The calculated q_e ($q_{e,\ calc}$) values estimated by the PFO are not close to the experimental q_e values ($q_{e,\ exp}$), indicating that the model is not suitable to describe the Cs^+ and Sr^{2+} sorption kinetics on NZ1 and NZ2. However, the $q_{e,\ calc}$ from PSO are close to the $q_{e,\ exp}$, and the R^2 values are higher than those obtained for the PFO. Thereby, the Cs^+ and Sr^{2+} sorption kinetics on NZ1 and NZ2 followed PSO rather than PFO. This indicates that the sorption process of Cs^+ on NZ1 and Sr^{2+} on NZ2 occurs as chemisorption.

El-Rahman et al. [48] studied the kinetics of Cs^+ and Sr^{2+} on synthetic zeolite A using linear forms of PFO and PSO models and found that PFO was not adequate to predict the sorption kinetics of both ions. Also, similar results were found for the sorption of Sr^{2+} and Cs^+ on zeolitic materials synthesized from Jeju volcanic rocks [39] and natural zeolites [5]. Moreover, it can be observed that the k_2 constants obtained for Cs^+ sorption are higher than those obtained for the sorption of Sr^{2+} on both zeolites. This agrees with the findings of Kwon et al. [1] and is ascribed to the strong hydration of Sr^{2+} [1].

The two cations are sorbed mainly by strong coordinate bonds with an aluminosilicate zeolite skeleton. Thus, they are difficult to extract back. However, the production of secondary solid waste is a key challenge in the sorption process. To be deposited after sorption without the risk of release to the environment, Abdollahi et al. [5] developed a stabilization procedure by heating the resulting secondary solid waste at temperatures ranging between 700 and 1100 °C for 2, 6, and 12 h. The leaching tests showed that Cs^+ and Sr^{2+} were well stabilized in the natural zeolite structure, and this represents a practical solution to manage the zeolite that retained these cations.

Considering that natural zeolite is a widespread and cheap material with good adsorption capacity, it represents a promising alternative for the treatment of water contaminated with radioactive Cs^+ and Sr^{2+} after a minimum pretreatment procedure. Therefore, it can be used for the decontamination of large amounts of contaminated water.

Table 4. Kinetic parameters for Cs^+ and Sr^{2+} sorption on NZ1 and NZ2 at different initial concentrations.

Cation	Zeolite	Model	Parameters	10 mg/L	50 mg/L	100 mg/L
Cs^+	NZ1	PFO	k_1 (1 min^{-1})	0.18	0.21	0.23
			$q_{e,\,calc}$ (mg g^{-1})	0.92	4.65	8.59
			R^2	0.9544	0.9498	0.81558
		PSO	k_2 (g mg·min^{-1})	0.28	0.07	0.04
			$q_{e,\,calc}$ (mg g^{-1})	0.97	4.9	9.02
			R^2	0.9795	0.9897	0.9326
			$q_{e,\,exp}$	0.96	4.81	9.25
	NZ2	PFO	k_1 (1 min^{-1})	1.01	0.38	0.57
			$q_{e,\,calc}$ (mg g^{-1})	0.91	4.72	8.70
			R^2	0.510	0.890	0.73
		PSO	k_2 (g mg·min^{-1})	1.47	0.13	0.10
			$q_{e,\,calc}$ (mg g^{-1})	0.96	4.95	9.21
			R^2	0.854	0.989	0.948
			$q_{e,\,exp}$	0.99	4.89	9.34
Sr^{2+}	NZ1	PFO	k_1 (1 min^{-1})	0.15	0.11	0.15
			$q_{e,\,calc}$ (mg g^{-1})	0.86	4.27	8.14
			R^2	0.9532	0.9257	0.7846
		PSO	k_2 (g mg·min^{-1})	0.21	0.03	0.02
			$q_{e,\,calc}$ (mg g^{-1})	0.93	4.69	8.79
			R^2	0.9786	0.9676	0.8926
			$q_{e,\,exp}$	0.93	4.66	9.26
	NZ2	PFO	k_1 (1 min^{-1})	0.30	0.20	0.22
			$q_{e,\,calc}$ (mg g^{-1})	0.89	4.64	8.55
			R^2	0.8072	0.9901	0.9217
		PSO	k_2 (g mg·min^{-1})	0.58	0.06	0.04
			$q_{e,\,calc}$ (mg g^{-1})	0.93	4.90	9.01
			R^2	0.9423	0.9875	0.9774
			$q_{e,\,exp}$	0.94	4.72	8.93

4. Conclusions

In this study, the use of thermally treated natural zeolite from a quarry from NW of Romania showed a high potential for the removal of Cs^+ and Sr^{2+} from aqueous solutions. Two different particle sizes of (0.5–1.25 and 0.1–0.5 mm) thermally treated zeolite at 200 °C were characterized and tested in batch experiments. The experiments were carried out using different NZ1 and NZ2 dosages (0.5, 1, 2 g) which were contacted with 50 mL of Cs^+ and Sr^{2+} solution (10, 50, 100 mg L^{-1} initial concentrations) for 180 min. A good sorption capacity is due to the high cation-exchange capacity, surface area, and porosity. The removal efficiency is influenced by zeolite dosage, initial concentration, contact time, and grain sizes. Maximum removal efficiency of 99.3% for Cs^+ when a dosage of 0.5 g zeolite with a grain size of 0.1–0.5 mm was contacted for 180 min with a solution with an initial concentration of 10 mg L^{-1}. For Sr^{2+} removal from solution, a maximum removal efficiency of 92.1% when a dosage of 0.5 g zeolite with a grain size of 0.1–0.5 mm was contacted for 180 min with a solution with an initial concentration of 10 mg L^{-1}. The competing ions of Na^+, K^+, Ca^{2+}, and Mg^{2+} from a multi-component solution with concentrations of 100 mg L^{-1} each decreased the maximum adsorption capacities to 73.5% for Cs^+ and 64.7% for Sr^{2+}. The sorption mechanism was investigated based on the adsorption and desorption studies correlated with FTIR spectroscopy, and it was found that the chemisorption dominates the retention of both Cs^+ and Sr^{2+}. Thus, mainly strong coordinate bonds with an aluminosilicate sorbent skeleton are formed. At the same time, some cations are retained by ion exchange and can be replaced from the zeolite structure by other cations. Therefore, the natural zeolite from Macicasu (Romania) is efficient and has a high adsorption capacity for Cs^+ and Sr^{2+} from aqueous solutions. Since zeolite is a natural and inexpensive material, it should be considered a viable solution for the treatment of wastewater contaminated with Cs^+ and Sr^{2+}.

Author Contributions: Conceptualization, M.Ș.; methodology, M.Ș. and E.N.; validation, M.Ș. and E.N.; formal analysis, M.Ș. and L.Ș.; investigation, M.Ș., C.T., E.N. and L.Ș.; resources, M.Ș.; data curation, E.N.; writing—original draft preparation, M.Ș. and E.N.; writing—review and editing, C.T. and L.Ș.; visualization, M.Ș.; supervision, M.Ș.; project administration, M.Ș.; funding acquisition, M.Ș., E.N., C.T. and L.Ș. All authors have read and agreed to the published version of the manuscript.

Funding: This research was funded by the Ministry of Research, Innovation and Digitization through Program 1—Development of the national research &development system, Subprogram 1.2—Institutional performance—Projects that finance the RDI excellence, Contract no. 18PFE/30 December 2021 and was carried out through the Core Program within the National Research Development and Innovation Plan 2022–2027, carried out with the support of MCID, project no. PN 23 05.

Institutional Review Board Statement: Not applicable.

Informed Consent Statement: Not applicable.

Data Availability Statement: The relevant data from this research are available in the authors' repositories.

Conflicts of Interest: The authors declare no conflict of interest. The funders had no role in the design of the study; in the collection, analyses, or interpretation of data; in the writing of the manuscript, or in the decision to publish the results.

References

1. Kwon, S.; Kim, C.; Han, E.; Lee, H.; Cho, H.S.; Choi, M. Relationship between zeolite structure and capture capability for radioactive cesium and strontium. *J. Hazard. Mater.* **2021**, *408*, 124419. [CrossRef]
2. Inan, S. Inorganic ion-exchangers for strontium removal from radioactive waste: A review. *J. Radioanal. Nucl. Chem.* **2022**, *331*, 1137–1154. [CrossRef]
3. Yang, H.M.; Jeon, H.; Lee, Y.; Choi, M. Sulfur-modified zeolite A as a low-cost strontium remover with improved selectivity for radioactive strontium. *Chemosphere* **2022**, *299*, 134309. [CrossRef] [PubMed]
4. Lonin, A.Y.; Levenets, V.V.; Omelnik, O.P.; Shchur, A.O. Removal of a mixture of Cs, Sr and Co cations from an aqueous solution using composite sorbents based on natural and synthetic zeolites. *J. Radioanal. Nucl. Chem.* **2022**, *331*, 5517–5523. [CrossRef]
5. Abdollahi, T.; Towfighi, J.; Rezaei-Vahidian, H. Sorption of cesium and strontium ions by natural zeolite and management of produced secondary waste. *Environ. Technol. Innov.* **2020**, *17*, 100592. [CrossRef]
6. Li, H.; Han, K.; Shang, J.; Cai, W.; Pan, M.; Xu, D.; Du, C.; Zuo, R. Comparison of adsorption capacity and removal efficiency of strontium by six typical adsorption materials. *Sustainability* **2022**, *14*, 7723. [CrossRef]
7. Zhang, X.Y.; Liu, Y. Ultrafast removal of radioactive strontium ions from contaminated water by nanostructured layered sodium vanadosilicate with high adsorption capacity and selectivity. *J. Hazard. Mater.* **2020**, *398*, 122907. [CrossRef]
8. Senila, M.; Cadar, O.; Senila, L.; Angyus, B.S. Simulated bioavailability of heavy metals (Cd, Cr, Cu, Pb, Zn) in contaminated soil amended with natural zeolite using diffusive gradients in thin-films (DGT) technique. *Agriculture* **2022**, *12*, 321. [CrossRef]
9. Painer, F.; Baldermann, A.; Gallien, F.; Eichinger, S.; Steindl, F.; Dohrmann, R.; Dietzel, M. Synthesis of zeolites from fine-Grained Perlite and Their Application as Sorbents. *Materials* **2022**, *15*, 4474. [CrossRef]
10. Neag, E.; Torok, A.I.; Tanaselia, C.; Aschilean, I.; Senila, M. Kinetics and equilibrium studies for the removal of Mn and Fe from binary metal solution systems using a Romanian thermally activated natural zeolite. *Water* **2020**, *12*, 1614. [CrossRef]

11. Ferreira, D.R.; Phillips, G.D.; Baruah, B. A comparison of the adsorption of cesium on zeolite minerals vs vermiculite. *Clays Clay Miner.* **2021**, *69*, 663–671. [CrossRef]
12. Senila, M.; Neag, E.; Cadar, O.; Kovacs, E.D.; Aschilean, I.; Kovacs, M.H. Simultaneous removal of heavy metals (Cu, Cd, Cr, Ni, Zn and Pb) from aqueous solutions using thermally treated Romanian zeolitic volcanic tuff. *Molecules* **2022**, *27*, 3938. [CrossRef] [PubMed]
13. Abdel-Karim, A.A.M.; Zaki, A.A.; Elwan, W.; El-Naggar, M.R.; Gouda, M.M. Experimental and modeling investigations of cesium and strontium adsorption onto clay of radioactive waste disposal. *Appl. Clay Sci.* **2016**, *132–133*, 391–401. [CrossRef]
14. Margeta, K.; Zabukovec, N.; Siljeg, M.; Farkas, A. Natural zeolites in water treatment—How effective is their use. In *Water Treatment*; Elshorbagy, W., Ed.; InTech: London, UK, 2013; pp. 84–85.
15. Krol, M.; Mozgawa, W.; Jastrzebski, W. Theoretical and experimental study of ion-exchange process on zeolites from 5-1 structural group. *J. Porous. Mater.* **2016**, *23*, 1–9. [CrossRef]
16. Zou, W.; Bai, H.; Zhao, L.; Li, K.; Han, R. Characterization and properties of zeolite as adsorbent for removal of uranium (VI) from solution in fixed bed column. *J. Radioanal. Nucl. Chem.* **2011**, *288*, 779–788. [CrossRef]
17. Fan, Y.; Huang, R.; Liu, Q.; Cao, Q.; Guo, R. Synthesis of zeolite A from fly ash and its application in the slow release of urea. *Waste Manag.* **2023**, *158*, 47–55. [CrossRef]
18. Pabis-Mazgaj, E.; Gawenda, T.; Pichniarczyk, P.; Stempkowska, A. Mineral composition and structural characterization of the clinoptilolite powders obtained from zeolite-rich tuffs. *Minerals* **2021**, *11*, 103. [CrossRef]
19. Mubarak, M.F.; Mohamed, A.M.G.; Keshawy, M.; elMoghny, T.A.; Shehata, N. Adsorption of heavy metals and hardness ions from groundwater onto modified zeolite: Batch and column studies. *Alex. Eng. J.* **2022**, *61*, 4189–4207. [CrossRef]
20. Sharma, P.; Sutar, P.P.; Xiao, H.; Zhang, Q. The untapped potential of zeolites in techno-augmentation of the biomaterials and food industrial processing operations: A review. *J. Future Foods* **2023**, *3*, 127–141. [CrossRef]
21. Inglezakis, V.J.; Kudarova, A.; Guney, A.; Kinayat, N.; Tauanov, Z. Efficient mercury removal from water by using modified natural zeolites and comparison to commercial adsorbents. *Sustain. Chem. Pharm.* **2023**, *32*, 101017. [CrossRef]
22. Kubota, T.; Fukutani, S.; Ohta, T.; Mahara, Y. Removal of radioactive cesium, strontium, and iodine from natural waters using bentonite, zeolite, and activated carbon. *J. Radioanal. Nucl. Chem.* **2013**, *296*, 981–984. [CrossRef]
23. Araissi, M.; Ayed, I.; Elaloui, E.; Moussaoui, Y. Removal of barium and strontium from aqueous solution using zeolite 4A. *Water Sci. Technol.* **2016**, *73*, 1628–1636. [CrossRef] [PubMed]
24. Baek, W.; Ha, S.; Hong, S.; Kim, S.; Kim, Y. Cation-exchange of cesium and cation selectivity of natural zeolites: Chabazite, stilbite, and heulandite. *Microporous Mesoporous Mater.* **2018**, *264*, 159–166. [CrossRef]
25. European Comission. Zero Pollution Action Plan. Available online: https://environment.ec.europa.eu/strategy/zero-pollution-action-plan_en (accessed on 31 March 2023).
26. Senila, M.; Cadar, O.; Miu, I. Mercury determination in natural zeolites by thermal decomposition atomic absorption spectrometry: Method validation in compliance with requirements for use as dietary supplements. *Molecules* **2019**, *24*, 4023. [CrossRef] [PubMed]
27. Chicinas, P.R.; Bedelean, H.; Maicaneanu, A. Romanian (Măcicaş) zeolitic volcanic tuff for malachite green removal. *Studia UBB Chemia* **2016**, *61*, 243–254.
28. Adekola, F.A.; Hodonou, D.S.S.; Adegoke, H.I. Thermodynamic and kinetic studies of biosorption of iron and manganese from aqueous medium using rice husk ash. *Appl. Water Sci.* **2016**, *6*, 319–330. [CrossRef]
29. Mozgawa, W.; Bajda, T. Spectroscopic study of heavy metals sorption on clinoptilolite. *Phys. Chem. Minerals* **2005**, *31*, 709–713. [CrossRef]
30. Mansouri, N.; Rikhtegar, N.; Panahi, H.A.; Atabi, F.; Shahraki, B.K. Porosity, characterization and structural properties of natural zeolite—Clinoptilolite—As a sorbent. *Environ. Prot. Eng.* **2013**, *39*, 139–152.
31. Senila, M.; Neag, E.; Cadar, O.; Hoaghia, M.A.; Roman, M.; Moldovan, A.; Hosu, A.; Lupas, A.; Kovacs, E.D. Characteristics of volcanic tuff from Macicasu (Romania) and its capacity to remove ammonia from contaminated air. *Molecules* **2022**, *27*, 3503. [CrossRef]
32. Thommes, M.; Kaneko, K.; Niemark, A.; Olivier, J.; Rodriguez-Reinoso, F.; Rouquerol, J.; Sing, K. Physisoprtion of gases, with special reference to the evaluation of surface area and pore size distribution. *Pure Appl. Chem.* **2015**, *87*, 1051–1064. [CrossRef]
33. Trgo, M.; Peric, J.; Medvidovic, V. Investigations of different kinetic models for zinc ions uptake by a natural zeolitic tuff. *J. Environ. Manag.* **2006**, *79*, 298–304. [CrossRef] [PubMed]
34. Pernyeszi, T.; Roland, F.; Janos, K. Methylene blue adsorption study on microcline particles in the function of particle size range and temperature. *Minerals* **2019**, *9*, 555. [CrossRef]
35. Ahmadi, S.J.; Akbari, N.; Shiri-Yekta, Z.; Mashhadizadeh, M.H.; Pourmatin, A. Adsorption of strontium ions from aqueous solution using hydrous, amorphous MnO_2–ZrO_2 composite: A new inorganic ion exchanger. *J. Radioanal. Nucl. Chem.* **2014**, *299*, 1701–1707. [CrossRef]
36. Luo, Q.; Jiang, D.; Hou, D.; Chen, W.; Hu, X.; He, Y. Effective adsorption of metal ions by modified clinoptilolite zeolite from simulated radioactive solution. *J. Radioanal. Nucl. Chem.* **2019**, *319*, 1069–1081. [CrossRef]
37. Inan, S.; Hicsonmez, U. Adsorption studies of radionuclides by Turkish minerals: A review. *J. Turk. Chem. Soc. A Chem.* **2022**, *9*, 579–600. [CrossRef]
38. Mimura, H.; Akiba, K. Adsorption behavior of cesium and strontium on synthetic zeolite P. *J. Nucl. Sci. Technol.* **1993**, *30*, 436–443. [CrossRef]

39. Lee, M.G.; Kam, S.-K.; Lee, C.-H. Kinetic and isothermal adsorption properties of strontium and cesium ions by zeolitic materials synthesized from Jeju volcanic rocks. *Environ. Eng. Res.* **2021**, *26*, 200127. [CrossRef]
40. Basuki, K.T.; Fatuzzahroh, M.; Ariyanti, D.; Saputra, A. Adsorption of strontium from an aqueous solution by TiO_2-pillared zeolite. *Int. J. Technol.* **2021**, *12*, 625–634. [CrossRef]
41. Belviso, C.; Abdolrahimi, M.; Peddis, D.; Gagliano, E.; Sgroi, M.; Lettino, A.; Roccaro, P.; Vagliasindi, F.; Falciglia, P.; Di Bella, G.; et al. Synthesis of zeolite from volcanic ash: Characterization and application for cesium removal. *Microporous Mesoporous Mater.* **2021**, *319*, 111045. [CrossRef]
42. Borai, E.H.; Harjula, R.; Malinen, L.; Paajanen, A. Efficient removal of cesium from low-level radioactive liquid waste using natural and impregnated zeolite minerals. *J. Hazard. Mater.* **2009**, *172*, 416–422. [CrossRef]
43. Lee, N.K.; Khalid, H.R.; Lee, H.K. Adsorption characteristics of cesium onto mesoporous geopolymers containing nano-crystalline zeolites. *Microporous Mesoporous Mater.* **2017**, *242*, 238–244. [CrossRef]
44. Abdel Moamen, O.A.; Ismail, I.M.; Abdelmonem, N.; Abdel Rahman, R.O. Factorial design analysis for optimizing the removal of cesium and strontium ions on synthetic nano-sized zeolite. *J. Taiwan. Inst. Chem. Eng.* **2015**, *55*, 133–144. [CrossRef]
45. Li, Y.; Simon, A.O.; Jiao, O.; Zhang, M.; Yan, W.; Rao, H.; Liu, J.; Zhang, J. Rapid removal of Sr^{2+}, Cs^{2+}, UO_2^{2+} from solution with surfactant and amino acid modified zeolite Y. *Microporous Mesoporous Mater.* **2021**, *302*, 110244. [CrossRef]
46. Shubair, T.; Tahara, A.; Khandaker, S. Optimizing the magnetic separation of strontium ion using modified zeolite with nano iron particles. *Case Stud. Therm. Eng.* **2022**, *6*, 100243. [CrossRef]
47. Sterba, J.H.; Sperrer, H.; Wallenko, F.; Welch, J.M. Adsorption characteristics of a clinoptilolite-rich zeolite compound for Sr and Cs. *J. Radioanal. Nucl. Chem.* **2018**, *318*, 267–270. [CrossRef] [PubMed]
48. El-Rahman, K.M.A.; El-Sourougy, M.R.; Abdel-Monem, N.M.; Ismail, I.M. Modeling the sorption kinetics of cesium and strontium ions on Zeolite A. *J. Nucl. Radiochem. Sci.* **2006**, *7*, 21–27. [CrossRef]

materials

Article

Modified Silica Nanoparticles from Rice Husk Supported on Polylactic Acid as Adsorptive Membranes for Dye Removal

João Otávio Donizette Malafatti [1], Francine Aline Tavares [2], Tainara Ramos Neves [2], Bruno Cano Mascarenhas [2], Simone Quaranta [3] and Elaine Cristina Paris [1,*]

1 Nanotechnology National Laboratory for Agriculture (LNNA), Embrapa Instrumentação, São Carlos 13560-970, Brazil
2 Department of Chemistry, Federal University of São Carlos, São Carlos 13565-905, Brazil
3 Institute for the Study of Nanostructured Materials, Italian National Research Council (ISMN–CNR), 00010 Rome, Italy; simone.quaranta@cnr.it
* Correspondence: elaine.paris@embrapa.br

Abstract: Industrial effluents and wastewater treatment have been a mainstay of environmental preservation and remediation for the last decade. Silica nanoparticles (SiO_2) obtained from rice husk (RH) are an alternative to producing low-cost adsorbent and agriculture waste recovery. One adsorption challenge is facilitating the adsorbate separation and reuse cycle from aqueous medium. Thus, the present work employs SiO_2 supported on polylactic acid (PLA) nanofibers obtained by the electrospinning method for Rhodamine B (RhB) dye adsorption. The silica surface was modified with trimethylsilyl chloride (TMCS) to increase affinity towards organic compounds. As a result, the silanized surface of the silica from rice husk (RHSil) promoted an increase in dye adsorption attributed to the hydrophobic properties. The PLA fibers containing 40% SiO_2 (w w^{-1}) showed about 85–95% capacity adsorption. The pseudo-first-order kinetic model was demonstrated to be the best model for PLA:SiO_2 RHSil nanocomposites, exhibiting a 1.2956 mg g^{-1} adsorption capacity and 0.01404 min^{-1} kinetic constant (k_1) value. In the reuse assay, PLA:SiO_2 membranes preserved their adsorption activity after three consecutive adsorption cycles, with a value superior to 60%. Therefore, PLA:SiO_2 nanocomposites from agricultural waste are an alternative to "low-cost/low-end" treatments and can be used in traditional treatment systems to improve dye removal from contaminated waters.

Keywords: silica nanoparticles; rice husk; PLA; electrospinning; nanocomposite; adsorption

Citation: Malafatti, J.O.D.; Tavares, F.A.; Neves, T.R.; Mascarenhas, B.C.; Quaranta, S.; Paris, E.C. Modified Silica Nanoparticles from Rice Husk Supported on Polylactic Acid as Adsorptive Membranes for Dye Removal. *Materials* **2023**, *16*, 2429. https://doi.org/10.3390/ma16062429

Academic Editors: Roberta G. Toro, Daniela Caschera and Abeer M. Adel

Received: 25 February 2023
Revised: 15 March 2023
Accepted: 16 March 2023
Published: 18 March 2023

1. Introduction

Environmental remediation processes have been applied in waste treatment to minimize the ecological impacts occasioned by human actions [1,2]. The adsorption of pollutants with low-cost adsorbents is controlled mainly by physical interactions, making this process more attractive for wastewater treatment since this characteristic allows the adsorbent to be reused in many cycles [3,4]. Among the desirable features of adsorbents are the high specific surface area (SSA), easy regeneration, and low-cost production [5]. In this context, activated carbon is the most used material due to its efficient adsorption capacity [6–8]. On the other hand, nanomaterials are a promising alternative due to the improvement of properties through surface and structural modifications [9–12]. Therefore, many researchers have studied alternative adsorbents that can efficiently substitute for activated carbon, including agricultural residues (sugar cane bagasse, coconut husk, corn cob) [13], zeolites [14,15], and silica [16–18], among others [19,20].

Silica (SiO_2) has a tetrahedral structure in which the silicon atom occupies the central position coordinated with four oxygen atoms [21]. As a result, crystalline forms, such as quartz, tridymite, and cristobalite, could be formed. On the other hand, the random arrangement of tetrahedra characterizes silica in the amorphous (non-crystalline) form [22]. SiO_2

nanoparticles are distinct compounds with efficient surface chemical reactivity, thermal stability, a high surface area, and a porous structure [23,24]. Therefore, silica nanostructures have been widely used for applications in catalysis, thermal insulators, sensors, and adsorption [25]. Although SiO_2 nanoparticles can be synthesized efficiently by the sol–gel process [26], rice husk extraction is an alternative to obtaining high-purity silica from an agribusiness residue. The main constituents of rice husks are lignin, cellulose, and hemicellulose, accounting for about 80% of the total mass, and 17% of the mineral ash [27–29]. Thus, the rice husk is a natural residue that, with the appropriate extraction method, can produce high-purity silica with a high SSA from 250 to 900 $m^2\ g^{-1}$ by decomposition of organic compounds [30,31].

The efficiency of removing organic pollutants from aqueous solutions can be increased by modifying the adsorbent surface, expanding the material hydrophobicity, and improving the interaction with the pollutant molecules [4,32]. The high chemical reactivity of silica originates from the efficient surface covering by silane groups (Si-O-H) from $R_nSiX_{(4-n)}$ compounds, where R is the organic part that can be an alkyl or organofunctional group and X is an alkoxy group (ethoxy or methoxy) or a halide [33,34]. Roe and Zang [35] studied the surface hydrophobization of silica nanoparticles for different silane sources. These authors obtained surfaces with maximum hydrophobicity using the combination of n-octyltrimethoxysilane (OTMS) and bis(triethoxysil)ethane (BTEOSE), producing a contact angle of 139.1°. Velmurugan et al. [36] obtained functionalized silica modified with amino groups from 3-aminopropyltriethoxysilane and evaluated the efficiency against the anionic dye Congo red adsorption. The pristine silica showed around 40% of dye removal and 98% after the surface modification. The authors attributed the high adsorption efficiency to the attractive forces between the adsorbent amine group and the dye molecules. In this context, silica with a hydrophobic nature and highly efficient adsorption of organic contaminants can be produced from functionalization reactions with suitable silane reagents. However, removing particulate solids from aqueous solutions, especially in nanometer scales, induce a challenge due to the complex, large-scale application of these materials [37].

The use of supports in heterogeneous adsorption methods promotes an adsorbent-accessible recovery and simple separation from the aqueous medium [38–41]. Additionally, nanocomposites combine individual characteristics to increase desirable properties [42–45]. Fibrous membranes are one alternative to immobilizing the nanoparticles due to the porosity and diameter characteristics improving the surface area [46,47]. Specifically for adsorption process applications, using nanoparticles immobilized on fibers aims to facilitate the adsorbent material's removal from the aqueous medium and promote its reuse for many cycles [48]. One technique to produce fibrous membranes is electrospinning, a simple and practical way to introduce nanoparticles into a polymer matrix [32,37,49]. Xu et al. [50] obtained polyacrylic acid (PAA) fibers with SiO_2 functionalized with mercapto silane groups with a maximum indigo carmine dye adsorption of 523 mg g^{-1}. In further work [47], this group used the PAA:SiO_2 membrane to adsorb malachite green dye. In the results, the membranes demonstrate stability after multiple cycles without significant loss of their adsorption capacity. Among the different polymers, polylactic acid (PLA) has gained special attention attributed to its non-toxicity, biodegradability, and flexibility [51,52]. Therefore, PLA has been used in applications such as biomedical [53], active packaging [54], and remediation supports [14,55]. From this perspective, a system based on PLA:SiO_2 fibers shows desirable characteristics for a membrane to be applied in adsorption processes.

The great potential stemming from joining SiO_2 nanoparticles has been investigated recently. For instance, PLA/silica composites have been successfully applied to packaging [56] and oil removal [57]. Nevertheless, studies concerning PLA:SiO_2 membranes for water remediation (i.e., dye elimination) are scarce. Furthermore, research has focused on high-cost, synthetic silica-based solutions for removing water contaminants. On the other hand, plenty of waste sources are available to synthesize silica nanoparticles. Therefore, the present study developed adsorbing membranes predicated on low-cost silica nanoparticles incorporated into silica particles. Specifically, mesoporous silica was attained from rice husk agri-waste by simple acid leaching and thermal annealing. Then, this material was supported on PLA fibers measuring up to both commercial and sol-gel synthesized silica nanoparticles for Rhodamine B removal. In addition, the SiO_2 external surface was modified with the trimethylsilyl chloride silane (TMSCl) to promote interaction with Rhodamine B (RhB). Thus, the proposed nanocomposite material can be regarded as a "low-end" platform for the adsorption of pollutants from aqueous solutions. In fact, rice-husk-derived PLA:SiO_2 membranes are suitable for recovery and multiple reuses.

2. Materials and Methods

2.1. Materials

Silica commercial (AC), tetraethyl orthosilicate (TEOS), trimethylsilyl chloride (TMSCl), ammonium hydroxide (NH_4OH), chloroform, and RhB were purchased from Sigma-Aldrich® (St. Louis, MO, USA). Dimethylsulfoxide (DMSO) was purchased from Labsynth® (Diadema, Brazil). Toluene was purchased from Dinâmica® (Indaiatuba, Brazil). All chemicals were used as received without further purification.

2.2. SiO_2 from Rice Husk Synthesis (RH)

Rice husk residues (RHs) were used to prepare the RH SiO_2 sample, as reported in the literature [31,58,59]. First, pristine RHs were washed with deionized water and dried overnight in an oven at 100 °C. Then, in sequence, 10 g of dried RHs were soaked into a 3M HCl solution (200 mL) and maintained under stirring and reflux at 80 °C. After 3 h, the RHs were removed from the leaching solution, washed to neutral pH with distilled water, and dried in an oven at 80 °C. Finally, the material was annealed in a muffle furnace at 600 °C for 3 h. The characterization results and RhB adsorption performances of the prepared silica nanoparticles were compared and contrasted to commercially available nanostructured Sigma-Aldrich silica (AC) and synthetic silica obtained by the sol–gel method (SG).

2.3. SiO_2 by Sol–Gel Synthesis (SG)

SG SiO_2 nanoparticles were synthesized by the sol–gel method according to the procedure reported by Stöber [60]. Briefly, 8 mL of NH_4OH and 7.5 mL of tetraethyl orthosilicate were added to an aqueous ethanol solution (1:1.5 V V^{-1}) in a round bottom flask. The mixture was maintained under vigorous stirring for 1 h at room temperature. After that, solvents were evaporated to attain a solid residue. Next, this material was thermally treated for 3 h at 350 °C and subsequently at 400 °C for 2 h.

2.4. SiO_2 Surface Modification with Trimethylsilyl Chloride (TMSCl)

The TMSCl was used to functionalize SiO_2 surface nanoparticles in a procedure adapted from Kulkarni et al. [33,61]. At room temperature, 1.0 g of dry SiO_2 was dispersed in 15 mL of toluene by constant stirring for 1 h. Then, 2 mL of TMSCl was added, and the mixture was kept under reflux for 48 h. The obtained material was washed with toluene and dried in an oven at 80 °C. Silanized samples were labeled XSil, where X is AC, SG, or RH.

2.5. PLA:SiO_2 Silanized Fibers

Silanized SiO_2 nanoparticles (Sil) were immobilized on PLA fibers by electrospinning. According to the methodology described above, PLA was initially solubilized in a mixture of chloroform and dimethylsulfoxide in a 3:1 (V V^{-1}) ratio. After complete PLA solubilization, modified SiO_2 was added, and the solution was continuously stirred at room temperature for 2 h. After homogenizing, the mixture was transferred to a glass syringe to electrospin the PLA:SiO_2 fibers. A 40 wt% silica load (concerning PLA mass) was chosen for the composite fabrication. Electrospinning parameters were set as follows: applied voltage at 20 kV, needle tip-to-collector distance equal to 5 cm, and flow rate at 3 mL h^{-1}.

2.6. Characterizations

Silica particles and PLA:SiO_2 Sil composite fibers were fully characterized (i.e., by structure, composition, morphology). Specifically, the crystallographic phase composition of SiO_2 particles was investigated by X-ray diffraction (XRD). A Lab-XRD-6000 Shimadzu® (Kyoto, Japan) diffractometer equipped with a Cu-Kα radiation source (λ = 1.5406 Å), 2θ angular range from 5° to 80°, and a scanning step of 1° min^{-1} was used. Chemical and structural features were obtained by ^{29}Si solid-state nuclear magnetic resonance (NMR) spectroscopy performed with a Bruker (Billerica, MA, USA) Avance III HD 400 MHz spectrometer operating at 10 kHz. Thermogravimetric analysis (TGA) was exploited to verify the complete removal of organic residues in the RH SiO_2 sample and quantify the composite fibers' silica amount. TGA was carried out using a Netzsch (Selb, Germany) model STA 409® equipment in an alumina pan, from room temperature to 900 °C under nitrogen flux and with a 10 °C min^{-1} scan rate. The morphology of SiO_2 particles and fibers was investigated by scanning electron microscopy (SEM) using a JEOL (Tokyo, Japan) JSM-6510 field emission electron microscope equipped with an EDS (electron dispersion spectroscopy) detector for compositional analysis. Finally, particle size was measured by transmission electron microscopy (TEM). Therefore, the samples were dispersed in acetone through ultrasonication and analyzed with a FEI® (Hillsboro, OR, USA) TECNAI G2 F20 microscope. The specific surface area (SAA), micropore area (A_{Pore}), external area (A_{Ext}), and pore diameter of the powder samples were determined by nitrogen adsorption/desorption through the Brunauer–Emmett–Teller (BET) method. Zeta potential and contact angle measurements were performed on pristine and TMSCl-functionalized SiO_2 particles (SG and RH) using a Malvern Instruments® (Malvern, UK), Zetasizer Nano ZS90 and an Optical contact angle and surface tension meter model CAM100 by KSV, respectively.

2.7. Rhodamine B Adsorption Assays

2.7.1. SiO_2 Nanoparticles

SiO_2-free unmodified and TMSCl-modified particles were evaluated to verify the efficiency of the different samples: SiO_2 extracted from rice husk, sol–gel synthesis, and commercial SiO_2. Thus, 40 mg of each adsorbent was added to an aqueous solution (40 mL) containing 5 mg L^{-1} of RhB dye and continuously stirred for 24 h at a neutral pH. At the end of the tests, powder adsorbents were removed by centrifugation. Finally, the residual RhB concentration was evaluated by UV-Vis absorption spectroscopy (with a Shimadzu model UV-1601PC).

2.7.2. PLA:SiO_2 Sil Membranes

The RhB dye adsorption was initially carried out with different adsorbent concentrations (1, 2, and 4 g L^{-1}) to determine the ideal amount to achieve high adsorption efficiencies. First, the test was performed in RhB solution (5 mg L^{-1}, 20 mL) under continuous stirring for 24 h at 25 °C. Then, the adsorbents were removed manually, and the RhB residual solution concentration was determined by UV-Vis spectroscopy.

Subsequently, the time required to reach equilibrium to remove the maximum percentage of the dye was evaluated. A kinetic study was performed for 2 and 4 g L^{-1} of PLA:SiO_2 SGSil and PLA:SiO_2 RHSil, respectively. The dye removal was monitored at different time points (30, 60, 120, 180, 240, and 1440 min). The kinetic study for the RhB dye adsorption with the membranes was performed by applying the pseudo-first-order (Equation (1)) and pseudo-second-order (Equation (2)) models according to the following equations:

Pseudo-first order:

$$\log(q_e - q_t) = \log q_e - \frac{k_1}{2.303} t \tag{1}$$

Pseudo-second order:

$$\frac{t}{q_t} = \frac{1}{k_2 q_e^2} + \frac{1}{q_e} t \tag{2}$$

where q_e (mg g^{-1}) is the RhB amount adsorbed at equilibrium, q_t (mg g^{-1}) is the amount adsorbed at time t (min), and k_1 (min^{-1}) and k_2 (g mg^{-1} min^{-1}) are the rate constants for the pseudo-first-order and pseudo-second-order reactions, respectively. According to the literature [62–64], these kinetic models are the ones most commonly applied for RhB adsorption systems, presenting the best fit for the experimental results.

A reuse assay was performed to verify the stability of the adsorbent membranes for RhB dye removal. The evaluation up to three reuse cycles was realized using the best response conditions for the adsorption process. In this stage, the reused fibers were subjected to the RhB adsorption/desorption process. After the recovery, the membranes were washed using ethanol and dried to undergo a new cycle. All the experiments were replicated thrice under agitation at 25 °C for 180 min.

3. Results and Discussion

The XRD patterns concerning SG and RH are reported in Figure 1. The commercial silica (AC) diffractogram is also inserted for comparison in Figure 1. All samples presented a broad diffraction peak at 2θ = 22.5°, a distinctive feature of amorphous (i.e., poorly crystalline) silica [25]. There is significant evidence that the RH sample showed no crystalline reflections of impurities from the original raw material. The AC nanoparticles were found to be made out of pure, mainly amorphous, silica. Furthermore, TMSCl functionalization did not affect the materials' crystal structure or composition (see Figure 1).

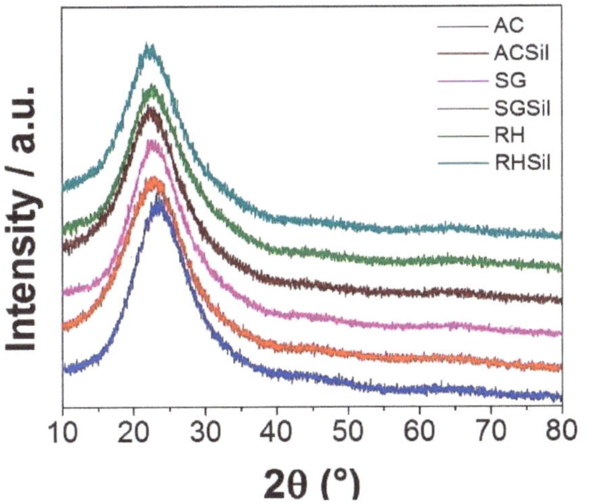

Figure 1. XRD diffractograms for SiO_2 with unmodified surface and modified surface with TMSCl silane.

Thermal analysis was performed on pristine and acid-treated rice husks (Figure 2) to confirm the RH purity. Moreover, TGA curves were also acquired for acid-treated rice husks before and after calcination at 600 °C.

Regardless of the specific treatment, the sample without heat treatment presented a mass loss of 2% at about 70 °C, corresponding to the release of physisorbed water. Additionally, a considerable mass loss of 6%, beginning at about 250 °C (inflection point at 330 °C), was attributed to the rapid cracking of cellulose, hemicellulose, and lignin [65]. An additional loss of 82% after 490 °C was attributed to the complete removal of cellulose, hemicellulose, and lignin degradation products [66]. As expected, the acid-treated rice husk's total mass loss was more significant than the pristine material. Indeed, acid leaching could remove rice husk's inorganic components (mainly potassium and calcium). In addition, the thermal treatment causes the crystallization of potassium and calcium oxides, acting as trapping centers for carbon residues [67]. Thus, HCl leaching can be considered instrumental in attaining high-purity SiO_2. Moreover, the acid-treated silica calcined at 600 °C presented a negligible mass loss indicating the complete elimination of any organic impurities, as was suggested by the visual inspection of the sample (white color). This fact was supported by the XRD results (Figure 1).

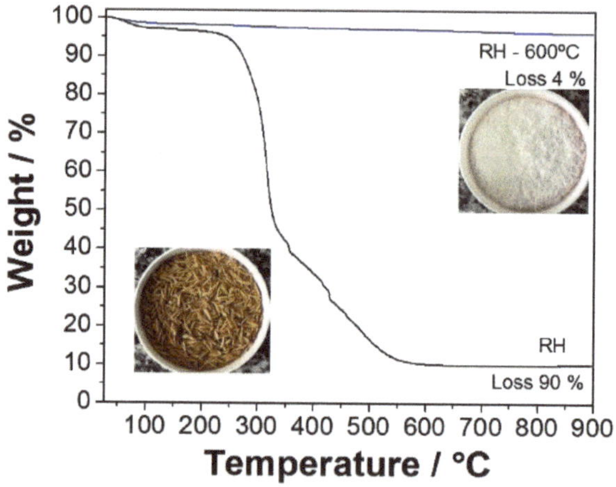

Figure 2. Thermogravimetric analysis of SiO_2 before and after rice husk (RH) treatment at 600 °C.

TEM images of commercial and synthetic SiO_2 are illustrated in Figure 3. The former comprises well-separated, quasi-spherical nanoparticles with an average diameter of around 15–16 nm. On the other hand, small particle sizes (5–10 nm) compared to AC were observed concerning SG and RH. As for SG, the formation of tiny particles was most likely favored by the large H_2O/NH_3 ratio used during the synthesis. In addition, water triggers TEOS hydrolysis, generating plenty of nuclei in the initial reaction stages. Conversely, the limited amount of ammonia keeps the nuclei from growing. Concerning RH, particle size is mainly determined by the annealing temperature. Usually, 600 °C is considered an efficient trade-off between high reactivity assured by small particle (i.e., large specific surface area) size and the complete removal of organic matter [30,31,58,66].

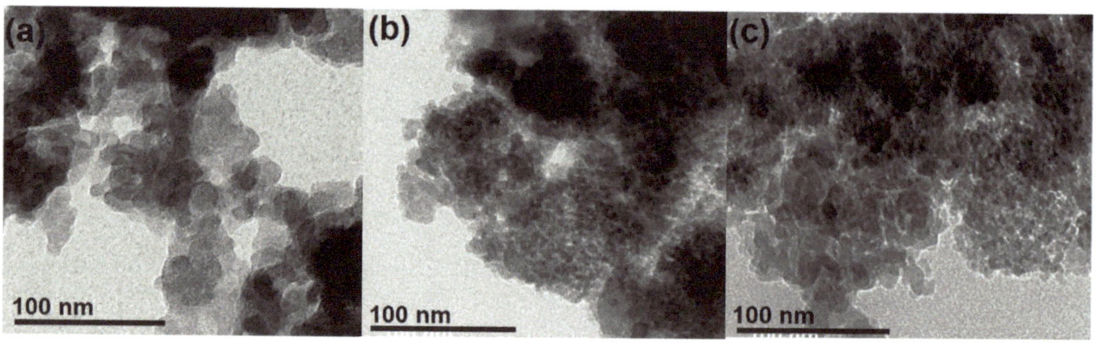

Figure 3. SEM images of SiO_2 nanoparticles: (**a**) AC, (**b**) SG, and (**c**) RH.

Specific surface area (SSA) data and pore size salient parameters for the silica samples (before and after TMSCl treatment) are reported in Table 1. In particular, the average particle size (estimated by TEM), micropore area (A_{Pore}), external area (A_{Ext}), total surface area ($A_{Pore} + A_{Ext} = SSA_{BET}$), and pore diameter are reported.

Table 1. N_2 adsorption/desorption results for the SiO_2 samples before and after silanization.

	Particle Diameter/nm	SSA_{BET} /m^2 g^{-1}	A_{Ext} /m^2 g^{-1}	A_{Pore} /m^2 g^{-1}	D_{pore} /nm
SiO_2					
AC	16	526.1373	249.6770	276.4603	3.26773
ACSil	16	278.2454	175.4967	102.7487	4.18450
SG	5	536.6674	527.2922	9.3751	4.22250
SGSil	5	384.5106	-	-	4.01733
RH	14	245.8613	225.6661	20.1952	5.12742
RHSil	14	202.5582	-	-	4.83049

SSA values partially reflected the particle size and the agglomeration degree of the samples. Therefore, RH had the lowest SSA_{BET} (246 m^2 g^{-1}) because of particle clustering brought about by the thermal treatment required to burn off the organics. On the other hand, SG SSA_{BET} (537 m^2 g^{-1}) benefited from the small particle size deriving from the excess water utilized in the material preparation. Despite larger particle sizes, AC SSA_{BET} (527 m^2 g^{-1}) proved to be similar to SG due to the low level of particle clustering. In addition, AC SSA_{BET} was almost equally shared between external and pore surface area.

On the contrary, the external surface area can be regarded as the only contribution to SSA_{BET} for RH and SG. The lack of porous structure for the two prepared samples is expected from the nature of the synthetic procedure. Indeed, the RH and SG preparation routes employed no templating agents (e.g., surfactants) to promote pore formation. Thus, their limited amount of mesopores probably resulted only from interparticle porosity. Interestingly, silanization affected each sample differently in terms of surface area. Although introducing organic groups from TMSCl decreased the SSA_{BET} of all samples, the synthetic nanoparticles retained most of the SSA. Hence, commercial silica (CA) SSA_{BET} dropped dramatically (from 526 to 278 m^2 g^{-1}). In contrast, surface area reduction was reasonably contained for RH and SG (17.6% and 28.4%) compared to the initial SSA_{BET} value. It is worth noting that AC wider pore size and significant surface accessibility might have allowed a higher amount of silyl groups to be anchored on the commercial silica, hindering nitrogen physisorption.

Besides influencing the SSA, silanization also altered the silica surface hydrophilic character by replacing Si-OH groups on the SiO_2 with Si-O-Si-$(CH_3)_3$ moieties. Since Cl^- is a good leaving group for nucleophilic substitution reactions, silica surface hydrophobization using TMSCl was strongly favored [33]. Figure 4 shows the contact angle measurements performed on pristine and TMSCl-functionalized silica powders. The surface properties of each silica sample changed dramatically, switching from a very hydrophilic to a remarkably hydrophobic behavior. Such a feature is supposed to improve silica affinity for RhB, enhancing the adsorption rate and promoting the incorporation of silica nanoparticles into the PLA fibers.

Since XRD could not detect any structural changes brought about by silanization, ^{29}Si NMR spectra were collected on the SiO_2 surface-modified nanoparticles. In both nanoparticles, we detected the silica surface species referred to as Q^2–Q^4 according to the number of siloxane bonds (Figure 5). Specifically, the non-functionalized AC (i.e., pure silica) showed three signals corresponding to chemical shift values of −110, −101, and −92 ppm and representing the Q^2, Q^3, and Q^4 units, respectively. On the other hand, functionalized silica presented a 13 ppm shift associated with the Si-O-Si-$(CH_3)_3$ unit, aside from the pure silica signals. In addition, an intensity decrease of the signals corresponding to Q^2 and Q^3 was observed in the silanized silica.

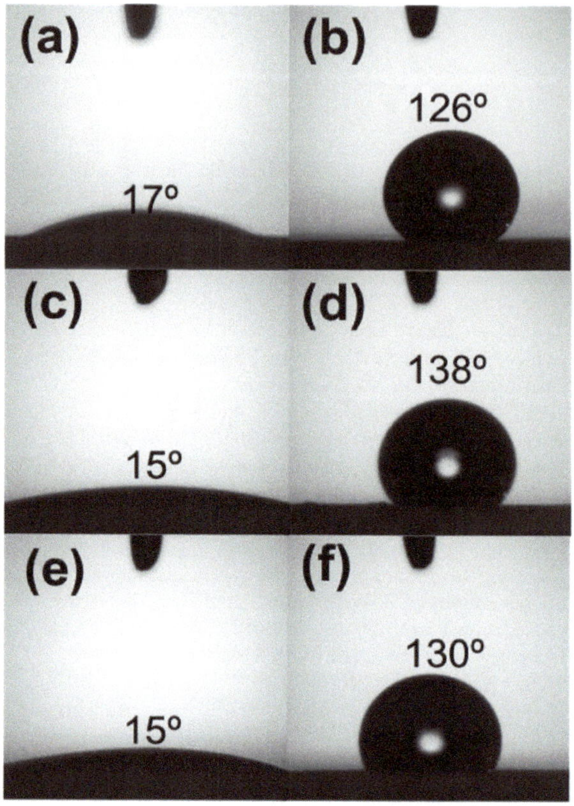

Figure 4. The contact angle for unmodified (left) and modified (right) SiO_2 surface: (**a**,**b**) AC, (**c**,**d**) SG, and (**e**,**f**) RH.

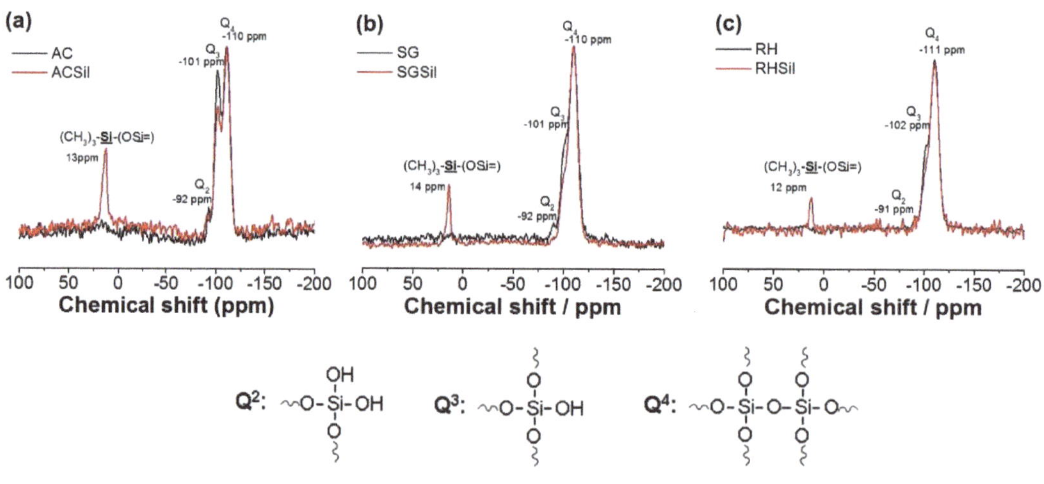

Figure 5. H[1] RMN for nanoparticles with unmodified and modified SiO_2 surfaces: (**a**) AC, (**b**) SG, and (**c**) RH.

After being characterized, all silica nanoparticles were tested for RhB adsorption. Figure 6 illustrates the UV-Vis absorption spectra of RhB solutions (5 ppm) depending on different adsorbents used for the dye removal. In addition, optical absorbance (i.e., color) attenuation was calculated from the 550 nm absorption peak of the initial RhB solution (blank).

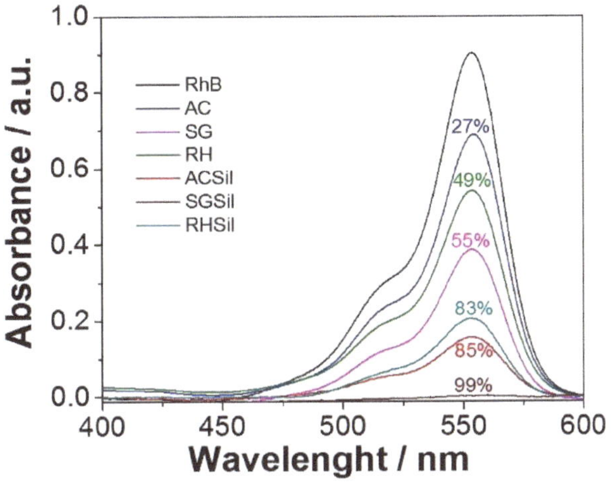

Figure 6. RhB dye adsorption by nanoparticles with unmodified and modified SiO_2 surfaces after 24 h.

The TMSCl-functionalized silica performed better than their unmodified counterparts in terms of RhB removal, confirming the crucial role played by surface hydrophobization. In particular, SGSil proved to be the most effective adsorbent, achieving a 99% color removal rate. The RHSil reached a dye removal rate (83%) similar to the CS one (85%). Not surprisingly, adsorption results for the silanized samples were perfectly in line with the SSA data (SGSil > ACSil > RHSil). Despite mesopores being large enough to accommodate the RhB molecule (length 1.8 nm, width 0.7 nm [68]), most adsorption took place on the external surface of RHS, with the dye removal percentage similar to the CS value. In addition, steric hindrance due to rhodamine aromatic groups reasonably hampered dye access to CS mesopores lowering the adsorption rate. Apart from the surface area, other phenomena such as electrostatic, hydrogen bond, and hydrophobic interactions must be considered for RhB adsorption on silica [69,70]. Regarding electrostatic interactions, dye removal can exploit the opposite charges on RhB molecules and SiO_2 surfaces. The former is a cationic dye, while the latter possesses a negatively charged surface as verified by zeta potential measurements (−35, −32, and −31 mV, for ACSil, SGSil, and RHSil, respectively). Consequently, even the non-functionalized samples demonstrated the ability to adsorb RhB. Nonetheless, electrostatic interactions are insufficient to ensure a high level of adsorption. Therefore, effects brought about by surface silanization, like van der Waals interactions mediated by the TMSCl methyl group, should be considered [69]. Finally, it has already been reported how the existence of hydrogen bonds between the hydroxyl groups on the silica surface and the carboxyl groups on RhB [70,71] may further enhance the adsorbent–adsorbate interaction.

RH and SG SiO_2-modified nanoparticles showed significant efficiency for RhB dye removal. Thus, both SiO_2 samples were deposited onto a PLA membrane using the electrospinning method. Figure 7 exhibits the SEM images for the pure PLA fibers and the PLA:SiO_2 nanocomposites. Initially, it can be seen that the SiO_2 insertion preserved the fiber shape with a rough appearance and diameter between 0.4 and 1.2 μm. However, regardless of the size distribution of the fiber diameters, it was verified from the histograms that the addition of SiO_2 nanoparticles promoted a decrease in the size and greater homogeneity of the fibers compared with the pure PLA (0.6–2 μm). This result may have been caused due to the intercalation of the nanoparticles between the PLA polymer chains, causing a decrease in polymer entanglement. Consequently, the electrospinning process favored the stretching of the polymer, resulting in more homogeneous fibers with smaller diameters [72]. Thus, the PLA:SiO_2 nanocomposite membranes in both cases allowed a gain of morphological properties of the fibers, indicating a good interaction between particle and polymer.

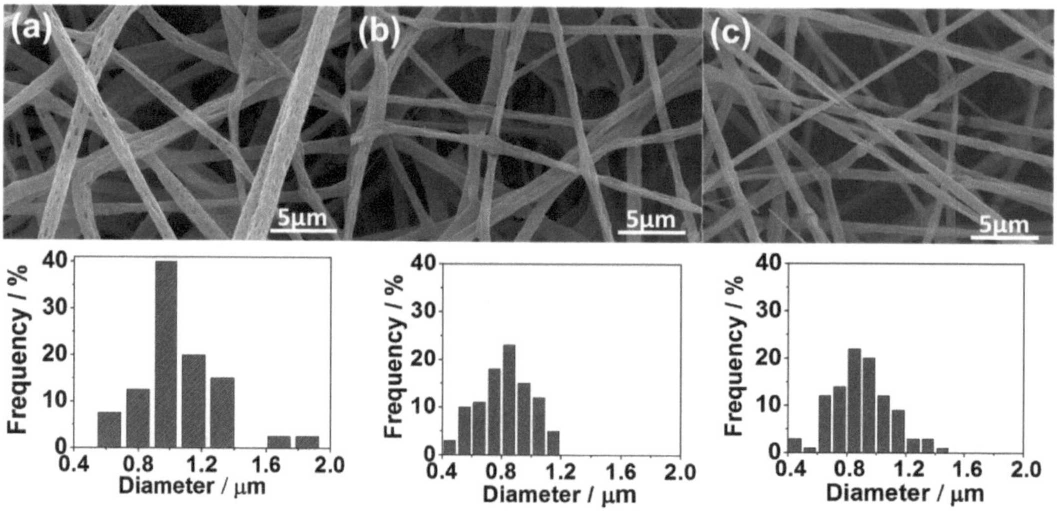

Figure 7. SEM images and fiber diameter histograms of PLA:SiO_2 membranes: (**a**) pure PLA, (**b**) SGSil, and (**c**) RHSil.

The presence of SiO_2 nanoparticles and their distribution in the PLA fibers were verified in the composites by energy dispersive X-ray spectroscopy (EDS), as shown in Figure 8. In the fibers of PLA:SiO_2 SGSil and PLA:SiO_2 RHSil, the detection of the element silicon (Si) in superposition with oxygen (O) demonstrated the presence of SiO_2 in the PLA fibers. In addition, the mappings obtained suggest a distribution in the membrane situated in separate regions, which indicates a certain degree of accumulation of the particulates. Such a result may be due to the high SiO_2 Sil concentration, 40% (w w^{-1}), favoring the interaction of particles with each other in the polymer solution. In this sense, both fibers showed a behavior, regardless of the SiO_2 agglomeration, of being more elucidated for the SiO_2 extracted from rice husks (Figure 8b).

After confirming the presence of the particulate materials in the PLA fibers, thermogravimetric analyses were performed (Figure 8) to quantify the silica mass present in the composites. The determination of particulates in the PLA:SiO_2 (SGSil and RHSil) fibers was obtained by comparing the PLA and the silica pure nanoparticles' loss events. The TGA curves show that the non-functionalized samples (SG and RH) had a slight loss in mass due to the adsorbed water loss. On the other hand, in the functionalized samples (SGSil and RHSil), a greater mass loss was observed (9% and 8%, respectively). These values were associated with the additional contribution related to the loss of organic material from surface modification (TMSCl). Concerning pure PLA, a considerable mass loss was observed at 350 °C due to the PLA thermal decomposition [51,55]. The remaining PLA mass probably corresponds to ash formation with increasing temperature. The composite materials of PLA:SiO_2 SGSil and PLA:SiO_2 RHSil showed a remaining group of about 27%. This percentage refers to the silica mass present in the fibers. Considering the polymeric solution added 40% (w w^{-1}) silica and the functionalized material nanoparticles lost about

9% (Figure 9), the silica mass is expected to be 31%. Thus, the value found in both TGA curves is within the desired values, demonstrating the complete incorporation of the 40% SiO_2 into the PLA fibers.

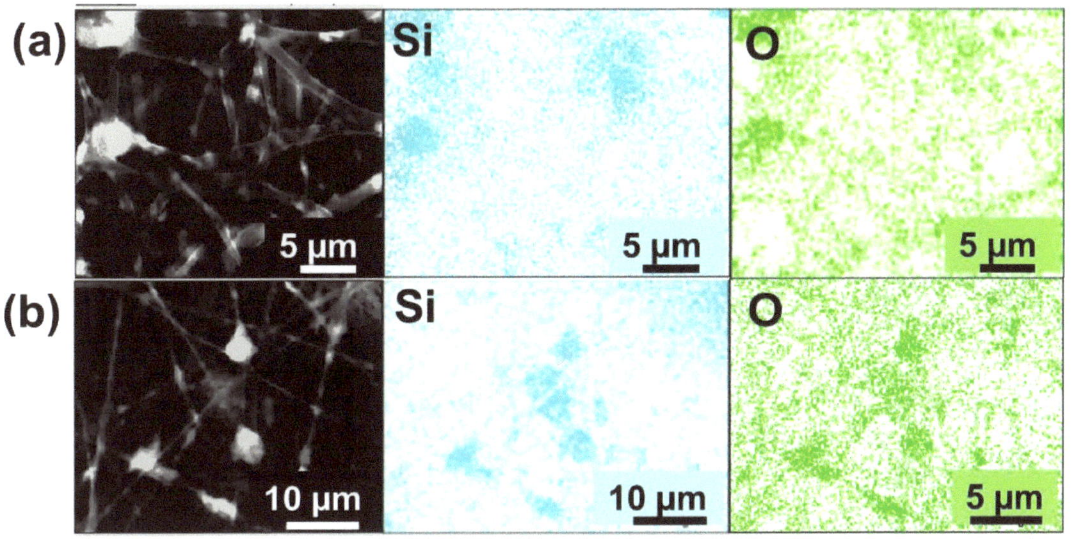

Figure 8. SEM-EDX images of PLA:SiO_2 membranes: (**a**) SGSil and (**b**) RHSil.

After characterization, PLA:SiO_2 SGSil and PLA:SiO_2 RHSil composite membranes were evaluated for RhB adsorption (5 mg L^{-1}). Preliminary experiments were conducted to establish the maximum capacity adsorption, as shown in Figure 10. Initially, PLA:SiO_2 RHSil fibers were shown to be effective in removing RhB over the entire range of adsorbent concentrations (1–4 g L^{-1}). A maximum adsorption capacity of 93% was achieved for 2 g L^{-1}. Conversely, the increase in RhB adsorption of PLA:SiO_2 RHSil was more sensitive to the adsorbent concentration. The highest capacity adsorption (83%) was reached with the maximum concentration evaluated (4 g L^{-1}). The difference in RhB dye adsorption by the membranes can be correlated to the characteristics of the SiO_2-modified nanoparticles. The result observed in Figure 10 corroborates with the free adsorbent particles in Figure 6. In that case, SGSil (99%) showed superior adsorption compared to RHSil (83%). This behavior is attributed to the RHSil's lower surface area (202 m^2 g^{-1}) after surface modification (Table 1). Thus, as expected, PLA:SiO_2 RHSil showed less effectiveness compared to PLA:SiO_2 SGSil in similar conditions. Moreover, silica immobilized in polymer fibers can reduce the number of SiO_2 active sites available for dye adsorption. Additionally, the distinct interaction of nanoparticles with the polymeric matrix influences the distribution of particles on the membrane surface, as demonstrated in Figure 8. Consequently, the surface area decreases, affecting RHSil adsorption performance in the filter membrane.

RhB adsorption kinetic curves are reported in Figure 11. In the observed results, the dye removal percentage increased steadily to 93% by 180 min for PLA:SiO_2 SGSil. On the other hand, the maximum capacity was slowly reached for PLA:SiO_2 RHSil up to 240 min. The longer time required for the PLA:SiO_2 RHSil membranes to achieve equilibrium may be related to its smaller surface area of nanoparticles and agglomeration effect in the fibers, resulting in fewer sites for adsorption. Thus, RhB dye removal for PLA:SiO_2 RHSil likely needs more diffusion through the fiber membranes resulting in a more extended time exposition for adsorption.

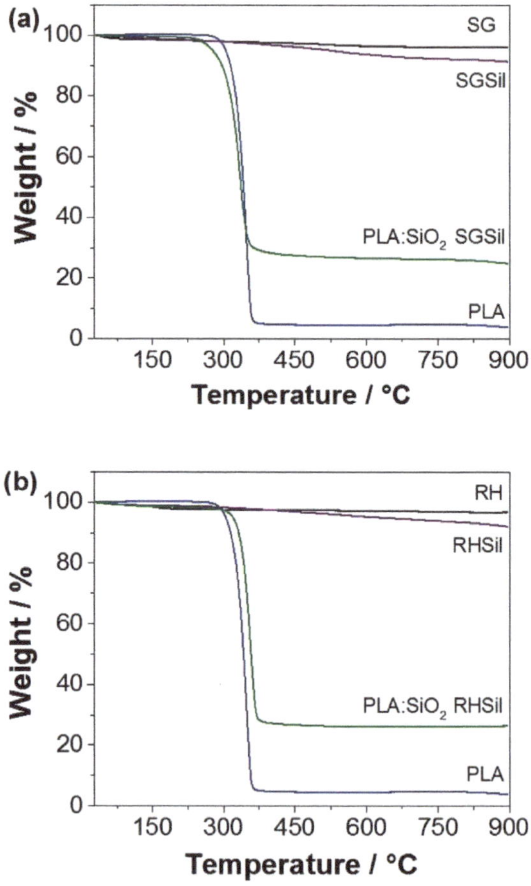

Figure 9. Thermogravimetric analysis of PLA:SiO_2 membranes: (**a**) SGSil and (**b**) RHSil.

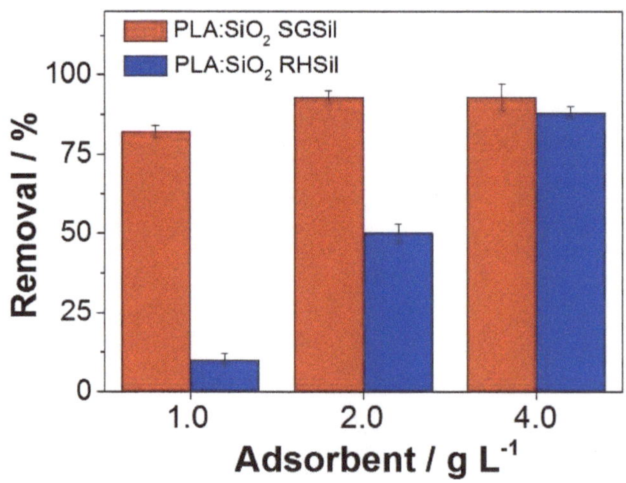

Figure 10. PLA:SiO_2 membrane adsorption removal against RhB dye.

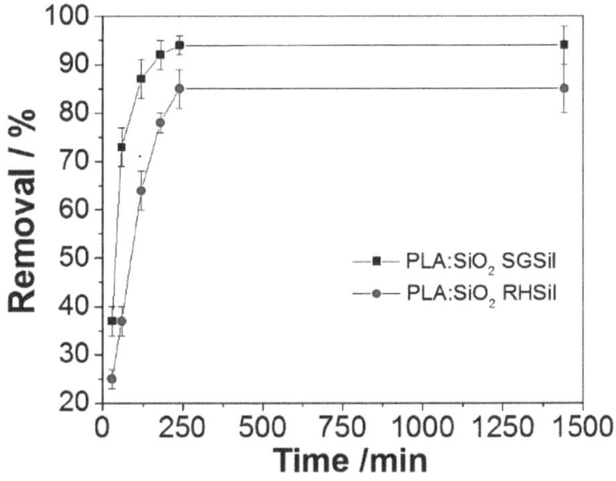

Figure 11. Kinetic study for RhB adsorption using PLA:SiO_2 nanocomposites.

The kinetic parameters were determined from linear regression (Figure 12) and are presented in Table 2. According to the correlation coefficient (R^2) analyses, the pseudo-first-order model best fits the experimental data, showing R^2 values of 0.9865 and 0.9842 for PLA:SiO_2 SGSil and PLA:SiO_2 RHSil, respectively. Based on this model, the calculated q_e values (2.7158 and 1.2956 mg g^{-1}) were close to the experimental q_e values (2.3012 and 1.0648 mg g^{-1}) established for dye adsorption using PLA:SiO_2 SGSil and PLA:SiO_2 RHSil, respectively. This result resembles those in the literature [73,74], which employed SiO_2-supported nanocomposites as adsorbents for RhB removal. In addition, PLA:SiO_2 SGSil fibers showed a speed constant (k_1) that was almost twice (0.02602 min^{-1}) as high as that of PLA:SiO_2 RHSil (0.01404 min^{-1}), confirming the more difficult diffusion access to adsorption sites.

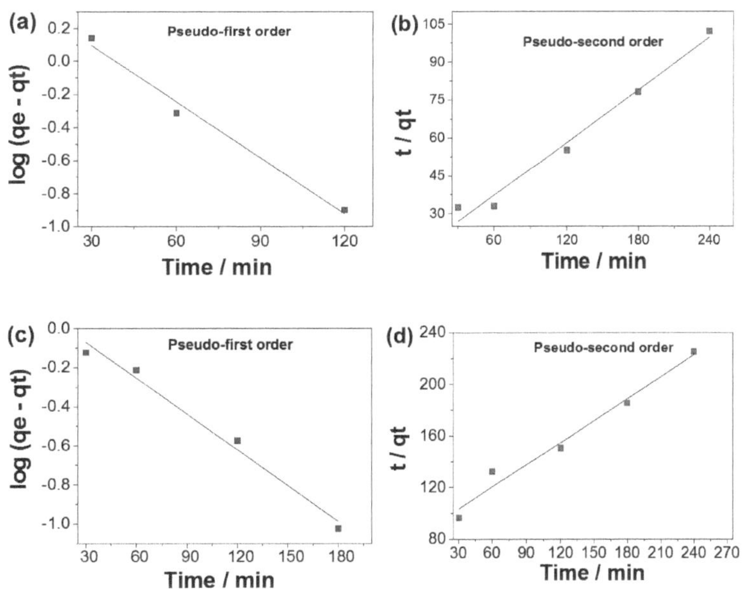

Figure 12. Kinetic assays of RhB dye adsorption (**a**,**b**) PLA:SiO_2 SGSil and (**c**,**d**) PLA:SiO_2 RHSil.

Table 2. Kinetic parameters of RhB dye adsorption from pseudo-first-order and pseudo-second-order models.

Adsorbent	Experimental	Pseudo-First Order			Pseudo-Second Order		
$PLA:SiO_2$	q_e mg g^{-1}	q_e mg g^{-1}	k_1 min^{-1}	R^2	q_e mg g^{-1}	k_2 g mg^{-1} min^{-1}	R^2
SGSil	2.3012	2.7158	0.02602	0.9865	2.8893	7.20×10^{-3}	0.9817
RHSil	1.0648	1.2956	0.01404	0.9842	1.758	3.74×10^{-3}	0.9775

RhB adsorption/desorption cycles were executed on $PLA:SiO_2$ SGSil and $PLA:SiO_2$ RHSil to evaluate nanocomposite regeneration/reuse capabilities. Figure 13 shows the RhB dye adsorption during three reuse cycles. PLA pure fibers as an adsorbent achieved a less than 20% removal rate in the 1° cycle. On the other hand, both filter membranes showed adsorptions greater than 80% (1° cycle). Hence, RhB removal was nearly entirely due to the SiO_2 nanoparticles supported on PLA fibers. In the reuse cycles, all materials demonstrated a decreased capacity for adsorption. In this way, both nanocomposites ($PLA:SiO_2$ SGSil and $PLA:SiO_2$ RHSil) maintained affinity interactions with RhB dye after desorption, promoting adsorption saturation. As a result, $PLA:SiO_2$ SGSil and $PLA:SiO_2$ RHSil had decreased adsorption capacities from 93% to 86% and 83% to 68%, respectively, after three reuse cycles.

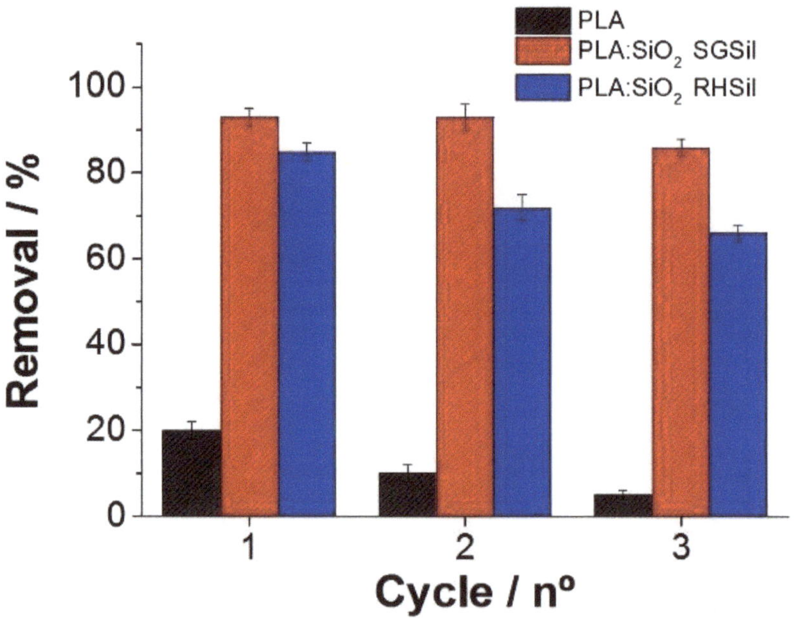

Figure 13. Adsorption cycles for $PLA:SiO_2$ nanocomposite fibers.

The current results are consistent with similar works involving dye adsorption on silica-polymer composite fibers. Wu et al. [75] investigated RhB dye adsorption on SiO_2 nanoparticles supported on polystyrene electrospun fibers. In this study, a 33% maximum dye adsorption was reported. Xu et al. [47] evaluated $PAA:SiO_2$ composite nanofiber membranes in the removal of malachite green dye. As a result, the membrane reached 98% dye removal after 240 min in the 1° cycle, maintaining about 70% dye removal after three adsorption/desorption cycles. Therefore, the system absorbent based on $PLA:SiO_2$ with TMSCl surface modification showed significant results in adsorption removal. Further, the present adsorptive membrane exhibited comparable results to the "low-cost/low-end" rice husk-based nanocomposite for wastewater treatment.

4. Conclusions

Mesoporous SiO_2 nanoparticles with diameters less than 20 nm, with high purity, and a surface area of 245 $m^2\ g^{-1}$ were obtained by extraction from rice husks (RH). The surface modification of SiO_2 nanoparticles with TMCS contributed to an increase in RhB dye adsorption from 49% to 83%. This result was comparable with that found for commercial and sol–gel synthesized silica. Furthermore, the modified SiO_2 nanoparticles [40% (w w^{-1})] were impregnated efficiently into PLA fibers using the electrospinning method. The PLA:SiO_2 nanocomposite membranes from rice husk showed more than 80% dye removal, maintaining 60% activity after three cycles. In this context, this paper offers a system with high RhB affinity, proving that the silica nanoparticles supported on PLA nanofibers promote easy removal and reuse cycles without significantly impairing adsorption. Silica (e.g., quartz) surface functionalization is already exploited on an industrial scale in direct and reverse minerals flotation. Furthermore, TMSCl is used mainly in organic synthesis as a trimethylsilyl group or anhydrous chloride source. The production fibers have been quickly improved to minimize economic costs for large-scale applications by solution blowing spinning from electrospinning knowledge. Finally, the development of adsorptive membranes from agri-wastes such as rice husks can minimize costs and enable their use in real applications involving the adsorption of organic pollutants in water to favor environmental remediation.

Author Contributions: Conceptualization, F.A.T. and E.C.P.; methodology, F.A.T. and B.C.M.; formal analysis, F.A.T.; investigation, F.A.T.; resources, E.C.P.; data curation, J.O.D.M.; writing—original draft preparation, F.A.T.; writing—review and editing, J.O.D.M., S.Q., T.R.N. and E.C.P.; visualization, S.Q.; supervision, E.C.P.; project administration, E.C.P.; funding acquisition, E.C.P. All authors have read and agreed to the published version of the manuscript.

Funding: This study was financed in part by the Coordenação de Aperfeiçoamento de Pessoal de Nível Superior-Brasil (CAPES) (Finance Code 001) and FAPESP (Grant Number 2021/14992-1). In addition, the authors acknowledge Embrapa (Grant Number 11.14.03.001.01.00), FINEP, SisNano, and AgroNano Network for financial support.

Institutional Review Board Statement: Not applicable.

Informed Consent Statement: Not applicable.

Data Availability Statement: Not applicable.

Conflicts of Interest: The authors declare no conflict of interest.

References

1. Rincón Joya, M.; Barba Ortega, J.; Malafatti, J.O.D.; Paris, E.C. Evaluation of Photocatalytic Activity in Water Pollutants and Cytotoxic Response of α-Fe_2O_3 Nanoparticles. *ACS Omega* **2019**, *4*, 17477–17486. [CrossRef]
2. Malafatti, J.O.D.; Moreira, A.J.; Sciena, C.R.; Silva, T.E.M.; Freschic, G.P.G.; Pereira, E.C.; Paris, E.C. Prozac®Removal Promoted by HAP:Nb_2O_5 Nanoparticles System: By Products, Mechanism, and Cytotoxicity Assessment. *J. Environ. Chem. Eng.* **2020**, *9*, 10482. [CrossRef]
3. Ali, I.; Asim, M.; Khan, T.A. Low Cost Adsorbents for the Removal of Organic Pollutants from Wastewater. *J. Environ. Manag.* **2012**, *113*, 170–183. [CrossRef]
4. Yagub, M.T.; Sen, T.K.; Afroze, S.; Ang, H.M. Dye and Its Removal from Aqueous Solution by Adsorption: A Review. *Adv. Colloid Interface Sci.* **2014**, *209*, 172–184. [CrossRef]
5. Mansa, R.F.; Sipaut, C.S.; Rahman, I.A.; Yusof, N.S.M.; Jafarzadeh, M. Preparation of Glycine–Modified Silica Nanoparticles for the Adsorption of Malachite Green Dye. *J. Porous Mater.* **2016**, *23*, 35–46. [CrossRef]
6. Srhir, B.; Rahali, A.; Elkhattabi, O.; Belhamidi, S.; Chhiti, Y.; Chlihi, K. Dielectric Study of the Adsorption of Cetylpyridinium and Phenol onto Activated Carbon. *Microporous Mesoporous Mater.* **2022**, *340*, 112036. [CrossRef]
7. Laksaci, H.; Belhamdi, B.; Khelifi, O.; Khelifi, A.; Trari, M. Elimination of Amoxicillin by Adsorption on Coffee Waste Based Activated Carbon. *J. Mol. Struct.* **2022**, *1274*, 134500. [CrossRef]

8. Boulika, H.; El Hajam, M.; Hajji Nabih, M.; Riffi Karim, I.; Idrissi Kandri, N.; Zerouale, A. Definitive Screening Design Applied to Cationic & Anionic Adsorption Dyes on Almond Shells Activated Carbon: Isotherm, Kinetic and Thermodynamic Studies. *Mater. Today Proc.* **2022**, *72*, 3336–3346. [CrossRef]
9. Paris, E.C.; Espinosa, J.W.M.; de Lazaro, S.; Lima, R.C.; Joya, M.R.; Pizani, P.S.; Leite, E.R.; Souza, A.G.; Varela, J.A.; Longo, E. Er3+ as Marker for Order-Disorder Determination in the $PbTiO_3$ System. *Chem. Phys.* **2007**, *335*, 7–14. [CrossRef]
10. Alves, M.C.F.; Souza, S.C.; Silva, M.R.S.; Paris, E.C.; Lima, S.J.G.; Gomes, R.M.; Longo, E.; De Souza, A.G.; Garcia Dos Santos, I.M. Thermal Analysis Applied in the Crystallization Study of $SrSnO_3$. *J. Therm. Anal. Calorim.* **2009**, *97*, 179–183. [CrossRef]
11. Raba-Páez, A.M.; Malafatti, J.O.D.; Parra-Vargas, C.A.; Paris, E.C.; Rincón-Joya, M. Effect of Tungsten Doping on the Structural, Morphological and Bactericidal Properties of Nanostructured CuO. *PLoS ONE* **2020**, *15*, e0239868. [CrossRef]
12. Malafatti, J.O.D.; Ruellas, T.M.D.O.; Meirelles, M.R.; Thomazi, A.C.; Renda, C.G.; Paris, E.C. Nanocarriers of Eu3+ Doped Silica Nanoparticles Modified by APTES for Luminescent Monitoring of Cloxacillin. *AIMS Mater. Sci.* **2021**, *8*, 760–775. [CrossRef]
13. Dos Santos, E.G.; De Alsina, O.L.S.; Da Silva, F.L.H. Desempenho de Biomassas Na Adsorção de Hidrocarbonetos Leves Em Efluentes Aquosos. *Quim. Nova* **2007**, *30*, 327–331. [CrossRef]
14. Mascarenhas, B.C.; Tavares, F.A.; Paris, E.C. Functionalized Faujasite Zeolite Immobilized on Poly(Lactic Acid) Composite Fibers to Remove Dyes from Aqueous Media. *J. Appl. Polym. Sci.* **2020**, *137*, 48561. [CrossRef]
15. Paris, E.C.; Malafatti, J.O.D.; Musetti, H.C.; Manzoli, A.; Zenatti, A.; Escote, M.T. Faujasite Zeolite Decorated with Cobalt Ferrite Nanoparticles for Improving Removal and Reuse in Pb^{2+} Ions Adsorption. *Chin. J. Chem. Eng.* **2020**, *28*, 1884–1890. [CrossRef]
16. Atia, A.A.; Donia, A.M.; Al-Amrani, W.A. Adsorption/Desorption Behavior of Acid Orange 10 on Magnetic Silica Modified with Amine Groups. *Chem. Eng. J.* **2009**, *150*, 55–62. [CrossRef]
17. Ebner, A.D.; Ritter, J.A.; Navratil, J.D. Adsorption of Cesium, Strontium, and Cobalt Ions on Magnetite and a Magnetite—Silica Composite. *Ind. Eng. Chem. Res.* **2001**, *40*, 1615–1623. [CrossRef]
18. Dutta, R.; Nagarjuna, T.V.; Mandavgane, S.A.; Ekhe, J.D. Ultrafast Removal of Cationic Dye Using Agrowaste-Derived Mesoporous Adsorbent. *Ind. Eng. Chem. Res.* **2014**, *53*, 18558–18567. [CrossRef]
19. Liu, C.; Zhu, C.; Wang, H.; Xie, S.; Zhou, J.; Fang, H. Synergistic Removal of Organic Pollutants by Co-Doped MIL-53(Al) Composite through the Integrated Adsorption/Photocatalysis. *J. Solid State Chem.* **2022**, *316*, 123582. [CrossRef]
20. Ahmadian, M.; Derakhshankhah, H.; Jaymand, M. Recent Advances in Adsorption of Environmental Pollutants Using Metal–Organic Frameworks-Based Hydrogels. *Int. J. Biol. Macromol.* **2023**, *231*, 123333. [CrossRef]
21. Chaves, M.R.M.; Dockal, E.R.; Souza, R.C.R.; Büchler, P.M. Biogenic Modified Silica as a Sorbent of Cadmium Ions: Preparation and Characterization. *Environ. Technol.* **2009**, *30*, 663–671. [CrossRef] [PubMed]
22. Rahman, I.A.; Padavettan, V. Synthesis of Silica Nanoparticles by Sol-Gel: Size-Dependent Properties, Surface Modification, and Applications in Silica-Polymer Nanocompositesa Review. *J. Nanomater.* **2012**, *2012*, 132424. [CrossRef]
23. Zhang, L.; Zhang, G.; Wang, S.; Peng, J.; Cui, W. Cation-Functionalized Silica Nanoparticle as an Adsorbent to Selectively Adsorb Anionic Dye from Aqueous Solutions. *Environ. Prog. Sustain. Energy* **2016**, *35*, 1070–1077. [CrossRef]
24. Rostamian, R.; Najafi, M.; Rafati, A.A. Synthesis and Characterization of Thiol-Functionalized Silica Nano Hollow Sphere as a Novel Adsorbent for Removal of Poisonous Heavy Metal Ions from Water: Kinetics, Isotherms and Error Analysis. *Chem. Eng. J.* **2011**, *171*, 1004–1011. [CrossRef]
25. Shen, Y.; Zhao, P.; Shao, Q. Porous Silica and Carbon Derived Materials from Rice Husk Pyrolysis Char. *Microporous Mesoporous Mater.* **2014**, *188*, 46–76. [CrossRef]
26. Singh, L.P.; Bhattacharyya, S.K.; Kumar, R.; Mishra, G.; Sharma, U.; Singh, G.; Ahalawat, S. Sol-Gel Processing of Silica Nanoparticles and Their Applications. *Adv. Colloid Interface Sci.* **2014**, *214*, 17–37. [CrossRef]
27. Ponce, J.; da Silva Andrade, J.G.; dos Santos, L.N.; Bulla, M.K.; Barros, B.C.B.; Favaro, S.L.; Hioka, N.; Caetano, W.; Batistela, V.R. Alkali Pretreated Sugarcane Bagasse, Rice Husk and Corn Husk Wastes as Lignocellulosic Biosorbents for Dyes. *Carbohydr. Polym. Technol. Appl.* **2021**, *2*, 100061. [CrossRef]
28. Soltani, N.; Bahrami, A.; Pech-Canul, M.I.; González, L.A. Review on the Physicochemical Treatments of Rice Husk for Production of Advanced Materials. *Chem. Eng. J.* **2015**, *264*, 899–935. [CrossRef]
29. Athinarayanan, J.; Periasamy, V.S.; Alhazmi, M.; Alatiah, K.A.; Alshatwi, A.A. Synthesis of Biogenic Silica Nanoparticles from Rice Husks for Biomedical Applications. *Ceram. Int.* **2015**, *41*, 275–281. [CrossRef]
30. Chumee, J.; Grisdanurak, N.; Neramittagapong, S.; Wittayakun, J. Characterization of AlMCM-41 Synthesized with Rice Husk Silica and Utilization as Supports for Platinum-Iron Catalysts. *Braz. J. Chem. Eng.* **2009**, *26*, 367–373. [CrossRef]
31. Xiong, L.; Sekiya, E.H.; Sujaridworakun, P.; Wada, S.; Saito, K. Burning Temperature Dependence of Rice Husk Ashes in Structure and Property. *J. Met. Mater. Miner.* **2009**, *19*, 95–99.
32. Taha, A.A.; Wu, Y.-N.; Wang, H.; Li, F. Preparation and Application of Functionalized Cellulose Acetate/Silica Composite Nanofibrous Membrane via Electrospinning for Cr(VI) Ion Removal from Aqueous Solution. *J. Environ. Manag.* **2012**, *112*, 10–16. [CrossRef] [PubMed]
33. Kulkarni, S.A.; Ogale, S.B.; Vijayamohanan, K.P. Tuning the Hydrophobic Properties of Silica Particles by Surface Silanization Using Mixed Self-Assembled Monolayers. *J. Colloid Interface Sci.* **2008**, *318*, 372–379. [CrossRef] [PubMed]
34. Wahyuningsih, K.; Yuliani, S. Hoerudin Characteristics of Silica Nanoparticles from Rice Husk as Influenced by Surface Modification with Used Solvent Containing Silane. *J. Eng. Technol. Sci.* **2021**, *53*, 210403. [CrossRef]

35. Roe, B.; Zhang, X. Durable Hydrophobic Textile Fabric Finishing Using Silica Nanoparticles and Mixed Silanes. *Text. Res. J.* **2009**, *79*, 1115–1122. [CrossRef]
36. Velmurugan, P.; Shim, J.; Oh, B.T. Removal of Anionic Dye Using Amine-Functionalized Mesoporous Hollow Shells Prepared from Corn Cob Silica. *Res. Chem. Intermed.* **2016**, *42*, 5937–5950. [CrossRef]
37. Hallaji, H.; Keshtkar, A.R.; Moosavian, M.A. A Novel Electrospun PVA/ZnO Nanofiber Adsorbent for U(VI), Cu(II) and Ni(II) Removal from Aqueous Solution. *J. Taiwan Inst. Chem. Eng.* **2015**, *46*, 109–118. [CrossRef]
38. Paris, E.C.; Malafatti, J.O.D.; Moreira, A.J.; Santos, L.C.; Sciena, C.R.; Zenatti, A.; Escote, M.T.; Mastelaro, V.R.; Joya, M.R. CuO Nanoparticles Decorated on Hydroxyapatite/Ferrite Magnetic Support: Photocatalysis, Cytotoxicity, and Antimicrobial Response. *Environ. Sci. Pollut. Res.* **2022**, *29*, 41505–41519. [CrossRef] [PubMed]
39. Paris, E.C.; Malafatti, J.O.D.; Sciena, C.R.; Junior, L.F.N.; Zenatti, A.; Escote, M.T.; Moreira, A.J.; Freschi, G.P.G. Nb_2O_5 Nanoparticles Decorated with Magnetic Ferrites for Wastewater Photocatalytic Remediation. *Environ. Sci. Pollut. Res.* **2021**, *28*, 23731–23741. [CrossRef]
40. Coutinho, T.C.; Malafatti, J.O.D.; Paris, E.C.; Tardioli, P.W.; Farinas, C.S. Hydroxyapatite-$CoFe_2O_4$ Magnetic Nanoparticle Composites for Industrial Enzyme Immobilization, Use, and Recovery. *ACS Appl. Nano Mater.* **2020**, *3*, 12334–12345. [CrossRef]
41. Lopes, M.M.; Coutinho, T.C.; Malafatti, J.O.D.; Paris, E.C.; de Sousa, C.P.; Farinas, C.S. Immobilization of Phytase on Zeolite Modified with Iron(II) for Use in the Animal Feed and Food Industry Sectors. *Process Biochem.* **2021**, *100*, 260–271. [CrossRef]
42. Manzoli, A.; Shimizu, F.M.; Mercante, L.A.; Paris, E.C.; Oliveira, O.N.; Correa, D.S.; Mattoso, L.H.C. Layer-by-Layer Fabrication of AgCl-PANI Hybrid Nanocomposite Films for Electronic Tongues. *Phys. Chem. Chem. Phys.* **2014**, *16*, 24275–24281. [CrossRef]
43. Fonseca, R.D.; Correa, D.S.; Paris, E.C.; Tribuzi, V.; Dev, A.; Voss, T.; Aoki, P.H.B.; Constantino, C.J.L.; Mendonca, C.R. Fabrication of Zinc Oxide Nanowires/Polymer Composites by Two-Photon Polymerization. *J. Polym. Sci. Part B Polym. Phys.* **2014**, *52*, 333–337. [CrossRef]
44. Abuhatab, S.; El-Qanni, A.; Marei, N.N.; Hmoudah, M.; El-Hamouz, A. Sustainable Competitive Adsorption of Methylene Blue and Acid Red 88 from Synthetic Wastewater Using NiO and/or MgO Silicate Based Nanosorbcats: Experimental and Computational Modeling Studies. *RSC Adv.* **2019**, *9*, 35483–35498. [CrossRef]
45. El-Qanni, A.; Nassar, N.N.; Vitale, G. Experimental and Computational Modeling Studies on Silica-Embedded NiO/MgO Nanoparticles for Adsorptive Removal of Organic Pollutants from Wastewater. *RSC Adv.* **2017**, *7*, 14021–14038. [CrossRef]
46. Darvishi Cheshmeh Soltani, R.; Khataee, A.R.; Safari, M.; Joo, S.W. Preparation of Bio-Silica/Chitosan Nanocomposite for Adsorption of a Textile Dye in Aqueous Solutions. *Int. Biodeterior. Biodegrad.* **2013**, *85*, 383–391. [CrossRef]
47. Xu, R.; Jia, M.; Zhang, Y.; Li, F. Sorption of Malachite Green on Vinyl-Modified Mesoporous Poly(Acrylic Acid)/SiO_2 Composite Nanofiber Membranes. *Microporous Mesoporous Mater.* **2012**, *149*, 111–118. [CrossRef]
48. Shen, J.; Li, Z.; Wu, Y.-N.; Zhang, B.; Li, F. Dendrimer-Based Preparation of Mesoporous Alumina Nanofibers by Electrospinning and Their Application in Dye Adsorption. *Chem. Eng. J.* **2015**, *264*, 48–55. [CrossRef]
49. Lee, J.J.L.; Ang, B.C.; Andriyana, A.; Shariful, M.I.; Amalina, M.A. Fabrication of PMMA/Zeolite Nanofibrous Membrane through Electrospinning and Its Adsorption Behavior. *J. Appl. Polym. Sci.* **2017**, *134*, 1–13. [CrossRef]
50. Xu, R.; Jia, M.; Li, F.; Wang, H.; Zhang, B.; Qiao, J. Preparation of Mesoporous Poly (Acrylic Acid)/SiO_2 Composite Nanofiber Membranes Having Adsorption Capacity for Indigo Carmine Dye. *Appl. Phys. A* **2012**, *106*, 747–755. [CrossRef]
51. Wu, G.; Liu, S.; Jia, H.; Dai, J. Preparation and Properties of Heat Resistant Polylactic Acid (PLA)/Nano-SiO_2 Composite Filament. *J. Wuhan Univ. Technol. Mater. Sci. Ed.* **2016**, *31*, 164–171. [CrossRef]
52. Kontou, E.; Georgiopoulos, P.; Niaounakis, M. The Role of Nanofillers on the Degradation Behavior of Polylactic Acid. *Polym. Compos.* **2012**, *33*, 282–294. [CrossRef]
53. Malafatti, J.O.D.; Bernardo, M.P.; Moreira, F.K.V.; Ciol, H.; Inada, N.M.; Mattoso, L.H.C.; Paris, E.C. Electrospun Poly(Lactic Acid) Nanofibers Loaded with Silver Sulfadiazine/[Mg–Al]-Layered Double Hydroxide as an Antimicrobial Wound Dressing. *Polym. Adv. Technol.* **2020**, *31*, 1377–1387. [CrossRef]
54. Malafatti, J.O.D.; Machado, T.; Ruellas, D.O.; Sciena, R.; Paris, E.C. PLA/Starch Biodegradable Fibers Obtained by the Electrospinning Method for Micronutrient Mineral Release. *AIMS Mater. Sci.* **2023**, *10*, 200–212. [CrossRef]
55. Lai, S.M.; Hsieh, Y.T. Preparation and Properties of Polylactic Acid (PLA)/Silica Nanocomposites. *J. Macromol. Sci. Part B Phys.* **2016**, *55*, 211–228. [CrossRef]
56. Cacciotti, I.; Nanni, F. Poly(Lactic) Acid Fibers Loaded with Mesoporous Silica for Potential Applications in the Active Food Packaging. *AIP Conf. Proc.* **2016**, *1738*, 270018. [CrossRef]
57. Qin, Y.; Shen, H.; Han, L.; Zhu, Z.; Pan, F.; Yang, S.; Yin, X. Mechanically Robust Janus Poly(Lactic Acid) Hybrid Fibrous Membranes toward Highly Efficient Switchable Separation of Surfactant-Stabilized Oil/Water Emulsions. *ACS Appl. Mater. Interfaces* **2020**, *12*, 50879–50888. [CrossRef]
58. Deshmukh, P.; Bhatt, J.; Peshwe, D.; Pathak, S. Determination of Silica Activity Index and XRD, SEM and EDS Studies of Amorphous SiO_2 Extracted from Rice Husk Ash. *Trans. Indian Inst. Met.* **2012**, *65*, 63–70. [CrossRef]
59. Wittayakun, J.; Khemthong, P.; Prayoonpokarach, S. Synthesis and Characterization of Zeolite NaY from Rice Husk Silica. *Korean J. Chem. Eng.* **2008**, *25*, 861–864. [CrossRef]
60. Stober, W.; Fink, A.; Bohn, E. Controlled Growth of Monodisperse Silica Spheres in the Micron Size Range. *J. Colloid Interface Sci.* **1968**, *26*, 62–69. [CrossRef]

61. Silva, A.L.P.; Nascimento, R.G.; Arakaki, L.N.H.; Arakaki, T.; Espínola, J.G.P.; Fonseca, M.G. Organofunctionalized Silica Gel as a Support for Lipase. *J. Non. Cryst. Solids* **2013**, *376*, 139–144. [CrossRef]
62. Zhang, Y.; Han, H.; Wang, X.; Zhang, M.; Chen, Y.; Zhai, C.; Song, H.; Deng, J.; Sun, J.; Zhang, C. Utilization of NaP Zeolite Synthesized with Different Silicon Species and $NaAlO_2$ from Coal Fly Ash for the Adsorption of Rhodamine B. *J. Hazard. Mater.* **2021**, *415*, 125627. [CrossRef] [PubMed]
63. Lv, Y.; Li, P.; Che, Y.; Hu, C.; Ran, S.; Shi, P.; Zhang, W. Facile Preparation and Characterization of Nanostructured Bioi Microspheres with Certain Adsorption-Photocatalytic Properties. *Mater. Res.* **2018**, 21. [CrossRef]
64. Gharbani, P.; Mehrizad, A. Preparation and Characterization of Graphitic Carbon Nitrides/Polyvinylidene Fluoride Adsorptive Membrane Modified with Chitosan for Rhodamine B Dye Removal from Water: Adsorption Isotherms, Kinetics and Thermodynamics. *Carbohydr. Polym.* **2022**, *277*, 118860. [CrossRef] [PubMed]
65. Ugheoke, I.B.; Mamat, O. A Critical Assessment and New Research Directions of Rice Husk Silica Processing Methods and Properties. *Maejo Int. J. Sci. Technol.* **2012**, *6*, 430–448.
66. Wang, W.; Martin, J.C.; Zhang, N.; Ma, C.; Han, A.; Sun, L. Harvesting Silica Nanoparticles from Rice Husks. *J. Nanoparticle Res.* **2011**, *13*, 6981–6990. [CrossRef]
67. Della, V.P.; Hotza, D.; Junkes, J.A.; De Oliveira, A.P.N. Estudo Comparativo Entre Sílica Obtida Por Lixívia Ácida Da Casca de Arroz e Sílica Obtida Por Tratamento Térmico Da Cinza de Casca de Arroz. *Quim. Nova* **2006**, *29*, 1175–1179. [CrossRef]
68. Pei, Y.; Wang, M.; Liu, Q.; Xu, X.; Yuan, L.; Zhao, Y. Novel SiO_2@MgxSiyOz Composite with High-Efficiency Adsorption of Rhodamine B in Water. *RSC Adv.* **2014**, *4*, 55237–55246. [CrossRef]
69. Ray, S.; Takafuji, M.; Ihara, H. Peptide-Based Surface Modified Silica Particles: Adsorption Materials for Dye-Loaded Wastewater Treatment. *RSC Adv.* **2013**, *3*, 23664–23672. [CrossRef]
70. Han, H.; Wei, W.; Jiang, Z.; Lu, J.; Zhu, J.; Xie, J. Removal of Cationic Dyes from Aqueous Solution by Adsorption onto Hydrophobic/Hydrophilic Silica Aerogel. *Colloids Surf. A Physicochem. Eng. Asp.* **2016**, *509*, 539–549. [CrossRef]
71. Shen, J.; Wu, Y.N.; Zhang, B.; Li, F. Adsorption of Rhodamine b Dye by Biomimetic Mesoporous SiO_2 Nanosheets. *Clean Technol. Environ. Policy* **2015**, *17*, 2289–2298. [CrossRef]
72. Wang, C.; Wang, Y.; Hashimoto, T. Impact of Entanglement Density on Solution Electrospinning: A Phenomenological Model for Fiber Diameter. *Macromolecules* **2016**, *49*, 7985–7996. [CrossRef]
73. Joshiba, G.J.; Kumar, P.S.; Govarthanan, M.; Ngueagni, P.T.; Abilarasu, A.; Carolin, C.F. Investigation of Magnetic Silica Nanocomposite Immobilized Pseudomonas Fluorescens as a Biosorbent for the Effective Sequestration of Rhodamine B from Aqueous Systems. *Environ. Pollut.* **2021**, *269*, 116173. [CrossRef] [PubMed]
74. Maruthapandi, M.; Eswaran, L.; Luong, J.H.T.; Gedanken, A. Sonochemical Preparation of Polyaniline@TiO_2 and Polyaniline@SiO_2 for the Removal of Anionic and Cationic Dyes. *Ultrason. Sonochem.* **2020**, *62*, 104864. [CrossRef] [PubMed]
75. Wu, C.; Yuan, W.; Al-Deyab, S.S.; Zhang, K.Q. Tuning Porous Silica Nanofibers by Colloid Electrospinning for Dye Adsorption. *Appl. Surf. Sci.* **2014**, *313*, 389–395. [CrossRef]

materials

Article

Molecular Sieve, Halloysite, Sepiolite and Expanded Clay as a Tool in Reducing the Content of Trace Elements in *Helianthus annuus* L. on Copper-Contaminated Soil

Mirosław Wyszkowski [1,*], Jadwiga Wyszkowska [2], Natalia Kordala [1] and Magdalena Zaborowska [2]

1 Department of Agricultural and Environmental Chemistry, University of Warmia and Mazury in Olsztyn, Łódzki 4 Sq., 10-727 Olsztyn, Poland
2 Department of Soil Science and Microbiology, University of Warmia and Mazury in Olsztyn, Łódzki 3 Sq., 10-727 Olsztyn, Poland
* Correspondence: miroslaw.wyszkowski@uwm.edu.pl

Abstract: The aim of this study was to determine the effect of copper soil contamination on the trace element content of sunflower aerial parts and in roots. Another aim was to assess whether the introduction of selected neutralizing substances (molecular sieve, halloysite, sepiolite and expanded clay) into the soil could reduce the impact of copper on the chemical composition of sunflower plants. Copper soil contamination with 150 mg Cu^{2+} kg^{-1} of soil and 10 g of each adsorbent per kg of soil were used. Soil contamination with copper caused a significant increase in the content of this element in the aerial parts (by 37%) and roots (by 144%) of sunflower. Enriching the soil with the mineral substances reduced the amount of copper in the aerial parts of sunflower. Halloysite had the greatest effect (35%), while expanded clay had the smallest effect (10%). An opposite relationship was found in the roots of this plant. In copper-contaminated objects, a decrease in the content of cadmium and iron and an increase in the concentrations of nickel, lead and cobalt in the aerial parts and roots of sunflower were observed. The applied materials reduced the content of the remaining trace elements more strongly in the aerial organs than in the roots of sunflower. Molecular sieve had the greatest reducing effect on the content of trace elements in sunflower aerial organs, followed by sepiolite, while expanded clay had the least impact. The molecular sieve also reduced the content of iron, nickel, cadmium, chromium, zinc and, especially, manganese, whereas sepiolite reduced the content of zinc, iron, cobalt, manganese and chromium in sunflower aerial parts. Molecular sieve contributed to a slight increase in the content of cobalt, while sepiolite had the same effect on the content of nickel, lead and cadmium in the aerial parts of sunflower. All materials decreased the content of chromium in sunflower roots, molecular sieve—zinc, halloysite—manganese, and sepiolite—manganese and nickel. The materials used in the experiment, especially the molecular sieve and to a lesser extent sepiolite, can be used effectively to reduce the content of copper and some other trace elements, particularly in the aerial parts of sunflower.

Keywords: copper contamination; materials; *Helianthus annuus* L.; trace elements

Citation: Wyszkowski, M.; Wyszkowska, J.; Kordala, N.; Zaborowska, M. Molecular Sieve, Halloysite, Sepiolite and Expanded Clay as a Tool in Reducing the Content of Trace Elements in *Helianthus annuus* L. on Copper-Contaminated Soil. *Materials* **2023**, *16*, 1827. https://doi.org/10.3390/ma16051827

Academic Editors: Roberta G. Toro, Abeer M. Adel and Daniela Caschera

Received: 18 January 2023
Revised: 20 February 2023
Accepted: 21 February 2023
Published: 23 February 2023

1. Introduction

As a consequence of man's agricultural and non-agricultural (industrial) activity, the properties and fertility of soil can worsen [1,2]. Particularly hazardous forms of soil degradation include those caused by the entry of foreign chemical substances to soil, both organic (e.g., crude oil chemicals, pesticides) [3] and inorganic ones (e.g., wastewater, municipal solid waste, deposited dust from the metallurgical industry) [2,4]. Trace elements accumulate in the environment, migrate within soil profiles, and are contained in various chemical compounds [2]. The development of industries and increasing mobility of human populations magnify the accumulation of trace elements in soil. There, they have persistent influence on both the soil dwelling organisms and the entire ecosystem because they do not undergo further transformations [5,6]. Soils used for food production that are contaminated with trace elements are therefore pose a direct threat to food safety (lower crop yields, poorer food/feed quality) [7,8] and to entire trophic chains [9]. Thus, it is necessary to implement sustainable measures aiming at the restoration of proper soil characteristics so that soils are able to perform their key economic and ecological functions [6].

Copper is an essential micronutrient in plants, where it plays an important role in such biological and physiological processes as photosynthesis, biosynthesis of proteins, transport of carbohydrates and mitochondrial respiration [10,11]. Copper is a co-factor of many enzymes (e.g., peroxidase dismutase, phenolic oxidase, plastocyanin) [12,13]; it also participates in conversions of nitrogen compounds and in plant tolerance mechanisms; in addition, it enhances the activity of nitrate reductase [12,14]. Copper also affects the permeability of cell membranes, metabolism of nucleic acids and the process of generative reproduction of plants (production of pollen and seeds) [15]. The copper content in plants typically ranges between 5 to 30 mg kg^{-1} [16]. Excess of this element has a negative effect on seed germination [17,18] and such metabolic process as respiration, enzyme activity [12,19] and photosynthesis (elevated photoinhibition, impaired structure of chloroplasts and composition of the membrane of thylakoids) [20]. Plants in which the physiological norms have been exceeded are observed to experience the following symptoms: distorted permeability of cell membranes [19], retarded growth, decreased mitotic activity of root cells, decreased size of conductive tissues [13,21] and chlorosis of leaves [22]. The toxic effect of copper, even towards tolerant plants, is observed when the soil content of copper exceeds 60–125 mg kg^{-1} [11]. Copper excess is also harmful to animals and people. It has been demonstrated that an excessively high concentration of copper can be cytotoxic, causing a number of health problems, e.g., stomachache, liver insufficiency [23], neurological disorders [24] and chronic anaemia [25].

Soil contamination with copper is a widespread problem in many countries [26], which dates back to the early days of industrial revolution [27]. The content of copper in soil continues to increase due to the application of copper-based fungicides and bactericides in horticulture and agriculture [16] as well as other pesticides [27]. Another source of metal-bearing contaminants is the disposal and recycling of waste, sewage sludge, copper-enriched animal manure in soil [27,28], as well as soil fertilisation and transport [26,29]. It has been demonstrated that in some areas occupied by special branches of agriculture, e.g., vineyards, the soil content of copper can exceed 500 mg kg^{-1} [18]. Excessive concentrations of copper are also detected in soils locally contaminated by mining or smelting (waste such as mine slag, slag from electric furnaces, solid waste from flue-gas desulfurization systems or floatation waste) and copper-bearing alloy processing [28–30].

There is ongoing search for remediation technologies applicable to contaminated soils, which would limit the negative impact of trace elements on a given biocenosis rather than remove such contaminants from the soil completely [29,31,32]. Attempts are made to achieve such an outcome through the immobilization of trace elements in soil by modifying the soil's properties, especially reaction, content of organic substance and sorption properties [33,34], but also by growing certain plants or through the precipitation of trace elements as insoluble salts [35]. The amendment of soil for the purpose of its remediation with mineral materials, rich in aluminosilicates and clay minerals, or zeolites as contaminant adsorbents decreases amounts of phytoavailable forms of trace elements, leading to the restoration of homeostasis in the soil [36,37]. Neutralizing substances applied as soil amendments (including bentonite, molecular sieve, halloysite, sepiolite, biochar) bind harmful trace elements into insoluble compounds or organic-mineral complexes [38,39], which can remain immobilised in the soil for long periods under favourable conditions [40]. Moreover, these substances contribute to the enhanced microbiological and biochemical activity of soil, supplement the soil's organic matter resources, and in consequence improve the soil's condition and fertility [41–43]. Nanotechnology seems to have a special role in combination with bioremediation, which is currently the most promising and economical method of removing contaminants from the soil [32]. In situ stabilisation limits the plants uptake of trace elements from the soil [44–46] by reducing their solubility and bioavailability [39,47]. In addition, this technique is inexpensive, does not destroy the soil's structure and biological activity [47], and prevents secondary contamination [48].

Effective, inexpensive and easily available soil amendments of natural origin are being sought, with view of using them broadly in in situ stabilisation of soils exposed to pressure of trace elements. Such materials include halloysite, sepiolite, molecular sieve and expanded clay. Sepiolite ($Mg_4Si_6O_{15}(OH)_2 \cdot 6H_2O$) is a natural hydrated magnesium silicate clay mineral with the structure similar to that of type 2:1 tri-octahedral silicates [49]. Some corner atoms on the surface of sepiolite sheets are bound to hydroxyl groups (Si-OH). These so-called silanol groups tend to be active sites accessible to ions or molecules, thereby playing the role of adsorption centres [50]. Molecular sieves are nano-porous materials, selective towards substances with specific molecules, which adsorb molecules with smaller dimensions [6]. The most common group of molecular sieves are zeolites, which are characterised by the crystal-like, uniform pore structure of [51]. They are also characterized by large specific surface, durability and the ability to regenerate [52]. Halloysite is a dioctahedral clay mineral, a 1:1 layer aluminosilicate [53]. It is characterized by high porosity (60–70%) and large specific area (56.2–58.0 $m^2\ g^{-1}$), high ion-exchange capacity and ease of chemical and mechanical processing [54]. As a natural mineral with a nanotubular structure, halloysite has many technological applications, as a mineral adsorbent (it adsorbs toxic substances and trace elements from water and wastewater) [55] and as support for catalysts [56]. The cation-exchange capacity of halloysite varies from 2 to 60 cmol(+) kg^{-1} [57]. Expanded clay is produced from natural materials processed in rotary kilns heated up to 1100–1200 °C. Expanded clay particles are nearly round, with high total porosity (up to 80%) [58,59]. Moreover, it is non-flammable, chemically stable, non-absorbent (up to 20%) and resistant to mould, fungi or rodents [58,60]. Expanded clay is broadly used in environmental engineering, civil engineering, geotechnology and horticulture [61]. The application of nanomaterials improves soil properties and plays an important role in stimulating plant growth and shaping their proper chemical composition [51]. In view of the significant areas of soil contaminated with trace elements, it is therefore necessary to find materials with a high efficiency in sorbing or binding them in immobilised forms that would not be taken up in excessive quantities by plants.

Therefore, we hypothesised that (1) copper contamination of the soil would increase the content of this element and some other metals in the aerial parts and roots of sunflower, (2) the application of materials such as molecular sieve, halloysite, sepiolite and expanded clay to the soil would reduce the potential negative effects of copper contamination on the chemical composition of sunflower plants.

2. Material and Methods

2.1. Methodology of the Plant Growing Trials

The experiment was conducted on soil collected from the humic topsoil. It was a soil with a texture composition of sandy loam (60.63% of 0.0–2.0 mm fractions, 35.99% of 0.02–0.05 mm fractions and 3.38% of <0.002 mm fractions), classified as a Eutric Cambisol [62]. It was a controlled plant growing experiment, set up in polyethylene pots of 3.0 dm^3 in capacity, conducted in the controlled environment of a greenhouse in Olsztyn (NE Poland). The basic properties of the soil used in the experiment were as follows: pH_{KCl}—soil reaction in 1 mol KCl dm^{-3}—6.00, hydrolytic acidity (HAC)—13.50 mmol (+) kg^{-1} DM, sum of exchangeable base cations (EBC)—145.00 mmol (+) kg^{-1} DM, cation exchange capacity (CEC)—158.50 mmol (+) kg^{-1} DM, base saturation (BS)—91.49%, content of: total nitrogen (N_{tot})—1.07 g kg^{-1} dry matter (DM), organic carbon (C_{org})—14.69 g kg^{-1} DM, phosphorus (P)—166.72 mg kg^{-1} DM, potassium (K)—171.31 mg kg^{-1} DM, magnesium (Mg)—443.21 mg kg^{-1} DM, copper (Cu)—4.20 mg kg^{-1} DM.

It was a typical, two-factorial experiment. The first factor was the copper soil contamination (without Cu—0 and with Cu—150 mg Cu^{2+} kg^{-1} of soil) used as $CuSO_4 \cdot 5H_2O$, and the second factor was the application of different materials (adsorbents) used to reduce the negative effect of copper on plants. Molecular sieve, halloysite, sepiolite, and expanded clay were used, all at a dose of 10 g kg^{-1} DM of soil.

The product called Silosiv A3, which contains molecular sieve, was tested. It is a crystalline aluminium silicate with mircopores of the size 0.3 nm (Grace, Enriching Lives, Everywhere, USA). Halloysite ($Al_2Si_2O_5(OH)_4$) is a mineral of high ion exchange capacity owing to its large specific surface and porosity (diameter of pores between 10–20 nm), most often used as a mineral adsorbent (Halosorb, Intermark, Poland) [63]. Sepiolite ($Mg_4[Si_6O_{15}(OH)_2]6H_2O$), or specifically the product named Sepiolite 60/100 (Sepiolsa Minersa Group, Spain) is hydrated magnesium silicate, with considerable adsorption properties, fibrous texture and the pore diameter of 8–1.6 mm [64,65]. Expanded clay is produced by baking loamy clay at a high temperature (1150–1200 °C). Same as molecular sieve, expanded clay is characterized by high porosity (GardenGURU, Poland) [66].

Additionally, in order to secure the nutritional requirements of the plants, each pot was supplied with fertilisers containing basic nutrients: nitrogen—110 mg N as $CO(NH_2)_2$, phosphorus—45 mg P as KH_2PO_4, potassium—110 mg K as KH_2PO_4 and KCl, and magnesium—20 mg as $MgSO_4 \cdot 7H_2O$. Prior to mixing with copper, adsorbents and mineral fertilisers, soil was passed through a 5 mm mesh net sieve. Next (3 June), it was transferred to polyethylene pots and seeds of the plant were sown. Sunflower (*Helianthus annuus* L.) was chosen to be the test phytoremediation plant, as it is highly tolerant to soil contamination with heavy metals [67] and simultaneously plays an important role in the global food economy. In terms of the size of area cultivated for oilseed crops, the sunflower occupies the fourth place [68,69], and its biggest producers are Ukraine, Russia and Argentina [68]. This is due to the relatively low translocation of trace elements from the roots to the aerial parts of sunflower, despite the significant content of their soluble forms in the soil rizosphere. The translocation coefficient for sunflower may even be <1 [67]. Sunflower biomass can also be used as a renewable energy resource [70]. The pots were watered with distilled water, maintaining the soil moisture content at 60% of maximum water capacity throughout the entire plant growing period. The pot experiment was conducted in triplicate. During the harvest of sunflower plants at BBCH 35 stage (Biologische Bundesanstalt, Bundessortenamt and Chemical Scale)—43 days after sowing (14 July), plant samples were taken for laboratory analysis.

2.2. Physicochemical and Chemical Analyses of Soil and Plants

Before the experiment, the basic parameters of the soil used in the experiment were determined: the grain-size distribution using a Malvern Mastersizer 2000 Laser Diffraction (Malvern, Worcestershire, UK) [71], soil's pH_{KCl} using a HI 2221 pH-meter (Hanna Instruments, Washington, UK) [72], hydrolytic acidity (HAC) and exchangeable base cations (EBC) using the Kappen method [73], total nitrogen content (N_{tot}) using Buchi B-324 a distillation apparatus (Buchi, Flawil, Switzerland) [74], organic carbon (C_{org}) on a Genesis 6 spectrophotometer (Thermo Electron Corporation, Switzerland, USA) [75], available phosphorus on a SQ118 spectrophotometer [76], potassium on a Jenway 6705 UV/VIS spectrophotometer (Jenway LTD, Staffordshire, UK) [76], magnesium on an atomic absorption spectrophotometer GBC 932AA (GBC Scientific Equipment, Braeside, Australia) [77], and copper by flame and electrothermal absorption spectrometry after extraction in aqua regia according to ISO 11047:1998 [78]. HAC and EBC values were used to calculate the soil's cation exchange capacity (CEC) and base saturation (BS) [73].

Samples of aerial parts and roots of sunflower plants were cut, dried at a constant temperature of 60 °C and ground to a flour-like texture. Then, the plant material was digested in concentrated 65% nitric acid (HNO_3 analytically pure, of the density of 1.40 g cm^{-3}) according to US-EPA3051 [79] in a microwave oven MARS 6—CEM Corporation, Matthews, NC, USA. Concentrations of trace elements (copper, cadmium, lead, chromium, nickel, zinc, manganese, iron and cobalt) in digested plant samples were determined by atomic absorption spectrometry (ASA) using a SpectrAA 240FS spectrophotometer (Varian Inc., Mulgrave, VIC, Australia) [80]. While performing these analyses, reference materials by Fluka were used (Cu 38996, Cd 51994, Pb 16595, Cr 02733, Ni 42242, Zn 188227, Mn 63534, Fe 16596, Co 119785.0100), and the quality of the analyses was monitored using the certified reference material NCS ZC 73030 (Chinese National Analysis Center for Iron and Steel, Beijing, China).

2.3. Statistical Analysis

The research results were statistically processed in Statistica 13.3 [81], using a two-factorial analysis of variance and the Tukey's test (HSD) to distinguish homogeneous groups at $p = 0.01$, and principal component analysis (PCA) and % of observed variation in order to determine the relationships between the content of trace elements in the aerial parts and roots of sunflower (dependent variables) versus the studied factors (soil contamination with copper and adsorbents—independent variables).

3. Results

Plant growth and development and the content of trace elements in sunflower depended on both the contamination of soil with copper and the application of adsorbents (Tables 1–3, Figure 1). The materials used in copper-contaminated sites (mainly sepiolite and then halloysite) had a beneficial effect on plant growth and development (Figure 1). In non-copper-contaminated objects, their effect was relatively small.

Table 1. Content of copper, cadmium and lead in sunflower—*Helianthus annuus* L. (mg kg^{-1} DM).

Object	Aerial Parts			Roots		
	Without Cu	**With Cu**	**% Change with Cu/without Cu**	**Without Cu**	**With Cu**	**% Change with Cu/without Cu**
			Copper content (mg kg^{-1} DM)			
Control	1.628 ± 0.017 ^{bc}	2.234 ± 0.124 ^{e}	37.2	4.388 ± 0.016 ^{c}	10.720 ± 0.044 ^{d}	144.3
Molecular sieve	1.486 ± 0.028 ^{ab}	1.936 ± 0.116 ^{d}	30.3	2.484 ± 0.024 ^{a}	15.570 ± 0.078 ^{e}	526.8
Halloysite	1.282 ± 0.010 ^{a}	1.449 ± 0.028 ^{ab}	13.0	3.611 ± 0.045 ^{b}	27.737 ± 0.026 ^{g}	668.1
Sepiolite	1.430 ± 0.003 ^{ab}	1.792 ± 0.025 ^{cd}	25.3	3.533 ± 0.070 ^{b}	19.576 ± 0.062 ^{f}	454.1
Expanded clay	1.501 ± 0.043 ^{ab}	2.012 ± 0.076 ^{de}	34.0	2.651 ± 0.074 ^{a}	29.419 ± 0.041 ^{h}	1009.7
Average	1.465 ^{A}	1.885 ^{B}	28.6	3.333 ^{A}	20.604 ^{B}	518.1
			Cadmium content (mg kg^{-1} DM)			
Control	0.013 ± 0.002 ^{cd}	0.011 ± 0.001 ^{bc}	−15.4	0.014 ± 0.001 ^{b}	0.006 ± 0.001 ^{a}	−57.1
Molecular sieve	0.009 ± 0.001 ^{ab}	0.009 ± 0.001 ^{ab}	0.0	0.007 ± 0.001 ^{a}	0.007 ± 0.001 ^{a}	0.0
Halloysite	0.006 ± 0.001 ^{a}	0.009 ± 0.001 ^{ab}	50.0	0.005 ± 0.001 ^{a}	0.032 ± 0.001 ^{d}	540.0
Sepiolite	0.016 ± 0.001 ^{d}	0.022 ± 0.001 ^{e}	37.5	0.006 ± 0.001 ^{a}	0.018 ± 0.001 ^{c}	200.0
Expanded clay	0.014 ± 0.001 ^{cd}	0.017 ± 0.001 ^{d}	21.4	0.012 ± 0.001 ^{b}	0.017 ± 0.001 ^{c}	41.7
Average	0.012 ^{A}	0.014 ^{B}	17.2	0.009 ^{A}	0.016 ^{B}	81.8
			Lead content (mg kg^{-1} DM)			
Control	0.174 ± 0.005 ^{a}	0.214 ± 0.030 ^{a}	23.0	0.128 ± 0.006 ^{b}	0.346 ± 0.001 ^{c}	170.3
Molecular sieve	0.197 ± 0.002 ^{a}	0.211 ± 0.018 ^{a}	7.1	0.099 ± 0.011 ^{a}	0.351 ± 0.002 ^{c}	254.5
Halloysite	0.388 ± 0.011 ^{d}	0.477 ± 0.007 ^{e}	22.9	0.408 ± 0.002 ^{d}	0.749 ± 0.003 ^{g}	83.6
Sepiolite	0.337 ± 0.009 ^{c}	0.316 ± 0.004 ^{c}	−6.2	0.439 ± 0.008 ^{e}	0.765 ± 0.010 ^{g}	74.3
Expanded clay	0.261 ± 0.001 ^{b}	0.264 ± 0.005 ^{b}	1.1	0.337 ± 0.008 ^{c}	0.697 ± 0.010 ^{f}	106.8
Average	0.271 ^{A}	0.296 ^{B}	9.2	0.282 ^{A}	0.582 ^{B}	106.1

Values are average ± standard deviation. Homogeneous groups indicated by different letters (^{A,B} for Cu contamination, $^{a–h}$ for interaction between Cu contamination and substance type) were calculated separately for each trace element. There are significant differences at $p \leq 0.01$ (Anova, Tukey's HSD test).

Table 2. Content of chromium, nickel and zinc in sunflower—*Helianthus annuus* L. (mg kg^{-1} DM).

Object	Aerial Parts			Roots		
	Without Cu	With Cu	% Change with Cu/without Cu	Without Cu	With Cu	% Change with Cu/without Cu
	Chromium content (mg kg^{-1} DM)					
Control	2.271 ± 0.090 ^{d}	1.844 ± 0.023 ^{c}	−18.8	1.825 ± 0.069 ^{a}	4.670 ± 0.081 ^{d}	155.9
Molecular sieve	3.491 ± 0.052 ^{g}	1.435 ± 0.012 ^{b}	−58.9	3.913 ± 0.070 ^{c}	3.992 ± 0.014 ^{cd}	2.0
Halloysite	2.799 ± 0.081 ^{e}	1.049 ± 0.001 ^{a}	−62.5	6.187 ± 0.064 ^{e}	2.996 ± 0.074 ^{b}	−51.6
Sepiolite	3.118 ± 0.069 ^{f}	0.877 ± 0.038 ^{a}	−71.9	2.503 ± 0.056 ^{ab}	3.052 ± 0.028 ^{b}	21.9
Expanded clay	1.715 ± 0.040 ^{c}	1.027 ± 0.056 ^{a}	−40.1	6.292 ± 0.023 ^{e}	3.137 ± 0.090 ^{b}	−50.1
Average	2.679 ^{B}	1.246 ^{A}	−53.5	4.144 ^{B}	3.569 ^{A}	−13.9
	Nickel content (mg kg^{-1} DM)					
Control	2.877 ± 0.041 ^{b}	3.130 ± 0.024 ^{c}	8.8	2.600 ± 0.063 ^{ab}	3.035 ± 0.081 ^{b}	16.7
Molecular sieve	2.485 ± 0.042 ^{a}	2.878 ± 0.045 ^{b}	15.8	2.377 ± 0.028 ^{a}	2.998 ± 0.080 ^{b}	26.1
Halloysite	2.343 ± 0.025 ^{a}	3.290 ± 0.077 ^{cd}	40.4	3.729 ± 0.054 ^{c}	2.821 ± 0.070 ^{ab}	−24.3
Sepiolite	3.415 ± 0.020 ^{d}	4.102 ± 0.078 ^{f}	20.1	2.610 ± 0.077 ^{ab}	2.752 ± 0.049 ^{ab}	5.4
Expanded clay	3.380 ± 0.041 ^{d}	3.727 ± 0.076 ^{e}	10.3	3.618 ± 0.088 ^{c}	4.463 ± 0.082 ^{d}	23.4
Average	2.900 ^{A}	3.425 ^{B}	18.1	2.987 ^{A}	3.214 ^{B}	7.6
	Zinc content (mg kg^{-1} DM)					
Control	28.32 ± 0.26 ^{d}	32.25 ± 0.14 ^{e}	13.9	110.77 ± 1.96 ^{g}	48.09 ± 0.31 ^{c}	−56.6
Molecular sieve	23.18 ± 0.05 ^{a}	25.11 ± 0.36 ^{bc}	8.3	38.31 ± 1.67 ^{b}	46.18 ± 1.04 ^{c}	20.5
Halloysite	28.22 ± 0.08 ^{d}	24.80 ± 0.10 ^{b}	−12.1	71.40 ± 0.05 ^{f}	61.51 ± 0.10 ^{e}	−13.9
Sepiolite	25.64 ± 0.30 ^{bc}	28.53 ± 0.22 ^{d}	11.3	29.18 ± 0.18 ^{a}	56.50 ± 0.05 ^{d}	93.6
Expanded clay	26.38 ± 0.06 ^{c}	32.25 ± 0.28 ^{e}	22.3	33.41 ± 0.47 ^{a}	70.30 ± 0.74 ^{f}	110.4
Average	26.35 ^{A}	28.59 ^{B}	8.5	56.61 ^{A}	56.52 ^{A}	−0.2

Values are average ± standard deviation. Homogeneous groups indicated by different letters (^{A,B} for Cu contamination, $^{a-g}$ for interaction between Cu contamination and substance type) were calculated separately for each trace element. There are significant differences at $p \leq 0.01$ (Anova, Tukey's HSD test).

The copper soil contamination (150 mg Cu^{2+} kg^{-1} of soil) caused a significant increase in the content of this element in aerial parts (by 37%) and in roots (by 144%) of sunflower grown in the control series, without the adsorbents application, relative to the uncontaminated pots (Table 1). The amendment of soil with the tested adsorbents had a positive effect by reducing the copper content in aerial parts of sunflower. Halloysite had the strongest effect while expanded clay produced the weakest influence, reducing the content of copper in sunflower's aerial parts by 35% and 10%, respectively, in comparison to the control. Molecular sieve and sepiolite had a smaller but also reducing effect (13% and 20%, respectively) on the content of copper in aerial parts of the sunflower. Contrary effects were observed in the roots of this plant. Molecular sieve, sepiolite, halloysite and expanded clay caused an increase by 45%, 83%, 159% and 174%, respectively, in the copper content in sunflower's roots relative to the pots where these substances were not added to soil. The increase in copper content in sunflower roots (as opposed to aerial parts) may be due to the positive effects of molecular sieve, sepiolite, halloysite and expanded clay on the physiological barrier, limiting excessive translocation of some elements from plant roots to aerial parts.

Table 3. Content of manganese, iron and cobalt in sunflower—*Helianthus annuus* L. (mg kg^{-1} DM).

Object	Aerial Parts			Roots		
	Without Cu	With Cu	% Change with Cu/without Cu	Without Cu	With Cu	% Change with Cu/without Cu
			Manganese content (mg kg^{-1} DM)			
Control	29.49 ± 0.14 ^{f}	32.11 ± 0.43 ^{g}	8.9	55.19 ± 0.61 ^{d}	22.61 ± 0.07 ^{a}	−59.0
Molecular sieve	15.74 ± 0.06 ^{a}	18.06 ± 0.09 ^{b}	14.7	33.79 ± 0.19 ^{c}	22.95 ± 0.73 ^{a}	−32.1
Halloysite	20.76 ± 0.09 ^{d}	23.86 ± 0.10 ^{e}	14.9	134.68 ± 0.77 ^{g}	99.66 ± 0.67 ^{f}	−26.0
Sepiolite	16.37 ± 0.26 ^{a}	19.29 ± 0.10 ^{c}	17.8	25.88 ± 0.07 ^{b}	21.85 ± 0.18 ^{a}	−15.6
Expanded clay	21.22 ± 0.04 ^{d}	32.85 ± 0.07 ^{g}	54.8	26.52 ± 0.08 ^{b}	67.41 ± 0.63 ^{e}	154.2
Average	20.72 ^{A}	25.23 ^{B}	21.8	55.21 ^{B}	46.90 ^{A}	−15.1
			Iron content (mg kg^{-1} DM)			
Control	24.94 ± 0.01 ^{f}	20.34 ± 0.37 ^{c}	−18.4	1911.71 ± 4.43 ^{g}	458.62 ± 7.31 ^{a}	−76.0
Molecular sieve	18.84 ± 0.09 ^{b}	19.31 ± 0.20 ^{b}	2.5	1444.87 ± 7.60 ^{e}	810.06 ± 7.41 ^{cd}	−43.9
Halloysite	23.50 ± 0.16 ^{e}	17.41 ± 0.02 ^{a}	−25.9	3195.57 ± 6.25 ^{h}	1437.98 ± 2.69 ^{e}	−55.0
Sepiolite	20.20 ± 0.04 ^{c}	17.48 ± 0.07 ^{a}	−13.5	672.19 ± 4.03 ^{b}	752.27 ± 6.58 ^{c}	11.9
Expanded clay	20.06 ± 0.05 ^{c}	21.76 ± 0.01 ^{d}	8.5	827.13 ± 8.16 ^{d}	1704.58 ± 10.39 ^{f}	106.1
Average	21.51 ^{B}	19.26 ^{A}	−10.5	1610.29 ^{B}	1032.70 ^{A}	−35.9
			Cobalt content (mg kg^{-1} DM)			
Control	0.037 ± 0.001 ^{a}	0.089 ± 0.004 ^{e}	140.5	0.111 ± 0.009 ^{a}	0.186 ± 0.002 ^{c}	67.6
Molecular sieve	0.040 ± 0.001 ^{ab}	0.093 ± 0.002 ^{e}	132.5	0.104 ± 0.001 ^{a}	0.215 ± 0.0052 ^{d}	106.7
Halloysite	0.093 ± 0.002 ^{e}	0.113 ± 0.002 ^{f}	21.5	0.155 ± 0.003 ^{b}	0.264 ± 0.012 ^{e}	70.3
Sepiolite	0.046 ± 0.004 ^{bc}	0.073 ± 0.002 ^{d}	58.7	0.154 ± 0.002 ^{b}	0.256 ± 0.007 ^{e}	66.2
Expanded clay	0.050 ± 0.001 ^{c}	0.090 ± 0.003 ^{e}	80.0	0.168 ± 0.005 ^{bc}	0.216 ± 0.003 ^{d}	28.6
Average	0.053 ^{A}	0.092 ^{B}	72.2	0.138 ^{A}	0.227 ^{B}	64.3

Values are average ± standard deviation. Homogeneous groups indicated by different letters (^{A,B} for Cu contamination, $^{a–h}$ for interaction between Cu contamination and substance type) were calculated separately for each trace element. There are significant differences at $p \leq 0.01$ (Anova, Tukey's HSD test).

Aerial parts of sunflower plants grown in pots with copper-contaminated soil not enriched with any of the tested substances were found to have increased concentrations of nickel and manganese by 9%, zinc by 14%, lead by 23% and cobalt by 141%, while the content of cadmium, iron and chromium was lower by 15%, 18% and 19%, respectively, in comparison with uncontaminated objects (Tables 1–3). The materials added to soil had a much larger reducing effect on the content of trace elements in aerial parts than in roots of sunflower plants. The strongest reducing effect on the content of trace elements in sunflower's aerial organs was produced by molecular sieve. In the series with copper-contaminated soil, molecular sieve reduced the content of iron by 5%, nickel by 8%, cadmium by 18%, chromium and zinc by 22% and manganese by as much as 44% in sunflower's aerial parts relative to the object where no adsorbent was applied. The application of this material led to a slight increase, by 4%, in the content of cobalt in sunflower's aerial parts. Halloysite, in turn, decreased the content of iron by 14%, cadmium by 18%, zinc by 23%, manganese by 26% and chromium by 43% while increasing the content of nickel by 5%, cobalt by 27% and lead by 123% in aerial parts of this plant grown in copper-contaminated soil. Sepiolite contributed to the reduction in the content of zinc by 12%, iron by 14%, cobalt by 18%, manganese by 40% and chromium by 52%, and to the elevation in the content of nickel by 31%, lead by 48% and cadmium by 100%. Expanded clay had the weakest positive and reducing effect on the content of trace elements in sunflower's aerial organs. Expanded clay only decreased the content of chromium, by 44%, while causing an increase in the content of iron by 7%, nickel by 19%, lead by 23% and cadmium by 55% in aerial parts of sunflower plants harvested from the pots with copper-contaminated soil relative to those obtained from soil not enriched with this material.

Figure 1. *Helianthus annuus* L. in the BBCH 34 phase. C—Control (uncontaminated soil); Cu—with Cu^{2+}; M—molecular sieve; H—halloysite; S—sepiolite; E—expanded clay.

In the series where no absorbents were applied, the copper soil contamination caused a decline in the content of cadmium and zinc by 57%, manganese by 59%, and iron by 76%, while leading to a higher content of nickel (by 17%), cobalt (by 68%), chromium (by as much as 156%) and lead (by 170%) in sunflower's roots relative to the uncontaminated object (Tables 1–3). All the tested materials reduced the content of chromium in sunflower's roots by 15% (molecular sieve) up to 33–36% (the other substances), while molecular sieve also reduced the root content of zinc by 4%, halloysite—the root content of manganese by 3%, and sepiolite—the root content of manganese and nickel by 3 and 9%, respectively, all relative to the objects without the application of these materials. In the remaining cases, the applied substances contributed to an increase or else had no effect on the trace elements content in sunflower's roots.

The PCA performed showed that the contribution of all analysed factors accounted for 64.29% of the total correlation of the set of data (Figure 2). Vectors of the content of chromium, nickel, zinc, manganese and iron in sunflower aerial parts were similar in length, meaning that they produced comparable effects, whereas the vectors corresponding to the content of cobalt, cadmium and lead were shorter, indicating that their effect was weaker. The content of zinc in sunflower's aerial parts was positively correlated with the content of manganese and copper, the content of nickel was positively correlated with cadmium and cobalt, that of copper with manganese, and cadmium with cobalt. Weaker positive correlations were determined between the content of chromium and iron, as well as nickel and lead in sunflower aerial parts. Negative correlations were detected between the content of chromium and nickel, and iron and lead, while weaker negative correlations occurred between cadmium and cobalt, and lead and manganese or zinc in a sunflower's aerial parts. Sepiolite expanded clay and molecular sieve had a stronger effect on the content of trace elements in sunflower's aerial parts than halloysite (Figure 3).

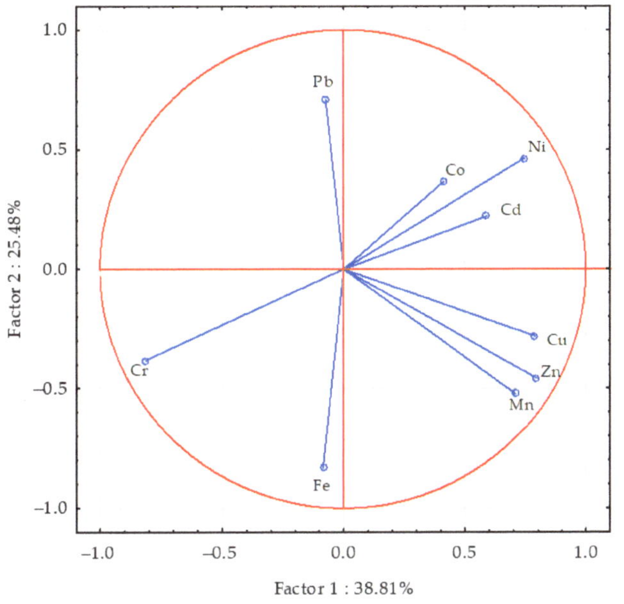

Figure 2. The trace elements content in the sunflower—*Helianthus annuus* L. aerial parts presented by a PCA analysis. Key: the vectors represent the analysed variables (content of Cu, Cd, Pb, Cr, Ni, Zn, Mn, Fe and Co).

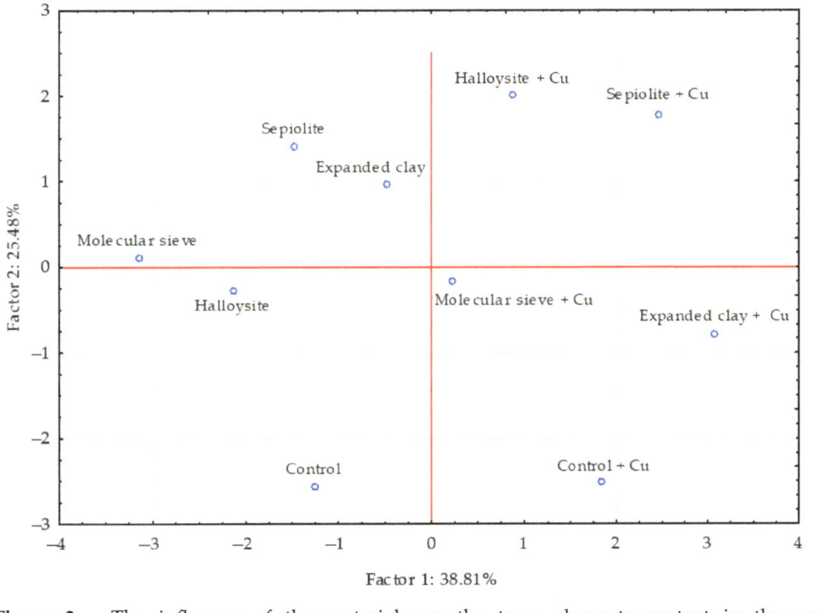

Figure 3. The influence of the materials on the trace elements content in the sunflower—*Helianthus annuus* L. aerial parts presented by a PCA analysis. Key: points indicate the samples with trace elements (uncontaminated: control—without materials, molecular sieve, halloysite, expanded clay; contaminated with copper: control + Cu —without materials + Cu, molecular sieve + Cu, halloysite + Cu, expanded clay + Cu).

The aggregated share of the analysed factors in the total correlation of the data sets was 67.54% (Figure 4). Quite large variation was demonstrated in the length of vectors representing the content of particular trace elements in sunflower roots. The zinc, nickel and especially chromium vectors were much shorter than those of the other elements, indicating that these three elements were less important. The content of copper and lead in sunflower's roots was positively correlated with cobalt and cadmium, the root content of cobalt was positively correlated with cadmium, that of manganese with iron and, less strongly, with zinc and nickel, and the root content of iron was correlated positively with zinc and nickel. Weak negative correlations were also determined between the content of chromium and iron versus concentrations of cobalt, copper, lead and cadmium in sunflower's roots. The scattering of points in Figure 5 suggests that the content of trace elements in the roots of sunflower grown in copper-contaminated soil was most strongly influenced by sepiolite and molecular sieve.

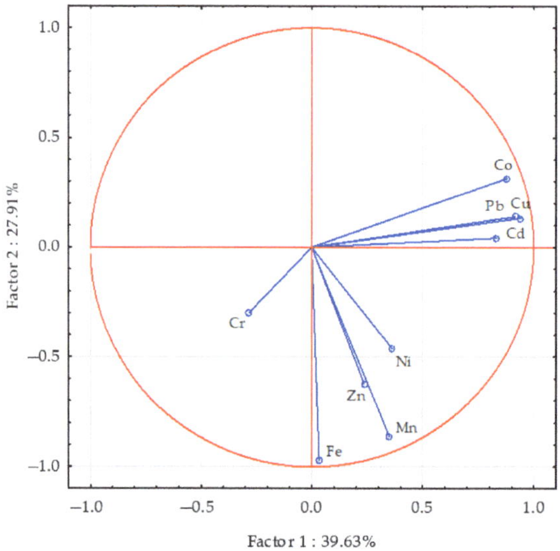

Figure 4. The trace elements content in the sunflower—*Helianthus annuus* L. roots presented by a PCA analysis. Key: the vectors represent the analysed variables (content of Cu, Cd, Pb, Cr, Ni, Zn, Mn, Fe and Co).

Figure 6 illustrates the percentage of observed variation, showing the cumulative impact of the analysed factors (soil contamination with copper and the application of different materials into soil) on the content of trace elements in sunflower's aerial organs and roots. Soil contamination with copper had a stronger effect only on the content of copper (50.5%), chromium (66.2%) and cobalt (55,7%) in sunflower's aerial parts, and on the content of copper (74.5%) and cobalt (72.1%) in sunflower's roots than the applied adsorbents did (39.8%, 15.8% and 37.0% in aerial parts and 11.7% and 21.7% in roots, respectively). The adsorbents applied to soil had a stronger impact than copper on the content of lead (92.7%), cadmium (84.4%), manganese (77.3%), nickel (66.9%), zinc (55.4%) and iron (35.4%) in aerial parts, and on the concentration of manganese (85.0%), nickel (69.5%), lead (54.4%), iron (47.1%), zinc (40.0%), chromium (28.0%) and cadmium (24.2%) in roots of sunflower plants. The soil contamination with copper had a considerable effect on the content of nickel (26.1%) and iron (23.6%) in aerial parts, and on the concentration of lead (44.1%) and cadmium (22.2%) in roots. It is worth emphasizing that the applied materials in the interaction with the soil contamination with copper had the strongest effect on the iron content in the aerial parts and on the concentrations of chromium, zinc and cadmium in roots of sunflower.

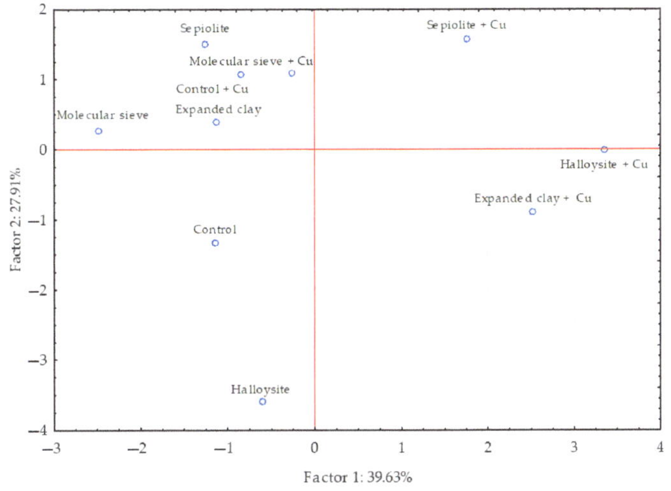

Figure 5. The influence of the materials on the trace elements content in the sunflower—*Helianthus annuus* L. roots presented by a PCA analysis. Key: points indicate the samples with trace elements (uncontaminated: control—without materials, molecular sieve, halloysite, expanded clay; contaminated with copper: control + Cu—without materials, molecular sieve + Cu, halloysite + Cu, expanded clay + Cu).

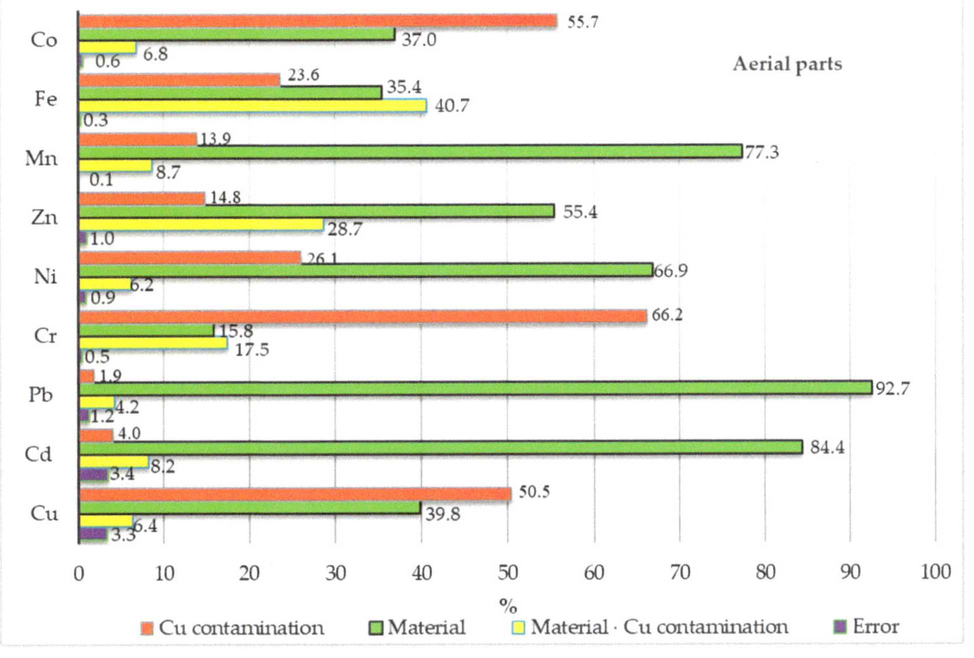

Figure 6. *Cont.*

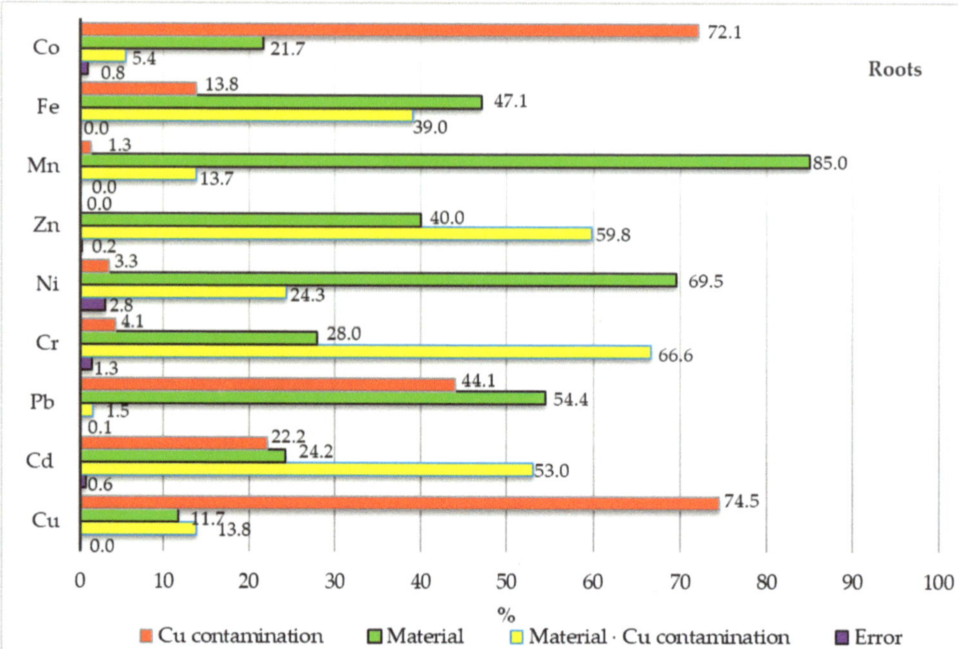

Figure 6. Relatively influence of factors on trace elements contents in the sunflower—*Helianthus annuus* L. (in percent).

4. Discussion

In this study, the contamination of soil with copper in a dose of 150 mg kg^{-1} was conducive to the accumulation of most trace elements in both aerial parts and roots of sunflower. The biggest changes were recorded for copper, whose content increased 1.4-fold in aerial biomass and 2.4-fold in roots of the test plant. A similar tendency was observed in their experiment by Wyszkowski and Brodowska [82]. These authors demonstrated that the soil contamination with copper in an amount of 200 mg kg^{-1} led to a 27-fold increase in the content of this element in aerial parts of maize as well as the greater accumulation of cobalt (by 75%), manganese (by 49%) and iron (by 30%) relative to the control. Copper may have had an antagonistic effect of on the translocation of iron and zinc from the stem to leaves of plants, e.g., rapeseed [83]. In the above experiment, the lowest applied dose of copper (50 mg Cu kg^{-1}) favoured the accumulation of manganese, iron and zinc increasing their content by 13%, 5% and 8%, respectively, compared to the control. This relationship was not observed in this experiment. These differences were most probably due to the different levels of copper contamination (here 150 mg kg^{-1} of soil) and the different plant species tested.

Copper contamination of soil influences soil physical properties, thereby modifying the translocation of trace elements in the soil–plant system [26]. As reported by Żołnowski and Wyszkowski [40], the copper contamination of soil resulted in a decreased soil pH and increased hydrolytic acidity (HAC). It also led to a decrease in total exchangeable bases (TEB), which resulted in the decreased cation exchange capacity (CEC) and base saturation (BS). Similar observations were reported by Guo et al. [84]. The decreased pH of soil exposed to copper pressure is most probably due to the replacement of H^+ by Cu^{2+} ions, and desorption of H^+ ions to the soil solution [26]. A decrease in the soil reaction to slightly acidic or acidic increased the solubility of chemical bonds of trace elements, thereby raising their mobility and content of phytoavailable forms [85]. In these conditions, trace elements migrate to the soil solution and become potentially available to plants. This may explain the higher content of copper, nickel, zinc, manganese and cobalt in the aerial parts and that of copper, lead, chromium and cobalt in the roots of sunflower found in this experiment.

Copper, same as other trace elements, does not undergo degradation [86], and due to the poor ability of soil for self-purification, copper can accumulate in the topsoil for many years, posing a local threat to groundwater, plants and animals [23]. As the actual uptake of trace elements by crops depends on the phytoavailability rather than concentration of these elements [49], chemical stabilisation of contaminants is an economically viable and environmentally reasonable solution. As reported by Wyszkowski and Brodowska [82], mineral amendments such as bentonite and zeolite change the soil pH and immobilize trace elements more effectively than organic soil amendments.

Nanomaterials application to soil plays an important role in limiting the contaminants uptake by plants [51]. In our experiment, all of the soil amendments tested, exception of expanded clay, were effective in reducing the negative effects of soil contamination with copper. The adsorbents introduced to soil decreased the bioavailability of copper, chromium, zinc, manganese and iron, limiting their translocation to aerial parts of sunflower. Sepiolite and molecular sieve proved to be most effective in the restoration of soil's homeostasis. The application of zeolite to soil has a beneficial effect on soils with low pH by raising their reaction [87,88], hence making it possible to immobilize contaminants effectively. As the soil pH raises, the solubility of chemical bonds of trace elements decreases while their adsorption on soil colloids increases, which leads to a decline in amounts of mobile and bioavailable forms of these elements [88]. High effectiveness of molecular sieves in the immobilization of cadmium and lead in contaminated soil has been confirmed by Huang-Ping and Shu-Hao [89]. Farooqi et al. [90] also demonstrated the usefulness of zeolite for the in situ stabilisation of agricultural soil fertilised for many years with wastewater. When this mineral material was applied to soil in an amount of 2% (w/w), the researchers noted reduced accumulation of copper (by 43%), cadmium (by 97%), nickel (by 70%) and lead (by 100%) in fruits of Solanum melongena. Under the same conditions, the content of copper, cadmium, nickel and lead in soil decreased by 55%, 47%, 55% and 58%, respectively, in comparison with the control series (without zeolite). Similar conclusions were reported by Chen et al. [91], who observed a decreased content of cadmium and lead in wheat following the application of zeolite to soil.

In our experiment, sepiolite statistically significantly reduced the content of copper, chromium, zinc, manganese, iron and cobalt in the aerial biomass and chromium in the root of sunflower. This effect can be attributed to the large specific surface area of sepiolite (230–320 $m^2\ g^{-1}$) [92], cation exchange capacity and soil pH regulation [49,93]. This is confirmed by other authors [94–96]. Liang et al. [95] proved that combined application of sepiolite and phosphorus fertiliser resulted in a decrease in the content of cadmium and lead in *Lactuca sativa* L. by 52% and 55%, respectively, relative to the control. The usefulness of sepiolite for in situ stabilisation of soil polluted with contaminants from the mining industry was also demonstrated by Lin et al. [94]. These researchers used a 4% addition of this mineral product, which enabled the reduction in the water-soluble fractions of cadmium and zinc by 57% and 41%, and their extractable fractions by 43% and 25%, respectively. The mobility of elements in the soil analysed also decreased by 69% (for cadmium) and by 52% (for zinc). The reduced translocation of cadmium and lead from soil to roots and shoots of spinach following the soil application of sepiolite was also shown by Sun et al. [97]. When this material was added to contaminated soil in a dose of 5%, the content of cadmium and lead in spinach roots decreased by 51% and 46%, respectively, while their concentration in spinach shoots was lower by 46% and 66%, respectively, in comparison with the control (without any addition of the neutralizing substance). These authors also determined that increasing doses of sepiolite resulted in the soil pH rising from 7.72 to 8.03.

An important property of the clay materials applied in our experiment is their ability to interact with inorganic substances. As a result of such interactions, the soil-borne, bioavailable forms of trace elements can transform into less readily available forms. This occurs through the adsorption on the surface of mineral grains, cation exchange or intercalation [98]. Once a mineral supplement is introduced to the contaminated soil, certain isomorphic substitutions occur in the tetrahedral network layers of the mineral, which leads to the formation of negatively charged adsorption sites. Such sites are then occupied by exchangeable cations, which compensate negative charges [93,99] and become immobilized. Soil amendments in the form of minerals, by inducing the immobilization of trace elements, can be therefore useful for the remediation of copper contaminated soils.

From the discussion of the research results obtained, it is difficult to find research on this topic with sunflower and many other plant species. The promising results obtained with these sorbents indicate that there is a need to continue research on the topic addressed in this paper. Possible future research should focus on plant species other than sunflower. These should mainly be energy crops whose biomass can be used for energy production. It would be advisable to carry out research under environmental conditions on a larger scale and with more trace elements. It would also be interesting to perform studies on the possibility of neutralising heavy metals in various interactions.

5. Conclusions

Soil contamination with copper caused a significant increase in the content of this element in the aerial parts (by 37%) and roots (by 144%) of sunflower. The enrichment of soil with the mineral substances reduced the amount of copper in the aerial parts of sunflower. The strongest 35% effect was achieved by halloysite, while the weakest 10% was obtained with expanded clay. Reverse relationships were found in the roots of this crop. In copper-contaminated objects in the series without adsorbents, a decrease in the content of cadmium and iron and an increase in the concentrations of nickel, lead and cobalt in aerial parts and roots of sunflower were observed. Concerning the content of the other analysed elements, either a reverse direction of change or no significant change was observed in both aerial parts and roots of sunflower. The applied materials reduced the content of the remaining trace elements more strongly in the aerial organs than in the roots of sunflowers. The molecular sieve had the highest reducing effect on the content of trace elements in sunflower aerial parts, followed by sepiolite, while expanded clay had the least effect. The molecular sieve also reduced the content of iron, nickel, cadmium, chromium, zinc and, especially, manganese, whereas sepiolite reduced the content of zinc, iron, cobalt, manganese and chromium in the aerial parts of the sunflower. The molecular sieve contributed to a slight increase in the content of cobalt, while sepiolite had the same effect on the content of nickel, lead and cadmium in aerial parts of sunflower. All materials decreased the content of chromium in the roots of the sunflower, molecular sieve—zinc, halloysite—manganese, and sepiolite—manganese and nickel. The tested materials either contributed to an increase or had no effect on the content of the other trace elements in sunflower roots.

Author Contributions: Conceptualization and methodology, J.W. and M.Z.; analysis, N.K.; writing—review and editing, M.W., N.K., J.W. and M.Z.; supervision, M.W. and J.W.; M.W., corresponding author. All authors have read and agreed to the published version of the manuscript.

Funding: The results presented in this paper were obtained as part of a comprehensive study financed by the University of Warmia and Mazury in Olsztyn, Faculty of Agriculture and Forestry, Department of Agricultural and Environmental Chemistry (grant No. 30.610.004-110) and Department of Soil Science and Microbiology (grant No. 30.610.006-110).

Institutional Review Board Statement: Not applicable.

Informed Consent Statement: Not applicable.

Data Availability Statement: All data are available in the manuscripts and from the authors.

Conflicts of Interest: The authors declare no conflict of interest. The funders had no role in the design of the study; in the collection, analyses, or interpretation of data; in the writing of the manuscript, or in the decision to publish the results.

References

1. Kopittke, P.M.; Menzies, N.W.; Wang, P.; McKenna, B.A.; Lombi, E. Soil and the intensification of agriculture for global food security. *Environ. Int.* **2019**, *132*, 105078. [CrossRef] [PubMed]
2. Alengebawy, A.; Abdelkhalek, S.T.; Qureshi, S.R.; Wang, M.Q. Heavy metals and pesticides toxicity in agricultural soil and plants: Ecological risks and human health implications. *Toxics* **2021**, *9*, 42. [CrossRef] [PubMed]
3. El-Shahawi, M.S.; Hamza, A.; Bashammakhb, A.S.; Al-Saggaf, W.T. An overview on the accumulation, distribution, transformations, toxicity and analytical methods for the monitoring of persistent organic pollutants. *Talanta* **2010**, *80*, 1587–1597. [CrossRef] [PubMed]
4. Riek, W.; Russ, A.; Marx, M. Concentrations of inorganic and organic pollutants in forest soils as an archive of anthropogenic inputs in the State of Brandenburg, Germany. *Appl. Sci.* **2021**, *11*, 1189. [CrossRef]
5. Garaiyurrebaso, O.; Garbisu, C.; Blanco, F.; Lanzén, A.; Martín, I.; Epelde, L.; Becerril, J.M.; Jechalke, S.; Smalla, K.; Grohmann, E.; et al. Long-term effects of aided phytostabilisation on microbial communities of metal-contaminated mine soil. *FEMS Microbiol. Ecol.* **2017**, *93*, fiw252. [CrossRef]
6. Boros-Lajszner, E.; Wyszkowska, J.; Kucharski, J. Use of a zeolite and molecular sieve to restore homeostasis of soil contaminated with cobalt. *Minerals* **2020**, *10*, 53. [CrossRef]
7. Chibuike, G.U.; Obiora, S.C. Heavy metal polluted soils: Effect on plants and bioremediation methods. *Appl. Environ. Soil Sci.* **2014**, *60*, 101–130. [CrossRef]
8. Ashraf, I.; Ahmad, F.; Sharif, A.; Altaf, A.R.; Teng, H. Heavy metals assessment in water, soil, vegetables and their associated health risks via consumption of vegetables, District Kasur, Pakistan. *SN Appl. Sci.* **2021**, *3*, 552. [CrossRef]
9. Rai, P.K.; Lee, S.S.; Zhang, M.; Tsang, Y.F.; Kim, K.-H. Heavy metals in food crops: Health risks, fate, mechanisms, and management. *Environ. Int.* **2019**, *125*, 365–385. [CrossRef]
10. Kabata-Pendias, A. *Trace Elements in Soil and Plants*, 4th ed.; CRS Press, Taylor and Francis Group: Boca Raton, FL, USA, 2010; 548p. [CrossRef]
11. Napoli, M.; Cecchi, S.; Grassi, C.; Baldi, A.; Zanchi, C.A.; Orlandini, S. Phytoextraction of copper from a contaminated soil using arable and vegetable crops. *Chemosphere* **2019**, *219*, 122–129. [CrossRef] [PubMed]
12. Yruela, I. Copper in plants. *Braz. J. Plant Physiol.* **2005**, *17*, 145–156. [CrossRef]
13. Kumar, V.; Pandita, S.; Singh Sidhu, G.P.; Sharma, A.; Khanna, K.; Kaur, P.; Bali, A.S.; Setia, R. Copper bioavailability, uptake, toxicity and tolerance in plants: A comprehensive review. *Chemosphere* **2021**, *262*, 127810. [CrossRef] [PubMed]
14. Hänsch, R.; Mendel, R.R. Physiological functions of mineral micronutrients (Cu, Zn, Mn, Fe, Ni, Mo, B, Cl). *Curr. Opin. Plant Biol.* **2009**, *12*, 259–266. [CrossRef]
15. Alhasany, A.R.; Noaema, A.H.; Alhmadi, H.B. The role of spraying copper and zinc on the growth and yield of *Vicia faba* L. *IOP Conf. Ser. Mater. Sci. Eng.* **2019**, *571*, 012048. [CrossRef]
16. Ballabio, C.; Panagos, P.; Lugato, E.; Huang, J.; Orgiazzi, A.; Jones, A.; Montanarella, L. Science of the total environment copper distribution in European topsoils: An assessment based on LUCAS soil survey. *Sci. Total Environ.* **2018**, *636*, 282–298. [CrossRef]
17. Nagajyoti, P.C.; Lee, K.D.; Sreekanth, T.V.M. Heavy metals, occurrence and toxicity for plants: A review. *Environ. Chem. Lett.* **2010**, *8*, 199–216. [CrossRef]
18. Adrees, M.; Ali, S.; Rizwan, M.; Zia-ur-Rehman, M.; Ibrahim, M.; Abbas, F.; Farid, M.; Qayyum, M.F.; Irshad, M.K. Mechanisms of silicon-mediated alleviation of heavy metal toxicity in plants: A review. *Ecotoxicol. Environ. Saf.* **2015**, *119*, 186–197. [CrossRef]
19. Wairich, A.; De Conti, L.; Lamb, T.I.; Keil, R.; Neves, L.O.; Brunetto, G.; Sperotto, R.A.; Ricachenevsky, F.K. Throwing copper around: How plants control uptake, distribution, and accumulation of copper. *Agronomy* **2022**, *12*, 994. [CrossRef]
20. Sharma, P.; Sirhindi, G.; Kumar, A.; Harpreet, S. Consequences of copper treatment on pigeon pea photosynthesis, osmolytes and antioxidants defense. *Physiol. Mol. Biol. Plants* **2017**, *23*, 809–816. [CrossRef]
21. Gowayed, S.M.H.; Almaghrabi, O.A. Effect of copper and cadmium on germination and anatomical structure of leaf and root seedling in maize (*Zea mays* L.). *Aust. J. Basic Appl. Sci.* **2013**, *7*, 548–555.
22. Shabbir, Z.; Sardar, A.; Shabbir, A.; Abbas, G.; Shamshad, S. Copper uptake, essentiality, toxicity, detoxification and risk assessment in soil-plant environment. *Chemosphere* **2020**, *259*, 127436. [CrossRef] [PubMed]
23. Operacz, A.; Bigaj, A.; Hap, K.; Kotowski, T. The Effect of sample preparation and measurement techniques on heavy metals concentrations in soil: Case study from Kraków, Poland, Europe. *Appl. Sci.* **2022**, *12*, 2137. [CrossRef]
24. Sadiq, S.; Ghazala, Z.; Chowdhury, A.; Büsselberg, D. Metal toxicity at the synapse: Presynaptic, postsynaptic, and long-term effects. *J. Toxicol.* **2012**, *2012*, 132671. [CrossRef] [PubMed]
25. Gaetke, L.M.; Chow-Johnson, H.S.; Chow, C.K. Copper: Toxicological relevance and mechanisms. *Arch. Toxicol.* **2014**, *88*, 1929–1938. [CrossRef] [PubMed]
26. Htwe, T.; Onthong, J.; Duangpan, S.; Techato, K.; Chotikarn, P.; Sinutok, S. Effect of copper contamination on plant growth and metal contents in rice plant (*Oryza sativa* L.). *Commun. Soil Sci. Plant Anal.* **2020**, *51*, 2349–2360. [CrossRef]

27. Apori, O.S.; Hanyabui, E.; Asiamah, Y.J. Remediation technology for copper contaminated Soil: A review. *Asian Soil Res. J.* **2018**, *1*, 1–7. [CrossRef]
28. Rehman, M.; Liu, L.; Wang, Q.; Saleem, M.H.; Bashir, S.; Ullah, S.; Peng, D. Copper environmental toxicology, recent advances, and future outlook: A review. *Environ. Sci. Pollut. Res.* **2019**, *26*, 18003–18016. [CrossRef]
29. Żołnowski, A.C.; Wyszkowski, M.; Rolka, E.; Sawicka, M. Mineral materials as a neutralizing agent used on soil contaminated with copper. *Materials* **2021**, *14*, 6830. [CrossRef]
30. Izydorczyk, G.; Mikula, K.; Skrzypczak, D.; Moustakas, K.; Witek-Krowiak, A.; Chojnacka, K. Potential environmental pollution from copper metallurgy and methods of management. *Environ. Res.* **2021**, *197*, 111050. [CrossRef]
31. Wyszkowski, M. Soil Contamination with copper and its effect on selected soil properties after applying neutralizing substances. *Pol. J. Environ. Stud.* **2019**, *28*, 2465–2471. [CrossRef]
32. Bhatt, P.; Pandey, S.C.; Joshi, S.; Chaudhary, P.; Pathak, V.M.; Huang, Y.; Wu, X.; Zhou, Z.; Chen, S. Nanobioremediation: A sustainable approach for the removal of toxic pollutants from the environment. *J. Hazard. Mater.* **2022**, *427*, 128033. [CrossRef] [PubMed]
33. Feng, N.; Dagan, R.; Bitton, G. Toxicological approach for assessing the heavy metal binding capacity of soils. *Soil Sediment Contam.* **2007**, *16*, 451–458. [CrossRef]
34. Czaban, J.; Siebielec, G. Effects of bentonite on sandy soil chemistry in a long-term plot experiment (II); Effect on pH, CEC, and macro- and micronutrients. *Pol. J. Environ. Stud.* **2013**, *22*, 1669–1676.
35. Yan, A.; Wang, Y.; Tan, S.N.; Mohd Yusof, M.L.; Ghosh, S.; Chen, Z. Phytoremediation: A promising approach for revegetation of heavy metal-polluted land. *Front. Plant Sci.* **2020**, *11*, 359. [CrossRef] [PubMed]
36. Inazumi, S.; Shishido, K.I.; Nontananandh, S.; Moriiwa, K. Remediation of heavy metals polluted soil using metal insolubilizing materials. *J. Environ. Prot.* **2018**, *9*, 770–789. [CrossRef]
37. Ukalska-Jaruga, A.; Siebielec, G.; Siebielec, S.; Pecio, M. The effect of soil amendments on trace elements' bioavailability and toxicity to earthworms in contaminated soils. *Appl. Sci.* **2022**, *12*, 6280. [CrossRef]
38. Nwachukwu, O.I.; Pulford, I.D. Comparative effectiveness of selected adsorbant materials as potential amendments for the remediation of lead-, copper- and zinc-contaminated soil. *Soil Use Manag.* **2008**, *24*, 199–207. [CrossRef]
39. Kumpiene, J. Trace elements immobilization in soil using amendments. In *Trace Elements in Soil*; Hooda, P.S., Ed.; John Wiley and Sons, Ltd.: London, UK, 2010; pp. 353–379. [CrossRef]
40. Żołnowski, A.C.; Wyszkowski, M. Mineral neutralizers as a tool for improving the properties of soil contaminated with copper. *Minerals* **2022**, *12*, 895. [CrossRef]
41. Beesley, L.; Moreno-Jimenez, E.; Gomez-Eyles, J.L.; Harris, E.; Robinson, B.; Sizmur, T. A review of biochars' potential role in the remediation, revegetation and restoration of contaminated soils. *Environ. Pollut.* **2011**, *159*, 3269–3282. [CrossRef]
42. Gonzalez-Núnez, R.; Alba, M.D.; Orta, M.M.; Vidal, M.; Rigol, A. Remediation of metal-contaminated soils with the addition of materials—Part I: Characterization and viability studies for the selection of non-hazardous waste materials and silicates. *Chemosphere* **2011**, *85*, 1511–1517. [CrossRef]
43. Khati, P.; Bhatt, P.; Kumar, R.; Sharma, A. Effect of nanozeolite and plant growth promoting rhizobacteria on maize. *3 Biotech* **2018**, *8*, 141. [CrossRef] [PubMed]
44. Kucharski, R.; Sas-Nowosielska, A.; Małkowski, E.; Japenga, J.; Kuperberg, J.M.; Pogrzeba, M.; Krzyżak, J. The use of indigenous plant species and calcium phosphate for the stabilization of highly metal-polluted sites in southern Poland. *Plant Soil* **2005**, *273*, 291–305. [CrossRef]
45. Kosiorek, M.; Wyszkowski, M. Remediation of cobalt-contaminated soil using manure, clay, charcoal, zeolite, calcium oxide, main crop (*Hordeum vulgare* L.), and after-crop (*Synapis alba* L.). *Minerals* **2020**, *10*, 429. [CrossRef]
46. Wyszkowski, M.; Kordala, N. Role of different material amendments in shaping the content of heavy metals in maize (*Zea mays* L.) on soil polluted with petrol. *Materials* **2022**, *15*, 2623. [CrossRef] [PubMed]
47. Palansooriya, K.N.; Shaheen, S.M.; Chen, S.S.; Tsang, D.C.W.; Hashimoto, Y.; Hou, D.; Bolan, N.S.; Rinklebe, J.; Ok, Y.S. Soil amendments for immobilization of potentially toxic elements in contaminated soils: A critical review. *Environ. Int.* **2020**, *134*, 105046. [CrossRef]
48. Kim, K.R.; Park, J.S.; Kim, M.S.; Koo, N.; Lee, S.H.; Lee, J.S.; Lee, S.S.; Ok, Y.S.; Kim, J.-G. Changes in heavy metal phytoavailability by application of immobilizing agents and soil cover in the upland soil nearby abandoned mining area and subsequent metal uptake by red pepper. *Korean J. Soil. Sci. Fert.* **2010**, *43*, 864–871.
49. Liang, X.; Han, J.; Xu, Y.; Sun, Y.; Wang, L.; Tan, X. In situ field-scale remediation of Cd polluted paddy soil using sepiolite and palygorskite. *Geoderma* **2014**, *235–236*, 9–18. [CrossRef]
50. Tekin, N.; Dinçer, A.; Demirbaş, O.; Alkan, M. Adsorption of cationic polyacrylamide onto sepiolite. *J. Hazard Mater.* **2006**, *134*, 211–219. [CrossRef]
51. Kumari, S.; Sharma, A.; Chaudhary, P.; Khati, P. Management of plant vigor and soil health using two agriusable nanocompounds and plant growth promotory rhizobacteria in Fenugreek. *3 Biotech* **2020**, *10*, 461. [CrossRef] [PubMed]
52. Galamboš, M.; Suchánek, P.; Rosskopfová, O. Sorption of anthropogenic adionuclides on natural and synthetic inorganic sorbents. *J. Radioanal. Nucl. Chem.* **2012**, *293*, 613–633. [CrossRef]
53. Meng, Q.; Chen, H.; Lin, J.; Lin, Z.; Sun, J. Zeolite A synthesized from alkaline assisted pre-activated halloysite for efficient heavy metal removal in polluted river water and industrial wastewater. *J. Environ. Sci.* **2017**, *56*, 254–262. [CrossRef] [PubMed]

54. Levis, S.R.; Deasy, P.B. Characterisation of halloysite for use as a microtubular drug delivery system. *Int. J. Pharm.* **2002**, *243*, 125–134. [CrossRef] [PubMed]
55. Anastopoulos, I.; Mittal, A.; Usman, M.; Mittal, J.; Yu, G.; Núñez-Delgado, A.; Kornaros, M. A review on halloysite-based adsorbents to remove pollutants in water and wastewater. *J. Mol. Liq.* **2018**, *269*, 855–868. [CrossRef]
56. Massaro, M.; Noto, R.; Riela, S. Halloysite nanotubes: Smart nanomaterials in catalysis. *Catalysts* **2022**, *12*, 149. [CrossRef]
57. Joussein, E.; Petit, S.; Churchman, J.; Theng, B.; Righi, D.; Delvaux, B. Halloysite clay minerals—A review. *Clay Minerals* **2005**, *40*, 383–426. [CrossRef]
58. Rashad, A.M. Lightweight expanded clay aggregate as a building material—An overview. *Constr. Build. Mater.* **2018**, *170*, 757–775. [CrossRef]
59. Zheng, W.; He, D.; Li, H.; Wang, F.; Yang, Y.; Zhang, J. The preparation and properties of a shell structure ceramsite. *Materials* **2020**, *13*, 1009. [CrossRef]
60. Boudaghpour, S.; Hashemi, S. A study on light expanded clay aggregate (LECA) in a geotechnical view and its application on greenhouse and greenroof cultivation. *Int. J. Geol.* **2008**, *4*, 59–63.
61. Latosińska, J.; Żygadło, M.; Czapik, P. The influence of sewage sludge content and sintering temperature on selected properties of lightweight expanded clay aggregate. *Materials* **2021**, *14*, 3363. [CrossRef]
62. IUSS Working Group WRB. World Reference Base for Soil Resources 2014. In *International Soil Classification System for Naming Soils and Creating Legends for Soil Maps. Update 2015*; World Soil Resources; Reports No. 106; FAO: Rome, Italy, 2015; p. 192. Available online: https://www.fao.org/3/i3794en/I3794en.pdf (accessed on 18 November 2022).
63. Keeling, J.L. *The Mineralogy, Geology and Occurrences of Halloysite*; Apple Academic Press: Oakville, ON, Canada, 2015. Available online: https://www.researchgate.net/publication/272481304_The_mineralogy_geology_and_occurrences_of_halloysite (accessed on 11 January 2023).
64. Sun, Y.; Sun, G.; Xu, Y.; Liu, W.; Liang, X.; Wang, L. Evaluation of the effectiveness of sepiolite, bentonite, and phosphate amendments on the stabilization remediation of cadmium contaminated soils. *J. Environ. Manag.* **2016**, *166*, 204–210. [CrossRef]
65. Hamid, Y.; Tang, L.; Hussain, B.; Usman, M.; Liu, L.; Ulhassan, Z.; He, Z.; Yang, X. Sepiolite clay: A review of its applications to immobilize toxic metals in contaminated soils and its implications in soil–plant system. *Environ. Technol. Innov.* **2021**, *23*, 101598. [CrossRef]
66. Wyszkowska, J.; Borowik, A.; Zaborowska, M.; Kucharski, J. Evaluation of the usefulness of sorbents in the remediation of soil exposed to the pressure of cadmium and cobalt. *Materials* **2022**, *15*, 5738. [CrossRef] [PubMed]
67. Kalyvas, G.; Bilias, F.; Gasparatos, D.; Zafeiriou, I.; Eissa, R.; Karamountzou, E.; Massas, I. Enhanced As, Pb and Zn uptake by *Helianthus annuus* from a heavily contaminated mining soil amended with EDTA and olive mill wastewater due to increased element mobilization, as verified by sequential extraction schemes. *Environments* **2022**, *9*, 61. [CrossRef]
68. FAO. *World Food and Agriculture—Statistical Yearbook*; FAO: Rome, Italy, 2021. Available online: https://www.fao.org/documents/card/en/c/cb4477en/ (accessed on 21 May 2022).
69. Hussain, M.; Shah, S.N.A.; Naeem, M.; Farooq, S.; Jabran, K.; Alfarraj, S. Impact of different mulching treatments on weed flora and productivity of maize (*Zea mays* L.) and sunflower (*Helianthus annuus* L.). *PLoS ONE* **2022**, *17*, e0266756. [CrossRef] [PubMed]
70. Iram, S.; Tariq, I.; Ahmad, K.S.; Jaffri, S.B. *Helianthus annuus* based biodiesel production from seed oil garnered from a phytoremediated terrain. *Int. J. Ambient. Energy* **2022**, *43*, 1763–1771. [CrossRef]
71. *PN-R-04032*; Soil and Mineral Materials—Sampling and Determination of Particle Size Distribution. Polish Committee for Standardization: Warsaw, Poland, 1998; pp. 1–12.
72. *ISO 10390*; Soil Quality—Determination of pH. International Organization for Standardization: Geneva, Switzerland, 2005.
73. Klute, A. *Methods of Soil Analysis. Part 3. Chemical Methods (Agronomy 9)*; Norman, A.G., Ed.; American Society of Agronomy: Madison, WI, USA, 1965; pp. 1201–1230.
74. *ISO 11261*; Soil Quality—Determination of Total Nitrogen—Modified Kjeldahl Method. ISO: Geneva, Switzerland, 1995.
75. Nelson, D.; Sommers, L. Total carbon, organic carbon, and organic matter. In *Method of Soil Analysis: Chemical Methods*; Sparks, D.L., Ed.; American Society of Agronomy: Madison, WI, USA, 1996; pp. 1201–1229.
76. Egner, H.; Riehm, H.; Domingo, W. Untersuchungen über die chemische bodenanalyse als grundlage für die beurteilung des nährstoffzustandes der böden. II. Chemische extractionsmethoden zur phospor- und kaliumbestimmung. *Ann. Royal Agricult. Coll. Swed.* **1960**, *26*, 199–215.
77. Schlichting, E.; Blume, H.; Stahr, K. *Bodenkundliches Praktikum*; Pareys studientexte 81; Blackwell Wissenschafts-Verlag: Berlin, Germany, 1995.
78. *ISO 11047*; Soil Quality—Determination of Cadmium, Chromium, Cobalt, Copper, Lead, Manganese, Nickel and Zinc—Flame and Electrothermal Atomic Absorption Spectrometric Methods. International Organization for Standardization: Geneva, Switzerland, 1998.
79. US-EPA Method 3051A. *Microwave Assisted acid Digestion of Sediment, Sludges, Soils, and Oils*; United States Environmental Protection Agency: Washington, DC, USA, 2007; pp. 1–30. Available online: https://www.epa.gov/sites/production/files/2015-12/documents/3051a.pdf (accessed on 20 December 2022).
80. Ostrowska, A.; Gawliński, S.; Szczubiałka, Z. *Methods for Analysis and Evaluation of Soil and Plant Properties*; Institute of Environmental Protection: Warsaw, Poland, 1991; pp. 1–334.

81. Tibco Software Inc. Statistica Version 13.3; Data Analysis Software System; Tibco Software Inc.: Palo Alto, CA, USA. 2021. Available online: http://statistica.io (accessed on 4 January 2023).
82. Wyszkowski, M.; Brodowska, M.S. Phytoextraction with maize of soil contaminated with copper after application of mineral and organic amendments. *Agronomy* **2020**, *10*, 1597. [CrossRef]
83. Wang, C.; Zhang, S.H.; Wang, P.F.; Hou, J.; Zhang, W.J.; Li, W.; Lin, Z.P. The effect of excess Zn on mineral nutrition and antioxidative response in rapeseed seedlings. *Chemosphere* **2009**, *75*, 1468–1476. [CrossRef]
84. Guo, X.Y.; Zuo, Y.B.; Wang, B.R.; Li, J.M.; Ma, Y.B. Toxicity and accumulation of copper and nickel in maize plants cropped on calcareous and acidic field soils. *Plant Soil* **2010**, *333*, 365–373. [CrossRef]
85. Kicińska, A.; Pomykała, R.; Izquierdo-Diaz, M. Changes in soil pH and mobility of heavy metals in contaminated soils. *Eur. J. Soil Sci.* **2022**, *73*, 13203. [CrossRef]
86. Vareda, J.P.; Valenta, A.J.M.; Durães, L. Heavy metals in Iberian soils: Removal by current adsorbents/amendments and prospective for aerogels. *Adv. Colloid. Interfac.* **2016**, *237*, 28–42. [CrossRef] [PubMed]
87. Shi, W.-Y.; Shao, H.-B.; Li, H.; Shao, M.-A.; Du, S. Progress in the remediation of hazardous heavy metal-polluted soils by natural zeolite. *J. Hazard. Mater.* **2009**, *170*, 1–6. [CrossRef] [PubMed]
88. Mondal, M.; Biswas, B.; Garai, S.; Sarkar, S.; Banerjee, H.; Brahmachari, K.; Bandyopadhyay, P.K.; Maitra, S.; Brestic, M.; Skalicky, M.; et al. Zeolites enhance soil health, crop productivity and environmental safety. *Agronomy* **2021**, *11*, 448. [CrossRef]
89. Huang-Ping, C.; Shu-Hao, C. Adsorption characteristics of both cationic and oxyanionic metal ions on hexadecyltrimethylammonium bromide-modified NaY zeolite. *Chem. Eng. J.* **2012**, *193–194*, 283–289. [CrossRef]
90. Farooqi, Z.U.R.; Ahmad, I.; Abdul Qadir, A.; Murtaza, G.; Rafiq, S.; Jamal, A.; Zeeshan, N.; Murtaza, B.; Javed, W.; Radicetti, E.; et al. Zeolite-assisted immobilization and health risks of potentially toxic elements in wastewater-irrigated soil under brinjal (*Solanum melongena*) cultivation. *Agronomy* **2022**, *12*, 2433. [CrossRef]
91. Chen, Z.S.; Lee, G.J.; Liu, J.C. The effects of chemical remediation treatments on the extractability and speciation of cadmium and lead in contaminated soils. *Chemosphere* **2000**, *41*, 235–242. [CrossRef] [PubMed]
92. Suárez, M.; García-Romero, E. Variability of the surface properties of sepiolite. *Appl. Clay Sci.* **2012**, *67–68*, 72–82. [CrossRef]
93. Abad-Valle, P.; Álvarez-Ayuso, E.; Murciego, A.; Pellitero, E. Assessment of the use of sepiolite amendment to restore heavy metal polluted mine soil. *Geoderma* **2016**, *280*, 57–66. [CrossRef]
94. Lin, D.S.; Liu, Y.; Xu, Y.M.; Zhou, Q.X.; Sun, G.H. Effects of sepiolite on the immobilization of cadmium and zinc in soil. *Acta Sci. Nat. Univ. Pekin.* **2010**, *46*, 346–350.
95. Liang, X.F.; Xu, Y.M.; Wang, L.; Su, G.H.; Qin, X.; Sun, Y. In-situ immobilization of cadmium and lead in a contaminated agricultural fields by adding natural clays combined with phosphate fertilizer. *Acta Sci. Circumstantiae* **2011**, *31*, 1011–1018.
96. Sun, Y.; Sun, G.; Xu, Y.; Wang, L.; Liang, X.; Lin, D. Assessment of sepiolite for immobilization of cadmium-contaminated soils. *Geoderma* **2013**, *193–194*, 149–155. [CrossRef]
97. Sun, Y.; Zhao, D.; Xu, Y.; Wang, L.; Liang, X.; Shen, Y. Effects of sepiolite on stabilization remediation of heavy metal-contaminated soil and its ecological evaluation. *Front. Environ. Sci. Eng.* **2016**, *10*, 85–92. [CrossRef]
98. Borisover, M.; Davis, J.A. Adsorption of inorganic and organic solutes by clay minerals. *Dev. Clay Sci.* **2015**, *6*, 33–70. [CrossRef]
99. Kumpiene, J.; Lagerkvist, A.; Maurice, C. Stabilization of As, Cr, Cu, Pb and Zn in soil using amendments—A review. *Waste Manag.* **2008**, *28*, 215–225. [CrossRef]

materials

MDPI

Article

Kinetics and Thermodynamics of Adsorption for Aromatic Hydrocarbon Model Systems via a Coagulation Process with a Ferric Sulfate–Lime Softening System

Deysi J. Venegas-García and Lee D. Wilson *

Department of Chemistry, University of Saskatchewan, 110 Science Place, Thorvaldson Building (Room 165), Saskatoon, SK S7N 5C9, Canada

* Correspondence: lee.wilson@usask.ca; Tel.: +1-306-966-2961; Fax: +1-306-966-4730

Abstract: The adsorption mechanisms for model hydrocarbons, 4-nitrophenol (PNP), and naphthalene were studied in a coagulation-based process using a ferric sulfate–lime softening system. Kinetic and thermodynamic adsorption parameters for this system were obtained under variable ionic strength and temperature. An in situ method was used to investigate kinetic adsorption profiles for PNP and naphthalene, where a pseudo-first order kinetic model adequately described the process. Thermodynamic parameters for the coagulation of PNP and naphthalene reveal an endothermic and spontaneous process. River water was compared against lab water samples at optimized conditions, where the results reveal that ions in the river water decrease the removal efficiency (RE; %) for PNP (RE = 28 to 20.3%) and naphthalene (RE = 89.0 to 80.2%). An aluminum sulfate (alum) coagulant was compared against the ferric system. The removal of PNP with alum decreased from RE = 20.5% in lab water and to RE = 16.8% in river water. Naphthalene removal decreased from RE = 89.0% with ferric sulfate to RE = 83.2% with alum in lab water and from RE = 80.2% for the ferric system to RE = 75.1% for alum in river water. Optical microscopy and dynamic light scattering of isolated flocs corroborated the role of ions in river water, according to variable RE and floc size distribution.

Keywords: *p*-nitrophenol; naphthalene; coagulation; kinetics; thermodynamics; ferric sulfate; lime-softening; aluminum sulfate; river water

Citation: Venegas-García, D.J.; Wilson, L.D. Kinetics and Thermodynamics of Adsorption for Aromatic Hydrocarbon Model Systems via a Coagulation Process with a Ferric Sulfate–Lime Softening System. *Materials* **2023**, *16*, 655. https://doi.org/10.3390/ma16020655

Academic Editors: Roberta G. Toro, Daniela Caschera and Abeer M. Adel

Received: 17 November 2022
Revised: 5 January 2023
Accepted: 8 January 2023
Published: 10 January 2023

1. Introduction

Many chemical process industries produce large volumes of oily wastewater that causes serious environmental problems when wastewater is not adequately treated [1]. In addition, oil spillage recurrently occurs during oil extraction, transportation, and storage processes, which leads to unintended soil and water contamination [2]. Hazardous chemicals are released from oil spills such as polycyclic aromatic hydrocarbons (PAHs), which are known to have harmful effects on ecosystems and human health [3]. Among the PAHs, naphthalene is relatively hydrophobic with relatively high melting/boiling points, electrochemical stability, and the propensity to reside in water, and in soils for long periods [4]. Among the various substituted arenes, *p*-nitrophenol (PNP) is regarded as a priority pollutant in industrial and agricultural wastewaters with high chemical stability, low water solubility, and high toxicity [5]. Methods such as filtration, adsorption, and coagulation are relevant for the treatment of effluent, apart from other water treatment methods [6]. Many researchers have reported the effectiveness of coagulation for the treatment of effluent with high levels of colloidal particles, organic matter, and suspended solids [7]. Coagulation is a suitable removal process for colloidal particles (negatively charged) that undergo neutralization by introduction of counter ions (positively charged) in the form of coagulants, which result in particle aggregation [8]. Chemical coagulants such as aluminum sulfate (alum), poly-aluminum chloride (PAC), alum potash, ferric sulfate or ferric chloride, among other coagulants, have been employed [9]. Four mechanisms that are either adsorption or non-adsorption processes may serve to drive coagulation [10]. Among them, inter-particle bridging and charge neutralization are the most significant factors that favour adsorption, while double layer compression and sweep flocculation are the non-adsorption factors [11]. In conventional industrial water treatment at the municipal level, lime-softening is a traditional and low-cost method used to reduce the hardness of water [12]. Lime addition to hard water (i.e., lime-softening) serves to increase the pH of water. In turn, bicarbonates transform into carbonates, where Ca^{2+} and Mg^{2+} are eliminated from water in the form of $CaCO_3$ and $Mg(OH)_2$ species, where such water softening is conceived for reducing the levels of calcium and magnesium ions [13]. Coagulation along with lime-softening has been proven to be an effective combination during water treatment processes for the removal of colloidal particles [14].

To reduce environmental damage to a minimum when accidental oil spills occur in aquatic environments, a scenario-specific response strategy was developed in a previous study [15]. Herein, the main aim of this work was to study the adsorption properties (kinetic and thermodynamic) of PNP and naphthalene that employs a ferric sulfate–lime softening system via a chemical coagulation process. The comparison of different coagulant systems, ferric vs. aluminum sulfate, and source water samples (laboratory vs. saline river water) was also studied for the removal of model contaminants (PNP and naphthalene). Finally, microscopy images and floc size were used to gain insight on the differences between the use of ferric sulfate over aluminum sulfate, and the effect of surface water on the properties of the flocs.

2. Materials and Methods

2.1. Materials

All of the chemicals were of analytical reagent (AR) grade. 4-nitrophenol (99%) was obtained from Alfa Aesar (Tewksbury, MA, USA). Naphthalene (99.7%) was acquired from Mallinckrodt Pharmaceuticals (Blanchardstown, Dublin, Ireland). Potassium phosphate dibasic (98%) and potassium phosphate monobasic (99%) were purchased from Fisher Scientific (New York, NY, USA). Calcium carbonate (99%) was obtained from BDH Chemicals (Mississauga, ON, Canada). All of the stock solutions were prepared using 18 MΩ cm Millipore water (Burlington, MA, USA), unless specified otherwise. Ferric sulfate (60% *w*/*v*), lime (10% *w*/*v*) solutions, and water samples were obtained from the South Saskatchewan River water samples after primary treatment at the City of Saskatoon Water Treatment Plant (Saskatoon, SK, Canada). Aluminum sulfate (48.5% *w*/*v*) was contributed by Chemtrade Logistics (Saskatoon, SK, Canada).

2.2. Kinetic Studies

Kinetic profiles for the coagulation process reflect the change in the adsorption rate and time according to the environmental conditions [16]. Kinetic adsorption equations such as the pseudo-first order (PFO) and pseudo-second order (PSO) models were used to examine the pollutant adsorption profiles [17]. The non-linear forms of the PFO and PSO kinetic models are expressed by Equations (1) and (2), respectively [18]:

$$q_t = q_e\left(1 - exp^{-k_1 t}\right) \tag{1}$$

$$q_t = \frac{{q_e}^2\, k_2 t}{1 + q_e k_2 t} \tag{2}$$

where q_t (mg g^{-1}) and q_e (mg g^{-1}) indicate the adsorption capacity of ferric sulfate towards pollutant removal at the time (t) and equilibrium. k_1 (min^{-1}) and k_2 (g mg^{-1} min^{-1}) are the respective rate constants for the PFO and PSO models.

Kinetic studies were performed using the one-pot method reported previously [19]. Briefly, a 600 mL beaker was filled with 500 mL of laboratory prepared water solution with a model pollutant that was mixed via magnetic stirring. A filter paper (Whatman no. 40) was folded into a cone and attached to the beaker that was immersed into the solution to a 2 cm depth (cf. Figure 1). An initial 2.5 mL aliquot was taken from the inner filter cone which is considered as the concentration at time (t) = 0 (initial concentration), where lime was then added to the solution with a 5 min mixing time. After adding the coagulant, the sampling continued at 1 min intervals for 3 min at 295 rpm, followed by 1 min intervals for 7 min additionally at 20 rpm, then for 10 min at 2 min intervals, and finally 40 min at 5 min intervals at a 20 rpm mixing speed. Then, stirring was stopped (t = 60 min) and sampling continued for a further 40 min at 10 min intervals. Sample aliquots were prepared for UV–vis analysis, PNP λ_{max} = 400 nm and naphthalene λ_{max} = 220 nm [15]. The adsorption capacity at different times was estimated by Equation (2). To measure adsorption kinetics at variable temperatures, the one-pot experiment was carried using an Endocal refrigerated circulating bath (−40 to 40 °C) with flow control (Neslab, Newington, NH, USA) at 22, 15, and 5 °C.

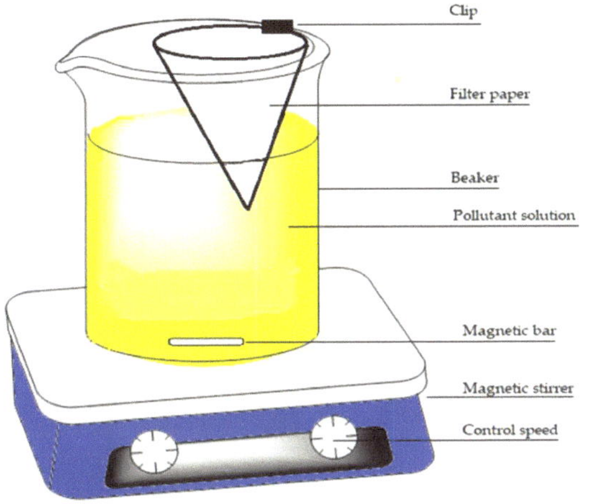

Figure 1. Diagram for the one-pot kinetic system.

2.3. Thermodynamic Study of Adsorption

Thermodynamic parameters play an important role in predicting adsorption behavior and they are strongly temperature dependent [20]. Thermodynamic parameters evaluated for PNP and naphthalene adsorption include the standard Gibbs energy change (ΔG°), standard enthalpy change (ΔH°), and the standard entropy change (ΔS°). These parameters were calculated using Equation (3), along with Equations (4) and (5) [21].

$$\Delta G^{\circ} = -RTlnK_e \tag{3}$$

where K_e is a thermodynamic equilibrium constant (L g^{-1}) for a liquid to solution partitioning process at equilibrium, as follows: $K_e = q_e/C_e$. The variable, q_e, refers to the amount or concentration of the adsorbate (PNP or naphthalene) adsorbed per unit mass of the adsorbent (mg g^{-1}) and C_e is the equilibrium concentration of PNP or naphthalene in solution (mg L^{-1}). R is the universal gas constant (8.314 J mol^{-1} K^{-1}) and T is the temperature in Kelvin (K). Equation (4) outlines the relationship between ΔG°, ΔH°, and ΔS°, according to T (K):

$$\Delta G^{\circ} = \Delta H^{\circ} - T\Delta S^{\circ} \tag{4}$$

Substituting Equation (3) into Equation (4), and re-arranging them gives Equation (5):

$$lnk_e = -\frac{\Delta H^{\circ}}{RT} + \frac{\Delta S^{\circ}}{R} \tag{5}$$

The values for the standard change for ΔH° and ΔS° were determined from the slope and intercept of the plot of lnK_e versus 1/T (slope = $-\Delta H^{\circ}/R$ and intercept = $\Delta S^{\circ}/R$) [22]. The standard change in Gibbs energy (ΔG°) for the adsorption process was calculated by employing Equation (4).

2.4. Coagulation Process

The coagulation process was performed using a program-controlled conventional jar test apparatus with six 2 L jars and stirrers, at room temperature (22 °C). Approximately 1 L of water containing the model pollutant was added to the jar test setup. For lab water samples, a simulation of hardness was carried out with the standard addition of $CaCO_3$. An aliquot of the simulated polluted water solution was sampled to measure the initial pollutant concentration. The coagulation experiment was carried out by adapting the procedure described by Agbovi and Wilson [10]. A predetermined amount of lime solution was added to the solution that was mixed for 20 min, where the coagulant was then added, followed by rapid stirring for 3 min at 295 rpm. Thereafter, the stirring rate was reduced to 25 rpm for 20 min. The stirring was stopped to allow the sample to settle overnight, where a 10 mL sample was centrifuged at 500 rpm for 30 min. The supernatant was analyzed by measurement of the UV−vis spectrum. PNP was buffered in order to measure concentration at pH 7. Figure 2 shows a diagram that illustrates the steps for the coagulation process and analysis.

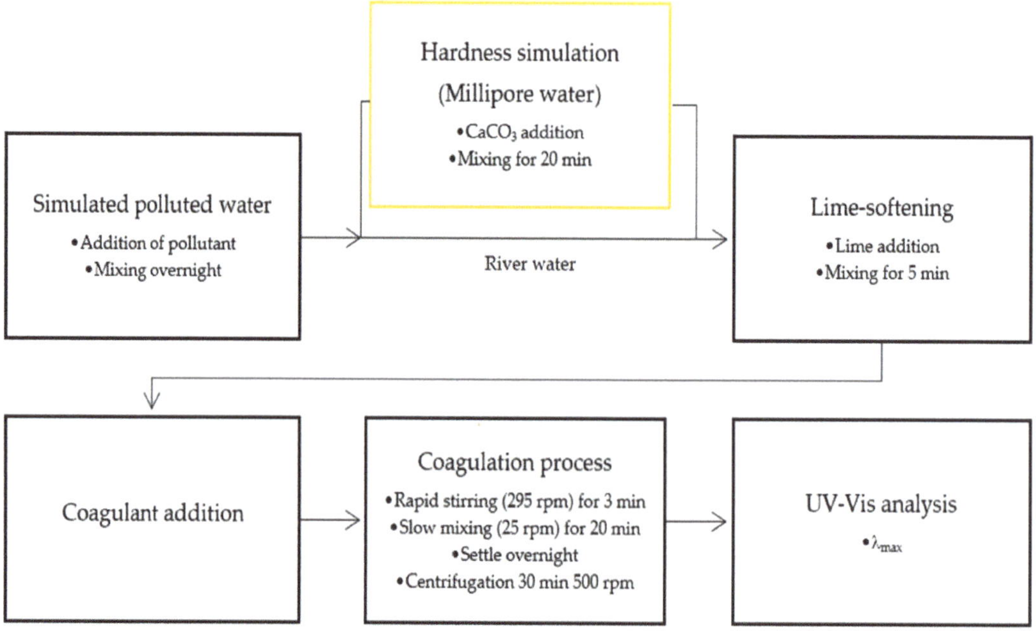

Figure 2. Block diagram for coagulation process for the removal of hydrocarbon oil models from lab (Millipore) versus river water samples.

Optimal coagulation conditions for PNP and naphthalene removal were obtained from a previous work and are listed in Table 1. Aluminum sulfate dosage was calculated with preliminary experiments considering optimal lime dosage and changing aluminum sulfate from 1 to 100 mg L^{-1}. The results reveal a higher RE (%) for PNP at 68 mg L^{-1} and for naphthalene at 75 mg L^{-1}.

Table 1. Optimal conditions for PNP and naphthalene removal [1].

Model	Initial Concentration (mg L^{-1})	Ferric Sulfate (mg L^{-1})	Lime (mg L^{-1})
PNP	2	74.5	101.6
Naphthalene	16.3	42	21

[1] Data obtained from a reported study [15].

The removal efficiency (RE, %) and the adsorption capacity (q_e) (mg·g^{-1}) were calculated by Equations (6) and (7), respectively [23].

$$\text{RE } (\%) = \frac{C_o - C_e}{C_o} * 100 \tag{6}$$

$$q_e = \frac{(C_o - C_e) * V}{m} \tag{7}$$

Here, C_o and C_e are the initial and equilibrium pollutant concentrations (mg L^{-1}), V is the volume (L), and m is the weight (g) of the coagulant metal salt system.

2.5. Optical Microscopy

Small samples of isolated wet flocs were deposited onto a microscope slide for analysis using optical microscopy. To avoid alteration of the floc structure due to compression effects, no cover slips were used during microscopy. Digital images of the flocs were captured on a Renishaw InVia Reflex Raman microscope (Renishaw plc, New Mills, UK) with 5× magnification with conventional backlighting.

2.6. Particle Size Distribution

The particle size of flocs that contain PNP and naphthalene were determined by dynamic light scattering (DLS) using a Malvern Zetasizer Nano ZS particle size analyzer (Worcestershire, UK). For this measurement, 0.02 mL floc suspension was diluted with 0.8 mL from the supernatant after centrifuging the floc suspension. A sample volume of 0.75 mL was used to measure the particle size in a disposable folded capillary cell (DTS1070) at 25 °C. Measured in triplicate, each measurement comprised an acquisition of 10 times and default settings for the refractive index of polystyrene particles dispersed in water were employed.

3. Results and Discussion

3.1. Kinetics

To study the effect of ferric sulfate on pollutant removal, PNP and naphthalene were employed as model hydrocarbon pollutants for this study. The adsorption process of PNP and naphthalene can occur into two stages: a fast initial adsorption stage, where ca. 90% of the equilibrium adsorption capacity was achieved during the subsequent slow adsorption phase, where the adsorption capacity gradually reached pseudo-equilibrium conditions. The kinetic behavior of ferric sulfate for the removal of PNP and naphthalene is shown in Figure 3, where the two kinetic models (PFO and PSO) were used to fit the kinetic adsorption profiles for removal of these model pollutants.

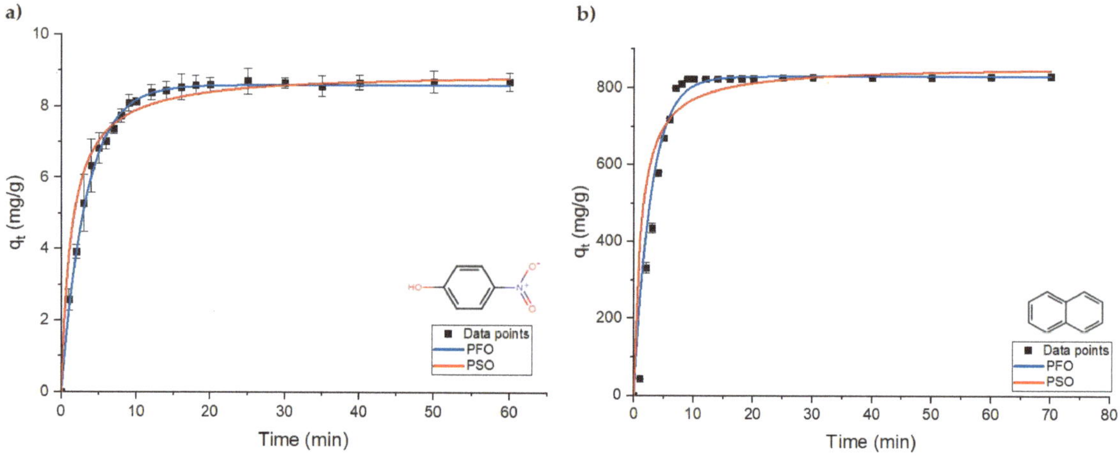

Figure 3. Kinetic profiles for the adsorption-based removal of model oil pollutants: (**a**) PNP and (**b**) naphthalene. The fitted lines correspond to the best-fit results for the PFO and PSO kinetic models.

The corresponding best-fit kinetic parameters are listed in Table 2. The high correlation with the PFO model provides insight on the dependence of model pollutant (PNP or naphthalene) removal for the coagulation process. Based on the R^2 values, the PNP and naphthalene uptake with ferric sulfate obeys the PFO model because of its higher R^2 values, in contrast with the lower R^2 values obtained by the PSO model.

Table 2. Kinetic adsorption parameters for ferric sulfate coagulation process for PNP and naphthalene.

Contaminant	Pseudo-First Order Model			Pseudo-Second Order Model		
	k_1 (min^{-1})	q_e (mg g^{-1})	R^2	k_2 (min^{-1})	q_e (mg g^{-1})	R^2
PNP	0.3091 ± 0.002	8.609 ± 0.037	0.997	0.0459 ± 0.0044	9.4503 ± 0.1366	0.981
Naphthalene	0.3086 ± 0.031	846.9 ± 18.4	0.943	$4.754 \times 10^{-4} \pm 1.2401 \times 10^{-4}$	932.5 ± 39.87	0.873

To evaluate the removal of PNP and naphthalene in real water samples, it is crucial to study the effect of temperature on the kinetics for the removal of the model compounds in the river water system. Figure 4 shows the kinetics for PNP and naphthalene removal at 5, 15, and 22 °C. Kinetic results for PNP show that by decreasing the temperature from 22 to 5 °C, the adsorption capacity decreased from 8 to 4 mg g^{-1}. Otherwise, naphthalene adsorption does not reveal a significant dependence with decreasing temperature (22 to 5 °C). Floc aggregation tends to be weaker when temperature decreases, because of the attenuated motional dynamics. This leads to fewer particle–particle collisions where the collision energy is low, with decreased coagulation efficiency [24].

3.2. Thermodynamics

The thermodynamic behavior for PNP and naphthalene removal using the ferric sulfate–lime softening coagulant system was determined by varying the temperature from 22 to 5 °C. The thermodynamic parameters such as the standard difference in Gibbs energy (ΔG°), enthalpy change (ΔH°), and entropy change (ΔS°) were determined from the van 't Hoff plot, as illustrated in Figure 5.

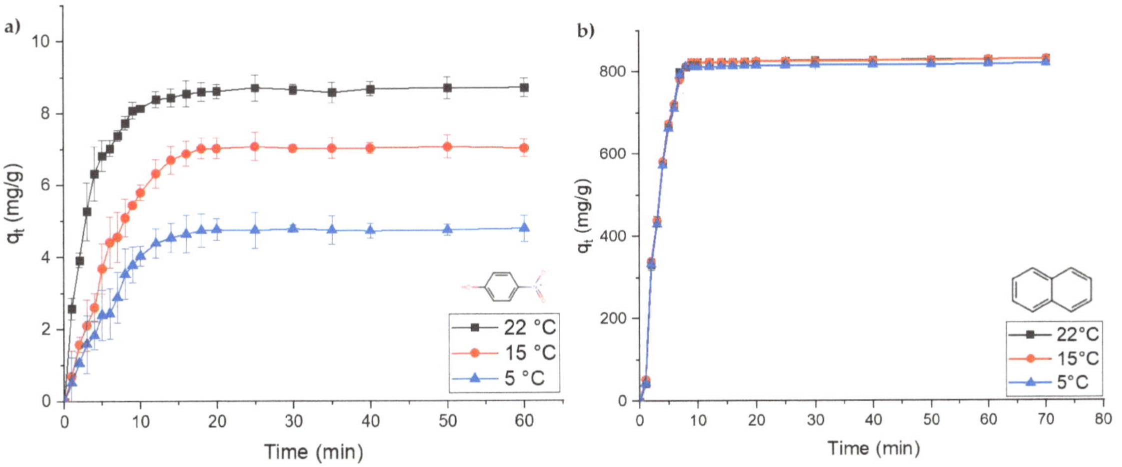

Figure 4. Kinetics for the adsorption of model pollutants at variable temperatures (5, 15, and 22 °C): (**a**) PNP and (**b**) naphthalene. The solid line represents a visual guide to illustrate the trend of the kinetic profile.

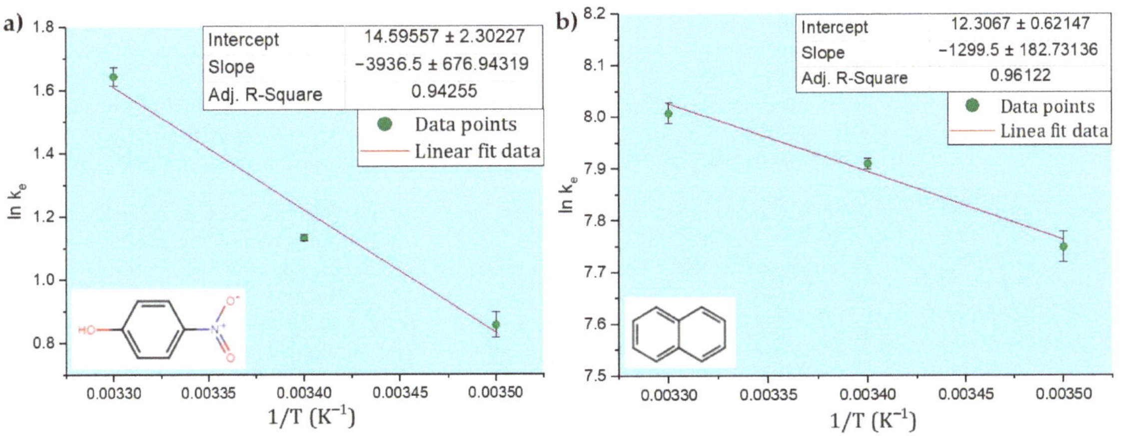

Figure 5. van 't Hoff plots for the model hydrocarbons in the adsorption process: (**a**) PNP and (**b**) naphthalene.

Table 3 shows the thermodynamic parameters $\Delta G°$, $\Delta H°$, and $\Delta S°$. The negative value of $\Delta G°$ indicates the spontaneity of the process and shows that the adsorption is favorable for PNP and naphthalene, which proceeds less favorably as the temperature decreases. Negative values of the Gibbs free energy ($\Delta G°$) with greater magnitude indicates a greater driving force for the adsorption, where a decrease in q_e with decreasing temperature indicates a lower adsorption at lower temperatures [25]. Positive values for the standard change in enthalpy ($\Delta H°$) confirmed that the process was endothermic for both systems [26]. The predominance of physical adsorption was verified by the magnitude of enthalpy change. According to Ohale et al. [27], an adsorptive process is physical in nature if the values of $\Delta H° < 80$ kJ mol^{-1}. The values of enthalpy obtained in this study meet this criterion: 32.7 kJ mol^{-1} for PNP and 10.8 kJ mol^{-1} for naphthalene. Positive values obtained for $\Delta S°$ show greater disorder (randomness) during the adsorption process.

Table 3. Thermodynamic parameters for ferric sulfate coagulation process for PNP and naphthalene.

Temp (K)	(1/K)	C_e	q_e (mg g^{-1})	K_e (L g^{-1})	ln K_e	$\Delta G°$ (kJ mol^{-1})	$\Delta H°$ (kJ mol^{-1})	$\Delta S°$ (J mol^{-1} K^{-1})
					PNP			
295.15	0.0033	1.631	8.44	5.177	1.6442	−3.08		
288.15	0.0034	1.700	5.28	3.106	1.1333	−2.23	32.72	121.4
278.15	0.0035	1.868	4.40	2.355	0.8569	−1.02		
					Naphthalene			
295.15	0.0033	0.279	836.8	3004.6	8.0078	−19.39		
288.15	0.0034	0.289	836.3	2891.7	7.9695	−18.72	10.80	102.5
278.15	0.0035	0.360	832.9	2316.9	7.7479	−17.70		

3.3. Coagulation Assay with South Saskatchewan River Water and Aluminum Sulfate

This section discusses the application of the coagulant systems studied, along with real water samples from the South Saskatchewan River and the alternative use of alum (aluminum sulfate) as a coagulant. Table 4 shows the removal of PNP and naphthalene at optimal conditions for river water samples and lab water samples (prepared in Millipore water). The PNP removal using ferric sulfate as a coagulant decreases from 28.0 to 20.3% when river water was used versus laboratory prepared samples. Considering the presence of different ions and organic compounds, there exists the possibility of competition between PNP and other ions present in the river water system [28]. For samples that employ alum as a coagulant, coagulation-based removal occurs in lab water at a lower level (20.5%) compared to ferric sulfate (28.0%). The effect of using river water with alum as the coagulant results in decreased removal of PNP from 20.5 to 16.8%, as compared with lab water. The size of flocs with alum as the coagulant are small when compared with flocs obtained using ferric sulfate. When naphthalene in a lab water sample undergoes coagulation with ferric sulfate, higher removal efficiency can be achieved (89.0%). This removal decreased to 80.2% when river water was used, assuming the effect of competition relates mainly to the presence of greater ionic strength of the media. Based on an analysis of river water, the amount of dissolved solids is estimated at 250 mg/L. By comparison, the total organic carbon (TOC) was 4.9 mg/L, which indicates that the major fraction of the dissolved solids is inorganic ($\approx$98%), in agreement with the role of prominent ionic strength effects in river water versus lab water. The use of alum as the coagulant for naphthalene removal has a lower RE (83.2%) when compared with ferric sulfate (89.0%). The same effect was observed when river water was used, as evidenced by decreasing naphthalene removal from 83.2 to 75.1%.

Table 4. Naphthalene and PNP removal efficiency (RE) with ferric and aluminum sulfate in river water samples.

Contaminant	Water	Coagulant	RE (%)
PNP	Lab	Ferric sulfate	28.0 ± 0.1
	River	Ferric sulfate	20.3 ± 0.1
	Lab	Aluminum sulfate	20.5 ± 0.3
	River	Aluminum sulfate	16.8 ± 0.4
Naphthalene	Lab	Ferric sulfate	89.0 ± 0.2
	River	Ferric sulfate	80.2 ± 0.1
	Lab	Aluminum sulfate	83.2 ± 0.3
	River	Aluminum sulfate	75.1 ± 0.5

3.4. Floc Characterization

3.4.1. Floc Images

To obtain further insight into the floc properties after the coagulation process, the flocs were extracted from the bottom of the jar test beaker after precipitation. Floc images (5× magnification) were captured on a Renishaw InVia Reflex Raman microscope (Renishaw plc, New Mills, UK). Some features can be appreciated from these images. PNP flocs formed with the ferric sulfate system in lab water (Figure 6a) reveal variable particle sizes (main floc at center and small flocs around). Figure 6b shows more compacted flocs (ferric sulfate–river water). The presence of the coagulant system (ferric-lime water) with different inorganic ions (CO_3^{2-}, Cl^-, F^-, SO_4^{2-}, etc.), along with trace organic compounds (ca. 2%; cf. Table S1 in the Supplementary Material) could create bigger and well-defined flocs [29]. PNP flocs formed under the alum system for lab (cf. Figure 6c) or river water (cf. Figure 6d) samples present similar trends.

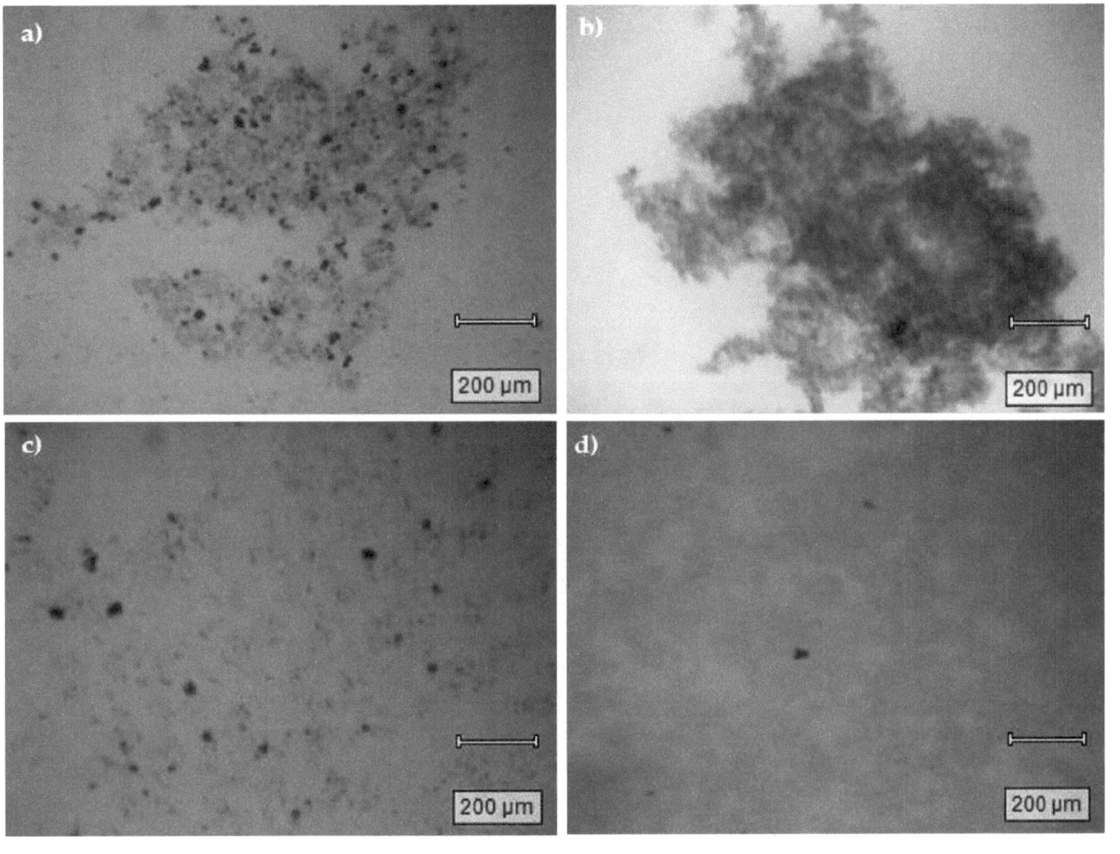

Figure 6. Floc sample images (5× magnification) obtained for variable coagulant systems: (**a**) PNP with lab water and ferric sulfate; (**b**) PNP with river water and ferric sulfate; (**c**) PNP with lab water and aluminum sulfate; (**d**) PNP with river water and aluminum sulfate.

Figure 7 shows flocs for naphthalene under different conditions where larger and well-defined flocs for systems using ferric sulfate as the coagulant was observed. The same behavior for PNP is observed for naphthalene, when the coagulation process was applied to river water samples (cf. Figure 7b), where bigger and denser flocs are formed. A similar behavior was observed for naphthalene flocs formed when employing the alum-based coagulant system in lab (cf. Figure 7c) or river water (cf. Figure 7d). The trend for naphthalene-based flocs is consistent with the absence of an ionic charge for this pollutant, in contrast with PNP-based flocs that possess a defined ionic charge due to ionization of PNP at the alkaline conditions employed (cf. RE (%) values in Table 4).

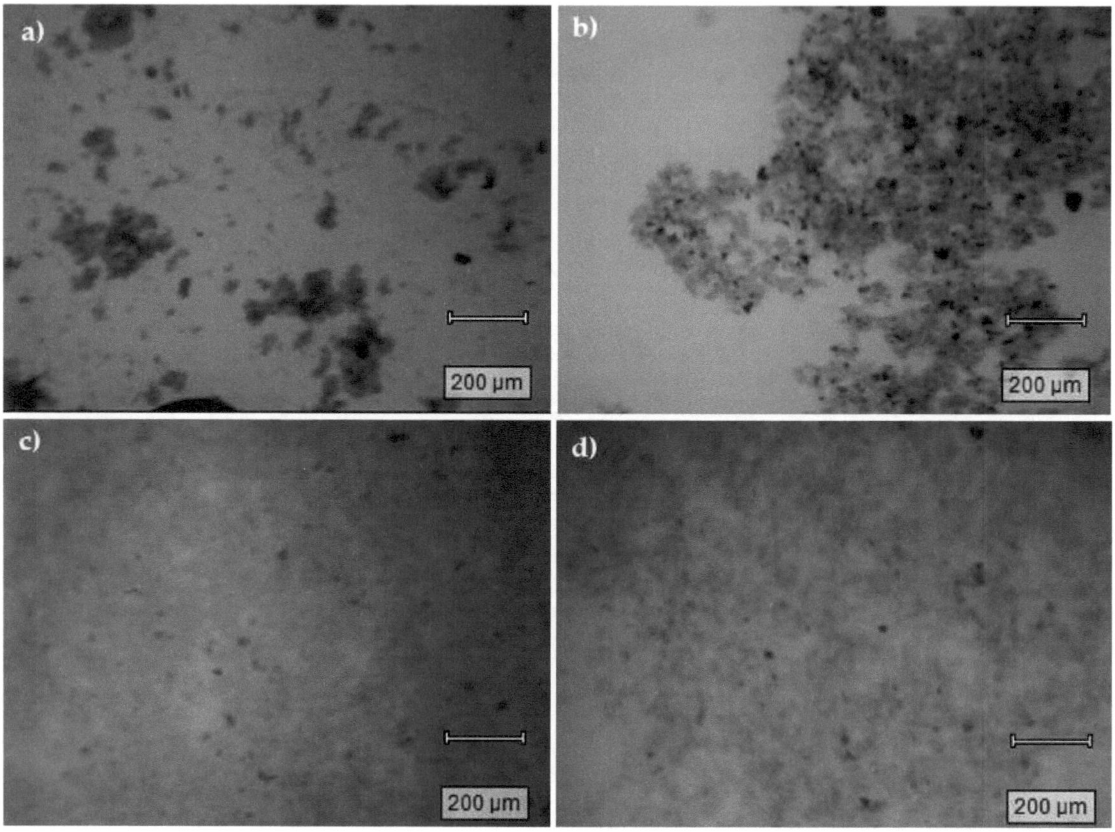

Figure 7. Floc sample images (5× magnification) obtained for variable coagulant systems: (**a**) naphthalene with lab water and ferric sulfate; (**b**) naphthalene with river water and ferric sulfate; (**c**) naphthalene with lab water and aluminum sulfate; (**d**) naphthalene with river water and aluminum sulfate.

Flocs are a complex mixture with FeOOH, PNP or naphthalene, and calcium carbonates (total alkalinity for river water = 113 mg/L, as $CaCO_3$), among other inorganic species. Hydrated iron salts can form different chemical species. The dark regions can be aggregates of ferric or calcium species due to their dominant contributions as cations, and their complexes with inorganic anions, which appeared as dark yellow under the microscope. The optical microscopy images were used to mainly identify differences in floc size and density to correlate with the kinetics of the coagulation process. In general, there appears to be a correlation between the floc density and the ionic strength upon comparison of ferric flocs in lab versus river water for PNP (cf. Figure 6a,b) and naphthalene (cf. Figure 7a,b). This trend is in accordance with the role of charge neutralization effects and the greater floc density observed for river water samples. This trend concurs with the greater ionic strength of river water and the reduced kinetic barrier for floc formation for either model pollutant (PNP or naphthalene).

3.4.2. Flocs Size

The results obtained related particle size (nm) and polydispersity index (PDI) in aqueous media are presented in Table 5. The PDI is a parameter used to describe the size particle range of flocs. The measured values could be any denomination between 0 and 1, where lower values indicate more homogenous dispersions than higher values. Usually, values < 0.5 are considered acceptable concerning the homogeneity of the prepared dispersions [30]. PNP flocs with lab water showed values for PDI higher than 0.5 (ferric sulfate PDI = 0.7 ± 0.1 and aluminum PDI = 0.8 ± 0.1). These flocs are considered as non-homogeneous dispersions. PNP flocs formed with river water present for PDI values of 0.5 ± 0.1 for ferric and aluminum sulfate. These results showed that size of the flocs are more homogeneous when river versus lab water was used. This trend was attributed to the presence of greater ion levels (ca. 250 mg/L) in river water (cf. Table S1, Supplementary Material). The trends for lab and river water corroborate the images for the floc results presented above (cf. Section 3.4.1).

Table 5. Particle size distribution of floc samples that contain PNP and naphthalene model pollutants.

Sample			Size (nm)	PDI
Contaminant	Water	Coagulant		
PNP	Lab	Ferric sulfate	1950.3 ± 424.8	0.7 ± 0.1
PNP	Lab	Aluminum sulfate	2200.0 ± 717.3	0.8 ± 0.1
PNP	River	Ferric sulfate	1795.0 ± 223.3	0.5 ± 0.1
PNP	River	Aluminum sulfate	2404.3 ± 242.7	0.5 ± 0.1
Naphthalene	Lab	Ferric sulfate	2059.3 ± 1323.3	1.0 ± 0.0
Naphthalene	Lab	Aluminum sulfate	3156.7 ± 368.5	0.6 ± 0.1
Naphthalene	River	Ferric sulfate	3274.3 ± 1202.6	1.0 ± 0.0
Naphthalene	River	Aluminum sulfate	2519.7 ± 615.1	0.4 ± 0.1

In lab water, the naphthalene flocs size presents a PDI = 1.0 for the system with ferric sulfate, but a lower value (PDI = 0.6) when alum was the coagulant. Even when both PDI values are above 0.5, naphthalene flocs formed with alum are apparently more homogenous than using ferric sulfate as the coagulant. When river water is used for naphthalene removal, the ferric sulfate PDI value is the same as that for lab water. Naphthalene uptake is lower for river water (RE% = 80.2) than lab water (RE% = 89.0) with ferric sulfate as coagulant. Based on these results, we can assume that the difference in the homogeneity of the flocs homogeneity strongly relates to the role of surface charge on the flocs and the ionic strength of the media. For the case of PNP (cf. Figure 6a,b), the ferric-based flocs in lab water tend to be more heterogeneous with lower density than those in river water. Similar trends are noted for naphthalene (cf. Figure 7a,b). Observed differences between alum and the hydrolysis products of ferric sulfate are anticipated for the adsorption mechanism of naphthalene due to the strong interactions between ferric species and carbonates. PDI differences for the alum systems concur with a naphthalene adsorption mechanism and effect of greater ion levels present in river water [31]. The greater density and homogeneity of the flocs formed in river can be related to the greater ionic strength of this media over lab water and the reduced nucleation barrier of flocs formed at higher ionic strength.

4. Conclusions

Coagulation of two types of model hydrocarbon oil-based compounds were studied as representative pollutants: *p*-nitrophenol (PNP) and naphthalene. Herein, an optimized ferric sulfate–lime softening coagulant system was adapted from previous work to characterize the kinetic and thermodynamic parameters of the coagulation-based removal process. The kinetic results were obtained using an in situ (one-pot) method, which showed that the PFO model provided a better fit over the PSO kinetic model for the two model pollutants. Kinetic removal (RE; %) profiles at different temperatures (5, 15, and 22 °C) were used to evaluate PNP and naphthalene. Upon decreasing the temperature from 22 to 5 °C, the adsorption capacity decreased from 8 to 4 mg g^{-1} for PNP, whereas temperature had a negligible effect on the naphthalene adsorption profile. Additionally, the coagulation process was spontaneous and endothermic in nature, based on the thermodynamic parameters. The prevalence of physical adsorption was verified by the enthalpy of the process: ΔH° = 32.7 kJ mol^{-1} for PNP and ΔH° = 10.8 kJ mol^{-1} for naphthalene. Aluminum sulfate was used as a reference coagulant to compare against the ferric sulfate system, for both river water and lab water samples, respectively. The removal efficiency (RE) of PNP with ferric sulfate system is lower for river water (RE = 20.3%) versus lab water (RE = 28.0%). By comparison, parallel trends are noted for alum, with the removal for PNP (RE = 16.8%) in river water versus PNP (RE = 20.5%) in lab water. Naphthalene coagulation follows parallel trends as noted for PNP, where the removal is generally higher overall, as follows: naphthalene with ferric sulfate as coagulant yields RE (80.2%) for river water versus lab water (RE = 89.0%). For the case of alum as the coagulant, the removal decreases from 83.2% (lab water) to 75.1% (river water). The floc images by optical microscopy and particle size analysis for PNP- and naphthalene-containing flocs show a corresponding trend related to the concentration of ions in the river water samples. The presence of greater ion concentration negatively affects the RE (%) for both ferric and aluminum systems, along with the floc size distribution, according to ionic strength effects in river water samples.

Supplementary Materials: The following supporting information can be downloaded at: https://www.mdpi.com/article/10.3390/ma16020655/s1. Table S1 outlines selected analytic data related to river water quality parameters that were used in this study.

Author Contributions: Conceptualization, L.D.W.; methodology, D.J.V.-G.; software, D.J.V.-G.; validation, D.J.V.-G. and L.D.W.; formal analysis, D.J.V.-G.; investigation, D.J.V.-G.; resources, L.D.W.; data curation, D.J.V.-G.; writing—original draft preparation, D.J.V.-G.; writing—review and editing, D.J.V.-G. and L.D.W.; visualization, D.J.V.-G.; supervision, L.D.W.; project administration, L.D.W.; funding acquisition, L.D.W. All authors have read and agreed to the published version of the manuscript.

Funding: This research was funded by the University of Saskatchewan and the City of Saskatoon, through the Research Junction Development Grant (Internal ID 354979) for the project entitled, "Addressing water security for the City of Saskatoon Water Treatment Plant—Evaluating the efficacy of a coagulant-flocculant system for addressing oil-based waterborne contaminants from oil spills". The APC was funded by the Materials Editorial office of MDPI.

Institutional Review Board Statement: Not applicable.

Informed Consent Statement: Not applicable.

Data Availability Statement: Not applicable.

Acknowledgments: The authors wish to make a land acknowledgement that this work was carried out in *Treaty 6 Territory and the Homeland of the Métis. As such, we pay our respect to the First Nations and Métis ancestors of this place and reaffirm our relationship with one another.* We acknowledge the support for this project through a Research Junction grant provided by the City of Saskatoon and the University of Saskatchewan. The provision of a research sample of aluminum sulfate by Chemtrade Logistics (Saskatoon, SK, Canada) is acknowledged by the authors. The technical support provided by Heng Yang, along with helpful discussions with Rob St. Pierre and Sunday Ibok, are gratefully acknowledged. Bernd G. K. Steiger is acknowledged for use of the watermark image in the graphical abstract of this contribution.

Conflicts of Interest: The authors declare no conflict of interest.

References

1. Travkin, V.M.; Solyanikova, I.P. Salicylate or Phthalate: The Main Intermediates in the Bacterial Degradation of Naphthalene. *Processes* **2021**, *9*, 1862. [CrossRef]
2. Setyawan, H.; Fauziyah, M.; Tomo, H.S.S.; Widiyastuti, W.; Nurtono, T. Fabrication of Hydrophobic Cellulose Aerogels from Renewable Biomass Coir Fibers for Oil Spillage Clean-Up. *J. Polym. Environ.* **2022**, *30*, 5228–5238. [CrossRef]

3. Anwana Abel, U.; Rhoda Habor, G.; Innocent Oseribho, O. Adsorption Studies of Oil Spill Clean-up Using Coconut Coir Activated Carbon (CCAC). *Am. J. Chem. Eng.* **2020**, *8*, 36. [CrossRef]
4. Kumar, J.A.; Krithiga, T.; Narendrakumar, G.; Prakash, P.; Balasankar, K.; Sathish, S.; Prabu, D.; Pushkala, D.P.; Marraiki, N.; Ramu, A.G.; et al. Effect of Ca2+ ions on naphthalene adsorption/desorption onto calcium oxide nanoparticle: Adsorption isotherm, kinetics and regeneration studies. *Environ. Res.* **2022**, *204*, 112070. [CrossRef] [PubMed]
5. He, X.-Q.; Cui, Y.-Y.; Zhang, Y.; Yang, C.-X. Fabrication of magnetic polydopamine@naphthyl microporous organic network nanosphere for efficient extraction of hydroxylated polycyclic aromatic hydrocarbons and p-nitrophenol from wastewater samples. *J. Chromatogr. A* **2021**, *1651*, 462347. [CrossRef]
6. Zhou, X.; Shi, L.; Moghaddam, T.B.; Chen, M.; Wu, S.; Yuan, X. Adsorption mechanism of polycyclic aromatic hydrocarbons using wood waste-derived biochar. *J. Hazard. Mater.* **2022**, *425*, 128003. [CrossRef]
7. Jaradat, A.Q.; Shtayat, A.R. A coagulation-flocculation process combined with continuous adsorption using eggshell waste materials for phenols and PAHs removal from landfill leachate. *Environ. Eng. Res.* **2021**, *27*, 210133. [CrossRef]
8. Agbovi, H.K.; Wilson, L.D. Optimisation of orthophosphate and turbidity removal using an amphoteric chitosan-based flocculant–ferric chloride coagulant system. *Environ. Chem.* **2019**, *16*, 599. [CrossRef]
9. Rosińska, A.; Dąbrowska, L. Influence of type and dose of coagulants on effectiveness of PAH removal in coagulation water treatment. *Water Sci. Eng.* **2021**, *14*, 193–200. [CrossRef]
10. Agbovi, H.K.; Wilson, L.D. Flocculation Optimization of Orthophosphate with FeCl 3 and Alginate Using the Box–Behnken Response Surface Methodology. *Ind. Eng. Chem. Res.* **2017**, *56*, 3145–3155. [CrossRef]
11. Gutierrez-Urbano, I.; Villen-Guzman, M.; Perez-Recuerda, R.; Rodriguez-Maroto, J.M. Removal of polycyclic aromatic hydrocarbons (PAHs) in conventional drinking water treatment processes. *J. Contam. Hydrol.* **2021**, *243*, 103888. [CrossRef]
12. Zhang, K.; Pernitsky, D.; Jafari, M.; Lu, Q. Effect of MgO Slaking on Silica Removal during Warm Lime Softening of SAGD Produced Water. *Ind. Eng. Chem. Res.* **2021**, *60*, 1839–1849. [CrossRef]
13. Zhang, L.; Mishra, D.; Zhang, K.; Perdicakis, B.; Pernitsky, D.; Lu, Q. Impact of influent deviations on polymer coagulant dose in warm lime softening of synthetic SAGD produced water. *Water Res.* **2021**, *200*, 117202. [CrossRef]
14. Dąbska, A.; Léthel, A. Swelling behaviours of compacted lime-softening sludge for application in landfill liners. *Sci. Rep.* **2021**, *11*, 15220. [CrossRef]
15. Venegas-García, D.J.; Wilson, L.D. Removal of Model Aromatic Hydrocarbons from Aqueous Media with a Ferric Sulfate–Lime Softening Coagulant System. *Surfaces* **2022**, *5*, 413–428. [CrossRef]
16. Kong, D.; Kusrini, E.; Wilson, L.D. Binary pectin-chitosan composites for the uptake of lanthanum and yttrium species in aqueous media. *Micromachines* **2021**, *12*, 478. [CrossRef]
17. Kristianto, H.; Manurung, N.; Wardhani, I.K.; Prasetyo, S.; Sugih, A.K.; Arbita, A.A. A kinetic, isotherm adsorption, and thermodynamic study of Congo red coagulation using Leucaena crude extract as natural coagulant. *Water Pract. Technol.* **2022**, *17*, 1332–1346. [CrossRef]
18. Inam, M.A.; Khan, R.; Lee, K.H.; Akram, M.; Ahmed, Z.; Lee, K.G.; Wie, Y.M. Adsorption Capacities of Iron Hydroxide for Arsenate and Arsenite Removal from Water by Chemical Coagulation: Kinetics, Thermodynamics and Equilibrium Studies. *Molecules* **2021**, *26*, 7046. [CrossRef] [PubMed]
19. Mohamed, M.H.; Wilson, L.D. Kinetic uptake studies of powdered materials in solution. *Nanomaterials* **2015**, *5*, 969–980. [CrossRef] [PubMed]
20. Mahaninia, M.H.; Wilson, L.D. A Kinetic Uptake Study of Roxarsone Using Cross-Linked Chitosan Beads. *Ind. Eng. Chem. Res.* **2017**, *56*, 1704–1712. [CrossRef]
21. Ejimofor, M.I.; Ezemagu, I.G.; Ugonabo, V.I.; Nnaji, P.C.; Anadebe, V.C.; Diyoke, C.; Menkiti, M.C. Adsorption kinetics, mechanistic, isotherm and thermodynamics study of petroleum produced water coagulation using novel Egeria radiate shell extract (ERSE). *J. Indian Chem. Soc.* **2022**, *99*, 100357. [CrossRef]
22. Edet, U.A.; Ifelebuegu, A.O. Kinetics, Isotherms, and Thermodynamic Modeling of the Adsorption of Phosphates from Model Wastewater Using Recycled Brick Waste. *Processes* **2020**, *8*, 665. [CrossRef]
23. Steiger, B.G.K.; Wilson, L.D. Modular chitosan-based adsorbents for tunable uptake of sulfate from water. *Int. J. Mol. Sci.* **2020**, *21*, 7130. [CrossRef] [PubMed]
24. Dayarathne, H.N.; Angove, M.J.; Aryal, R.; Abuel-Naga, H.; Mainali, B. Removal of natural organic matter from source water: Review on coagulants, dual coagulation, alternative coagulants, and mechanisms. *J. Water Process Eng.* **2021**, *40*, 101820. [CrossRef]
25. Abramian, L.; El-Rassy, H. Adsorption kinetics and thermodynamics of azo-dye Orange II onto highly porous titania aerogel. *Chem. Eng. J.* **2009**, *150*, 403–410. [CrossRef]
26. chaudhary, S.; Sharma, J.; Kaith, B.S.; yadav, S.; Sharma, A.K.; Goel, A. Gum xanthan-psyllium-cl-poly(acrylic acid-co-itaconic acid) based adsorbent for effective removal of cationic and anionic dyes: Adsorption isotherms, kinetics and thermodynamic studies. *Ecotoxicol. Environ. Saf.* **2018**, *149*, 150–158. [CrossRef]
27. Ohale, P.E.; Onu, C.E.; Ohale, N.J.; Oba, S.N. Adsorptive kinetics, isotherm and thermodynamic analysis of fishpond effluent coagulation using chitin derived coagulant from waste Brachyura shell. *Chem. Eng. J. Adv.* **2020**, *4*, 100036. [CrossRef]
28. Steiger, B.G.K.; Udoetok, I.A.; Faye, O.; Wilson, L.D. Counterion Effects in Metal Hybrid Biopolymer Materials for Sulfate Adsorption: An Experimental and Computational Study. *ACS Appl. Polym. Mater.* **2021**, *3*, 4595–4606. [CrossRef]

29. Castañeda, L.F.; Coreño, O.; Nava, J.L. Simultaneous elimination of hydrated silica, arsenic and phosphates from real groundwater by electrocoagulation using a cascade-shaped up-flow reactor. *Electrochim. Acta* **2020**, *331*, 135365. [CrossRef]
30. Hassan, D.H.; Shohdy, J.N.; El-Setouhy, D.A.; El-Nabarawi, M.; Naguib, M.J. Compritol-Based Nanostrucutured Lipid Carriers (NLCs) for Augmentation of Zolmitriptan Bioavailability via the Transdermal Route: In Vitro Optimization, Ex Vivo Permeation, In Vivo Pharmacokinetic Study. *Pharmaceutics* **2022**, *14*, 1484. [CrossRef]
31. Masdor, N.A.; Padrilah, S.N.; Rani, R.M.; Karim, M.S.A.; Bakar, M.Z.A.; Ismail, A.S. The Effect of Temperature and Incubation Period on the Size and Polydispersity Index of the Nano-Emulsion Containing Essential Oil from Cinnamomum zeylanicum. *Mater. Sci. Forum* **2022**, *1055*, 105–109. [CrossRef]

materials

MDPI

Article

Eco-Friendly Polysaccharide-Based Synthesis of Nanostructured MgO: Application in the Removal of Cu^{2+} in Wastewater

Nayara Balaba [1], Dienifer F. L. Horsth [1,2], Jamille de S. Correa [1], Julia de O. Primo [1], Silvia Jaerger [1], Helton J. Alves [3], Carla Bittencourt [2,*] and Fauze J. Anaissi [1]

1 Chemistry Department, Universidade Estadual Do Centro-Oeste, Guarapuava 85040-080, Brazil
2 Chimie des Interactions Plasma-Surface (ChIPS), Research Institute for Materials Science and Engineering, University of Mons, 7000 Mons, Belgium
3 Laboratório de Materiais e Energias Renováveis, LABMATER/UFPR, Universidade Federal do Paraná—UFPR, Palotina 85950-000, Brazil
* Correspondence: carla.bittencourt@umons.ac.be

Abstract: The present study described three synthesis routes using different natural polysaccharides as low-cost non-toxic fuels and complexing agents for obtaining MgO. Cassava starch, *Aloe vera* leaves (mainly acemannan) gel, and citric pectin powder were mixed with magnesium nitrate salt and calcined at 750 °C for 2 h. The samples were named according to the polysaccharide: cassava starch (MgO-St), citrus pectin (MgO-CP), and *Aloe vera* (MgO-Av). X-ray diffraction identified the formation of a monophasic periclase structure (FCC type) for the three samples. The N_2 adsorption/desorption isotherms (B.E.T. method) showed an important difference in textural properties, with a higher pore volume (V_{max} = 89.76 cc/g) and higher surface area (SA = 43.93 m^2/g) obtained for MgO-St, followed by MgO-CP (V_{max} = 11.01 cc/g; SA = 7.01 m^2/g) and MgO-Av (V_{max} = 6.44 cc/g; SA = 6.63 m^2/g). These data were consistent with the porous appearance observed in SEM images. Porous solids are interesting as adsorbents for removing metallic and molecular ions from wastewater. The removal of copper ions from water was evaluated, and the experimental data at equilibrium were adjusted according to the Freundlich, Langmuir, and Temkin isotherms. According to the Langmuir model, the maximum adsorption capacity (q_{max}) was 6331.117, 5831.244, and 6726.623 mg·g^{-1} for the adsorbents MgO-St, MgO-Av, and MgO-CP, respectively. The results of the adsorption isotherms indicated that the synthesized magnesium oxides could be used to decrease the amount of Cu^{2+} ions in wastewater.

Keywords: adsorption; heavy metals; magnesium oxide; copper; drinking water

Citation: Balaba, N.; Horsth, D.F.L.; Correa, J.d.S.; Primo, J.d.O.; Jaerger, S.; Alves, H.J.; Bittencourt, C.; Anaissi, F.J. Eco-Friendly Polysaccharide-Based Synthesis of Nanostructured MgO: Application in the Removal of Cu^{2+} in Wastewater. *Materials* **2023**, *16*, 693. https://doi.org/10.3390/ma16020693

Academic Editors: Roberta G. Toro, Daniela Caschera and Abeer M. Adel

Received: 10 December 2022
Revised: 28 December 2022
Accepted: 31 December 2022
Published: 10 January 2023

1. Introduction

The mining process presents many ores and reject concentrations, which contributes as a source of contamination to the environment through mine wastewater [1]. Rejects can be rich in toxic and bio-accumulative metallic elements, even far from the mining areas, as the contamination can be carried for miles [2]. In 2015, the Rio Doce estuary in Mariana (Minas Gerais, Brazil) was severely impacted by the breach of an iron mine tailing dam, causing major environmental damage. Although the tailing released from the breached dam was composed primarily of non-toxic minerals, the high Fe–oxy–hydroxide content may have promoted the chemical binding of metals and metalloids accumulated within the basin. As a result, after years, there is still evidence of trace metal contamination of As, Cd, Cr, Cu, Hg, Mn, Pb, Se, and Zn in the rivers and streams near the region. It is estimated that 663.2 km from the mining company, about 43 million m^3 of iron mine tailing were released [3,4]. Among the heavy metals, copper is one of the most abundant pollutants in mining wastewater [5]. Besides mines, it can be released into the environment through agricultural processes, solid waste disposal, welding and electroplating processes, sewage treatment, plumbing, and electrical wiring materials [5]. According to WHO (1993) standards, the permissible limit of Cu^{2+} in drinking water is 1.3–2.0 $mg{\cdot}L^{-1}$ because, in high concentrations, it can cause serious health effects such as damaging the central nervous system, increasing blood pressure, Alzheimer's disease, and Menkes syndrome [6].

Therefore, industrial wastewater must be treated to protect the environment before it is disposed of into water bodies. Contamination by water does not occur only through direct ingestion but by other means, such as consuming vegetables and fruits from areas irrigated with contaminated water, thus increasing health risks. Several methods for the decontamination of Cu^{2+} ions from water have been reported. Among them are flotation [6], electrocoagulation [7], ion exchange [8], precipitation [9], and adsorption [10,11]. Adsorption is one of the most common techniques due to its simplicity, high efficiency, and low cost [5]. In this context, magnesium oxide (MgO) has attracted interest due to its low cost, environmental friendliness, and high capacity to adsorb metal ions [12], dyes [13,14], carbon dioxide [15], and phosphates [16]. However, commercial MgO powder, often obtained by the thermal decomposition of $MgCO_3$ or $Mg(OH)_2$, as a metal adsorber has drawbacks, such as a low specific surface area, irregular morphology, and grain size [17]. Considerable methodologies for the synthesis of MgO have been reported in the literature such as precipitation [13], the calcination process [17], and hydrothermal- and microwave-assisted methods [18]. In light of this, the search for natural additives for materials synthesis has recently gained importance due to their reported low environmental impact, higher chemical reactivity, high combustion power, reduced calcination temperature, and action as a complexing gelling agent [19]. Synthesis routes using natural additives are considered eco-friendly. For example, polysaccharides from plants can be used as fuel, thus replacing other types of synthetic reagents such as complexing agents (oxalic, tartaric, and acetic acid) and surfactants (including cetyltrimethylammonium bromide, ethanol, ethylene glycol, and glycerin) [20] that are widely used as modifying agents to obtain MgO particles. Polysaccharides act as low-cost, non-toxic alternative additives produced from natural plants, acting as complexing agents for the central metal ions and can tailor the growth of MgO crystals [21].

This study used three different polysaccharides in the MgO synthesis: starch, citrus pectin, and *Aloe vera*. Synthesis 1 used starch from cassava (*Manihot esculenta*), an abundant plant in Brazil. This biodegradable polysaccharide consists of amylose and amylopectin molecules composed of D-Glucose units [22]. In synthesis 2, citrus pectin was used, a polysaccharide from the peels of citrus fruits. This polysaccharide consists of α-D-galacturonic acid units joined by glycosidic bonds (α-1,4) and esterified methyl carboxyl groups [23]. Finally, synthesis 3 used *Aloe vera* leaves (*Aloe Barbadensis Miller*), a succulent perennial. The main components of Aloe vera are glycoproteins, anthraquinones, saccharides, and low-molecular-weight substances [24]. The different natural additives used in the MgO syntheses resulted in distinct morphological and surface characteristics. These affected the adsorption of Cu^{2+} ions in wastewater.

2. Materials and Methods

2.1. Preparation of Magnesium Oxide

The MgO particles were prepared by three different synthesis routes using three sources of polysaccharides as fuel: cassava starch (synthesis 1), citrus pectin (synthesis 2), and *Aloe vera* (synthesis 3). In syntheses 1 and 3, starch extracted from cassava roots and *Aloe vera* leaves in natura were used, respectively. Both organic precursors were harvested from the city of Palmital-Paraná, Brazil. Synthesis 2 used commercial (Dynamic) citrus pectin. The magnesium salt used in the syntheses was magnesium nitrate hexahydrate ($Mg(NO_3)_2.6H_2O$, 98%, Dynamic). The analytical reagents were of high purity, and all solutions were prepared with deionized water.

The samples were labeled according to the polysaccharide used: starch (MgO-St), citrus pectin (MgO-CP), and *Aloe vera* (MgO-Av).

2.2. Synthesis 1—Cassava Starch (MgO-St)

The synthesis method was adapted from Primo et al. [25]. The first step of synthesis 1 was the extraction of starch from cassava roots. Initially, 500 g of natural cassava starch was extracted in 2500 mL of deionized water under mechanical stirring for 3 h. Then, the colloidal suspension was separated from the fibers and used as a base solution. Next, MgO-St oxide was synthesized from starch (300 g) and magnesium nitrate (64 g) and stirred for 20 min.

2.3. Synthesis 2—Citric Pectin (MgO-CP)

For synthesis 2, the first step was heating 1500 mL of deionized water to 80 °C and adding 15 g of commercial citrus pectin. This step was adapted from Dalpasquale et al. [23]. Next, this suspension was stirred in a mechanical shaker (800 rpm) for about 1 h to solubilize the citrus pectin. After homogenization of the colloidal suspension, 32 g of magnesium nitrate was added, keeping the agitation and temperature controlled for 3 h, thus forming the MgO-CP material.

2.4. Synthesis 3—Aloe vera (MgO-Av)

Synthesis 3 was adapted from Primo et al. (2020) [19]. To obtain the magnesium oxide, gel extracted from *Aloe vera* leaves was used. The gel from the leaves was first removed and processed in a blender (power: 1200 W). The gel was sieved and refrigerated (2 degrees Celsius). Thus, *Aloe vera* gel broth extract with a concentration of 90% was prepared in 100 mL of deionized water. Then, magnesium nitrate (24 g) was dissolved in the aloe extract solution while constantly being magnetically stirred (60 min).

The samples prepared in the different synthesis routes were calcined using the same calcination parameters. The suspensions were calcined in a muffle furnace at 750 °C with a heating ramp of 10 $°C \cdot min^{-1}$ for 60 min. The products obtained were pulverized, sieved using a 250 mm (60 mesh) sieve, and stored in a suitable container for characterization and application.

2.5. Adsorption Tests

Adsorption tests were performed to investigate the efficiency of the samples as adsorbents for removing copper from water. The initial and final concentrations of copper metal ions were evaluated for the adsorption test. The adsorption experiments were carried out in flasks of 100 mL containing 50 mL of an aqueous solution of copper sulfate pentahydrate ($CuSO_4.5H_2O$, 99%, Biotec Reagentes Analíticos, Curitiba, Brazil) with initial concentrations of 0.95, 1.6, 3.0, 5.4, 6.4 and 11 g. L^{-1}. To this end, 100 mg of each MgO (St, CP, and Av) sample was added to the solutions and kept under continuous shaking in a heating bath at 298 K for a contact time of 60 min. After the adsorption time, the solutions with adsorbent were centrifuged at 1200 rpm for 15 min and the aliquot was separated. Then, the copper ion concentration was monitored by UV–vis spectrophotometer (Shimadzu UV-1800, Kyoto, Japan) at 810 nm.

The amount of Cu^{2+} ions adsorbed at the end of the adsorption experiment and the ion percentage removal (%) by the MgO were calculated by applying Equations (1) and (2), respectively:

$$q = \frac{(C_o - C_f)}{m} \cdot V \tag{1}$$

$$\%\text{Removal} = 100 \cdot \frac{(C_o - C_f)}{C_o} \tag{2}$$

where q is the amount of ions adsorbed by the adsorbent in $mg \cdot g^{-1}$, C_o is the initial ion concentration in contact with the adsorbent ($mg \cdot L^{-1}$), C_f is the ion concentration ($mg \cdot L^{-1}$) after the batch adsorption process, m (g) is the mass of adsorbent, and *V* is the volume of ion solution (L).

2.6. Characterization Techniques

The crystalline structure and phase purity were characterized by X-ray diffractometry (XRD-D2 Phaser; Bruker, Billerica, MA, USA) with a copper cathode (λ = 1.5418 Å), 30 kV potential, 10 mA current, range between 10° and 80° (2θ), and 0.2 °/s increment. Phase Identification from Powder Diffraction, Version 3.4.2, was used with access to the COD database (Crystallography Open Database). The morphology of the MgO particulate samples was examined with a scanning electron microscope (SEM-VEGA 3; TESCAN Brun, Brno, Czech Republic); for the SEM analysis, each sample was dispersed in water, and a drop of dispersion was deposited on an Al sample holder. Before SEM analysis, the specimen surface was coated with a layer of gold. Nitrogen adsorption/desorption isotherms were obtained from the Quantachrome Instruments (Boynton Beach, FL, USA) gas Sorption Analyzer equipment, NOVA 2000 model, using the Quantachrome Novawin software, 11.03 version, the data were measured and analyzed in 24 June 2019. The samples were submitted to vacuum degassing at 200–250 °C for 2 h and 30 min, and the analyses were performed at liquid nitrogen temperature (−196.15 °C). The specific surface areas of the samples were calculated using the multi-point Brunauer–Emmet–Teller (BET) method. Pore volume and mean pore radius were calculated by desorption curve analysis using the Barrett–Joyner–Halenda (BJH) model. The zeta potentials (ζ) of the nanoparticles were determined from their electrophoretic mobilities according to Smoluchowski's equation [26]. The pH of these nanoparticles was adjusted between 2 and 12 using HCl or NaOH solutions. The measurements were performed on ZETASIZER Malvern equipment, NANO ZS90 model (Worcestershire, UK). For qualitative analysis of MgO synthesized before and after the adsorption of Cu^{2+} ions, energy dispersive spectroscopy (EDS) was performed. The measurements were conducted with an acquisition time of 20 s and an accelerating voltage of 15.0 kV, obtained from Oxford Instruments (Abingdon, United Kingdom), EDS XSTREAM2 model.

3. Results

3.1. Characterization of Synthesized MgO Particles

3.1.1. X-ray Diffractometry (XRD)

The XRD of the MgO-St, MgO-CP, and MgO-Av samples (Figure 1) showed diffraction peaks at 36.86°, 42.88°, 62.30°, 74.71°, and 78.73°, respectively. These diffraction peaks corresponded to the (111), (200), (220), (113), and (222) planes, respectively, which was in good agreement with the standard pattern of face-centered cubic (FCC) and periclase phase of MgO (COD: 9000505) with a space group of Fm-3 m [18]. Furthermore, the diffractograms (Figure 1) were similar, suggesting that the fuel used did not affect the crystal structure of the MgO particles. Nevertheless, the diffractogram peaks of the oxide synthesized with starch fuel (MgO-St) showed larger (57% wider, approx.) widths at their average height compared to the peaks of the MgO-CP and MgO-Av diffractograms, suggesting a smaller crystallite size [27,28].

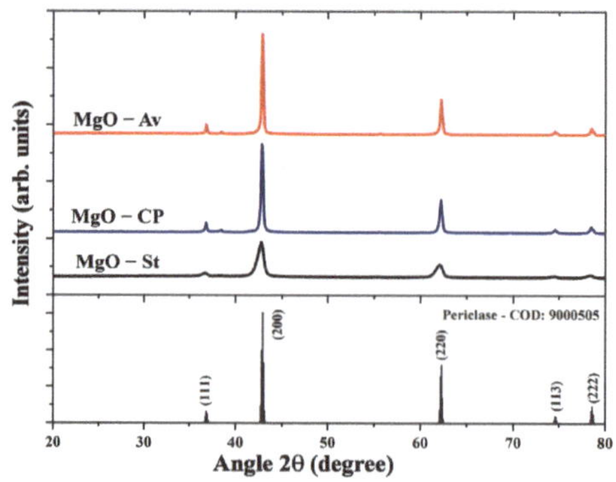

Figure 1. X-ray diffraction of the pattern Periclase [COD: 9000505] and the synthesized samples.

3.1.2. Morphological Analysis (SEM)

The scanning electron microscopy micrographs of the synthesized powders using various fuels are shown in Figure 2a–c. Figure 2a illustrates the morphology of the MgO-Av sample. It was composed of interconnected nanosheets [29]. Figure 2b shows the porous morphology of the MgO-CP sample. Finally, the morphology of the MgO-St sample (Figure 2c) showed sponge-like characteristics, with non-regular pores and holes along the entire surface. These pores were reported to be associated with releasing gases from the organic fuel when calcined [30].

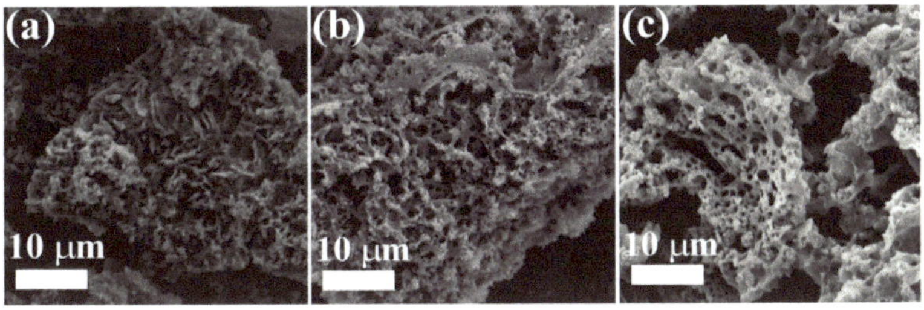

Figure 2. SEM images of the synthesized MgO samples MgO-Av (**a**), MgO-CP (**b**), and MgO-St (**c**).

3.1.3. Textural Properties (BET and BJH Isotherms)

The samples' N_2 adsorption–desorption isotherms (Figure 3) were calculated using the multi-point Brunauer–Emmet–Teller (BET) method. In addition, the pore volume and radius were calculated by analyzing the desorption curve using the Barrett–Joyner–Halenda (BJH) model. The isotherms were type IV according to the IUPAC classification, indicating completely reversible adsorption–desorption isotherms associated with mesopores for MgO-St sample (pore diameter between 20–50 Å) and micropores for MgO-Cp and MgO-Av samples (<20 Å) (Table 1) [31]. The MgO-St sample (Figure 3a) showed an H4-type hysteresis loop often related to materials with slit-like pores between particles, indicating that pore filling occurred [32].

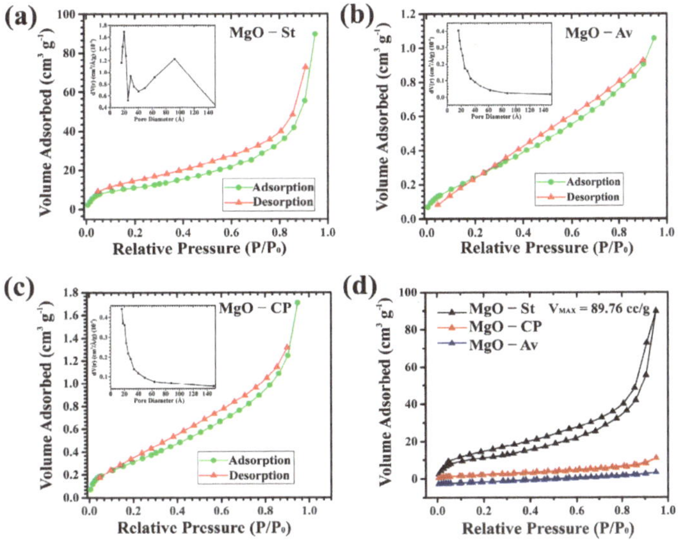

Figure 3. Adsorption and desorption isotherms with pore size distribution (inset) of MgO samples: (**a**) starch, (**b**) *Aloe vera*, (**c**) citrus pectin, and (**d**) comparison of the maximum volume of adsorbed N_2.

Table 1. Textural properties of the synthesized MgO.

Samples	Surface Area (m^2/g)	Pore Volume (cm^3/g)	Pore Diameter (Å)	Volume (cc/g)
MgO-St	43.93	0.130	20.06	89.76
MgO-CP	7.01	0.015	16.29	11.01
MgO-Av	6.63	0.009	16.29	6.44

The MgO-St surface area, pore volume, and diameter (Table 1) were larger than those of the other two materials, showing wider micropores and possibly narrow mesopores (between 20–500 Å) [31]. The MgO-Av (Figure 3b) and MgO-CP (Figure 3c) samples also showed an H4-type hysteresis loop, with a smaller surface area and pore volume, which implied large pores, however, this was not as profound as for the MgO-St sample. The maximum volume was due to N_2 being adsorbed for forming complete gas monolayers on the sample surface at partial pressure [32]. The MgO-St sample showed an increase of ~22.7% in the pore radius, suggesting a larger pore volume (0.131 cm^3/g) and a larger surface area (41 m^2/g). These textural aspects indicated a greater adsorption capacity of N_2 gas. The maximum volume adsorbed on the surface of each oxide presented values in cc/g (Figure 3d). The MgO-St sample showed the largest adsorbed volume (V_{MAX} = 89.76 cc/g), while the adsorbed volume for the MgO-Av (V_{MAX} = 6.44 cc/g) and MgO-CP (V_{MAX} = 11.01 cc/g) samples were much smaller (Table 1).

3.1.4. Zeta Potential (ζ) vs. pH

Figure 4 shows the changes in the zeta potential (ζ) of the samples during the NaOH titration for the synthesized oxides. The zeta potential allows the evaluation of the chemical stability of particles in a colloidal state [19]. Figure 4 shows that with the addition of acid solution, i.e., the addition of H^+ ions, the initial dissociation of MgO occurred, which results in the approximation of H^+ ions and the dissociation in Mg^{2+} and OH^-. This step decreased the zeta potential of the MgO due to the neutralization of the surface charge. At low pH, the number of positive ions increased, preferably at the magnesium oxide surface, increasing the zeta potential. The pH of the solution decreased with the increase in this charge, reducing the tendency of agglomeration [33].

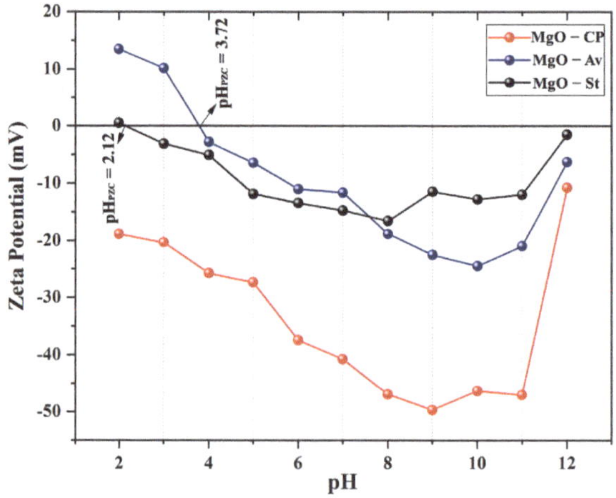

Figure 4. Zeta potential as a function of pH and point of zero charge (pH_{PZC}) for the MgO-St, MgO-CP, and MgO-Av samples.

The ζ potential of the MgO-CP sample during the pH titration range of 6–11 (Figure 4) was less than −30 mV, indicating that in this range, its particles were highly stable in the suspension, which prevented the particles from flocculating [34]. Therefore, the profile of the zeta potential for the MgO-CP sample standing out as the zero-charge point (pH_{PZC}) was not observed. This probably occurred due to the adsorption of negative ionizable groups on the MgO particles, which decreased the mobility, reduced the surface electric charge of MgO (positive), and consequently decreased the zeta potential and medium fluidity [33].

The samples synthesized with starch and *Aloe vera* showed positive values when the pH < 3; for the MgO-Av sample, the surface charge showed a positive ζ potential value equal to + 1.24 mV, and for the MgO-St sample, it was + 14 mV. With increasing the pH, the zero-charge point (pH_{PZC}) was reached at 2.12 for MgO-Av and 3.72 MgO-St.

The pH values of each solution were measured as soon as it was prepared, without any external acidic or basic modifications. The pH values of each solution are shown in Table 2.

Table 2. pH values of the solutions before adsorption.

Adsorption Solutions (g·L^{-1})	pH
0.95	3.77
1.6	3.54
3.0	3.27
5.4	3.11
6.4	3.07
11	2.88

The impact of the pH on Cu^{2+} removal was appreciable, while the interaction between negative and positive ions played a determining role in the adsorption on the adsorbent surface. This was due to changes not only to the active sites on the MgO surface, which could coordinate the copper metal but also to the solubility of Cu^{2+} in the aqueous solution [12]. Magnesium oxide is a highly alkaline particle, and therefore sorption of Cu^{2+} onto MgO surfaces can be achieved by chemical sorption and/or ion exchange [35]. In the studies by Ismail et al. (2022) [12], they also used MgO as an adsorbent to remove Cu^{2+} in a solution with a concentration of 90 g·L^{-1} in a pH range from 1 to 9, obtaining the best removal results in solutions at pH 3 and 5. This is because the minimal adsorption at pH 2.0 may be because there are more protons at lower pH available to protonate active groups on the surface and compete with the metal ions in the solution. At higher pH values, the lower number of H+ and higher number of negatively charged ligands results in greater copper adsorption [36]. As the pH increases, the adsorbent surface becomes more and more negatively charged, and therefore the adsorption of positively charged Cu^{2+} species is quite favorable. According to previous research [36,37], Cu^{2+} is only available in the divalent ion form at pH values below 5.0. At pH 6.0 and higher, various neutral hydrolysis species can be found.

4. Copper Ion Removal by MgO as Adsorbent

4.1. Adsorption Isotherms

The removal of Cu^{2+} ions onto the surfaces of the MgO-St, MgO-Av, and MgO-CP oxides was effectively reached, even at high initial concentrations. Figure 5 shows that the adsorption capacity of the synthesized MgO is related to the initial concentration of copper. The results reveal that the more is copper present in the solution, the stronger is the interaction between the copper ions with the adsorbent, and the more susceptible to exchange on the MgO surface [11]. Similar results were observed in [38], using mesoporous–macroporous MgO (HMMM) via a calcination process, to remove uranium from an aqueous solution in high concentration. The maximum adsorbed amount of all the magnesium oxides was obtained at a very close value of the equilibrium concentration of Cu^{2+} ion in solution, also observed in Figure 5; this can be explained by the fact that the structure of all the materials was the same as observed in the DRX results (Figure 1). The sites for the adsorption of Cu^{2+} ions were available in a very similar way; this made the amount of Cu^{2+} ions adsorbed practically equal in higher concentrations, as observed in Figure 5.

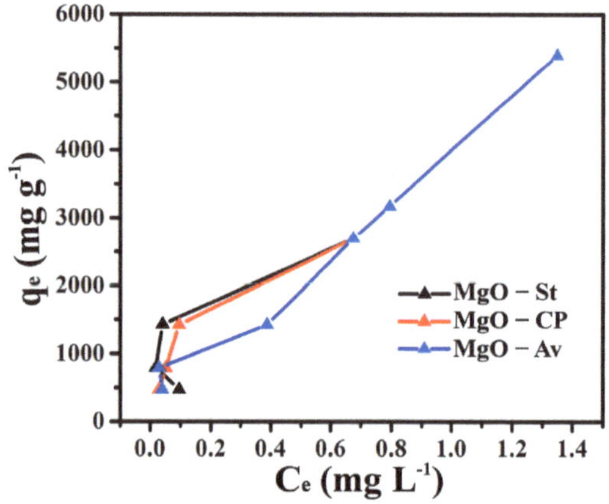

Figure 5. Experimental point curves for the adsorption of Cu^{2+} ions onto MgO-St, MgO-Av, and MgO-CP at 298 K.

The data for the adsorption of Cu^{2+} ions onto the MgO-St, MgO-Av, and MgO-CP oxides were adjusted according to the Langmuir, Freundlich, and Temkin isotherm models and their correlation parameters are presented in Table 2. For the practical design of adsorption systems, the correlation between theoretical data and empirical equations is significant [39]. The Langmuir model is applicable for systems with the assumption of ideal homogeneous surface adsorption. Therefore, the Freundlich model fits better for a heterogeneous system. The Langmuir isotherm in linear form is usually given as (Equation (3)):

$$\frac{C_e}{q_e} = \frac{1}{qL_{max}\frac{C_e}{q_{max}}} \tag{3}$$

where q_e ($mg \cdot g^{-1}$) is the amount of ions adsorbed per unit mass of MgO at equilibrium, K_L ($L \cdot mg^{-1}$) is the Langmuir constant related to the affinity of the binding sites, and q_{max} ($mg \cdot g^{-1}$) is a parameter associated with the maximum amount of Cu^{2+} per unit weight of MgO.

The linearized form of the Freundlich isotherm is given as (Equation (4)):

$$\ln q_e = \ln K_F + \frac{1}{n} \ln C_e \tag{4}$$

where K_F ($mg\ L^{-1}$) is the Freundlich constant, n is a parameter related to the adsorption intensity and system heterogeneity, and K_F and n are the Freundlich constants found from the intercept and slope of the straight line of the plot $\ln_{qe}$ versus $\ln_{Ce}$.

The Temkin model isotherm ignores extremely low and very high concentrations and assumes a linear rather than a logarithmic decrease in the heat of adsorption [40]. This model assumes a uniform distribution of the bonding energy up to some maximum bonding energy [40,41]. It can be expressed as (Equation (5)).

$$q_e = \frac{RT}{b} \ln C_e + \frac{RT}{b} \ln K_T \tag{5}$$

where b is the Temkin constant ($J \cdot mol^{-1}$), T is the absolute temperature (K), R is the gas constant (8.314 $J \cdot mol^{-1} \cdot K^{-1}$), and A is another Temkin isotherm constant ($L \cdot g^{-1}$).

The correlation coefficient (R^2) obtained by the Freundlich ($R_F{}^2$) isothermal model was better fitted than the Langmuir ($R_L{}^2$) and Temkin ($R_T{}^2$) models for the MgO-St and MgO-CP samples, as observed in Figure 6 and Table 3. However, the adsorption of ions onto the MgO-Av sample was well fitted for the Langmuir and Temkin models (Table 3). When observing the correlation coefficient (R^2), better-fit results were obtained for the Freundlich isotherm model ($R_F{}^2 = 0.989 > R_T{}^2 = 0.987$). Therefore, as the Freundlich was the best-fitted model, the adsorption process of Cu^{2+} ions onto the MgO-St, MgO-Av, and MgO-CP samples consisted of multilayer adsorption, as this isotherm model is an empirical model and is not limited to a monolayer process [40,41].

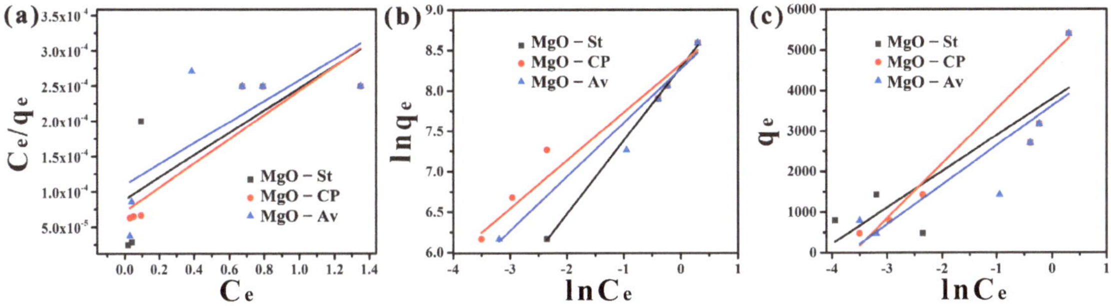

Figure 6. The linear form of the Langmuir (**a**), Freundlich (**b**), and Temkin (**c**) adsorption isotherms for the adsorption of Cu $^{2+}$ ions onto MgO-St, MgO-Av, and MgO-CP.

Table 3. Parameters of the Langmuir, Freundlich, and Temkin isotherms for adsorption of Cu $^{2+}$ ions onto MgO-St, MgO-Av, and MgO-CP.

Samples	Langmuir			Freundlich			Temkin		
	K_L ($L \cdot mg^{-1}$)	q_{max} ($mg \cdot g^{-1}$)	$R_L{}^2$	K_F ($L \cdot g^{-1}$)	1/n	$R_F{}^2$	K_T ($L \cdot g^{-1}$)	b ($J \cdot mol^{-1}$)	$R_T{}^2$
MgO-St	1.779	6331.117	0.589	3974.443	1.103	0.999	76.831	2.777	0.725
MgO-Av	2.386	5831.244	0.817	4125.036	1.683	0.989	40.953	1.840	0.987
MgO-CP	1.356	6726.623	0.547	3895.705	1.500	0.991	44.904	2.546	0.733

In the study of Huang et al. (2021) [41], a 2D carbon ribbon (MIL-53/C) and a nitrogen-doped carbon ribbon (MIL-53-N/C) were successfully synthesized using 2D-MOFs, and they were used in the removal of high fluoride concentrations in water, and the adsorption isotherm results also showed a better fit to the Freundlich model for the adsorption of fluoride by MIL-53/C. The adsorption was a multilayer process, and the maximum adsorption capacity was recorded at 299.17 $mg \cdot g^{-1}$ for the same sample.

The Freundlich isotherm was also best described in the study of Inyinbor et al. (2016) [40], in which the researchers investigated the uptake of Rhodamine B dye onto *Raphia hookeri* fruit epicarp, obtaining an R^2 value of 0.9969, suggesting an adsorption process that was not onto a uniform site but rather multilayer adsorption.

One of the other parameters that can be considered in the Freundlich model is the value of n. This factor determines the affinity of the process, that is, the process is considered to involve chemisorption ($n < 1$) or physisorption ($n > 1$). From Figure 6 and considering the parameters in Table 2, the Freundlich plot resulted in 1/n values of 1.103, 1.683, and 1.500 for the MgO-St, MgO-Av, and MgO-CP samples, respectively. These values indicated that physisorption occurred [42]. When a physisorption adsorption process is considered, in general, a van der Waals bond exists between the adsorbate and adsorbent, and the physisorption process is postulated to include the existence of multilayer adsorption onto the adsorbent, which is line with the Freundlich model.

The maximum adsorption capacity (q_{max}) obtained by the Langmuir isotherm model was calculated, and values of 6331.117, 5831.244, 6726.623 $mg \cdot \cdot g^{-1}$ were obtained for the

MgO–St, MgO–Av, and MgO–CP adsorbents, respectively, at a high concentration of Cu^{2+} ions. These values were comparable to the studies reported in the literature, and are significantly higher than those reported, as shown in Table 4. In addition, the maximum capacity of Cu^{2+} adsorption was highlighted when compared with the adsorption capacity of the other magnesium oxides due to the ion exchange process that occurs when MgO is added into different concentrations of Cu^{2+} ion solutions.

Table 4. Comparison of copper's adsorption capacities (mg·g^{-1}) using magnesium oxide as adsorbent.

Adsorbent	Adsorption Capacities q_{max} (mg·g^{-1})	References
Magnesium phyllosilicate modified with 2-aminophenyldisufide	208.44	(Moscofian and Airoldi, 2008) [43]
Lignin/MgO-SiO_2	83.98	(Ciesielczyk et al., 2017b) [44]
Electrospun SiO_2-MgO hybrid fibers	493.00	(Xu et al., 2019) [39]
MgO nanostructures	546.45	(Ismail et al., 2022) [12]

Moscofian and Airoldi (2008) [43] synthesized a magnesium phyllosilicate modified with 2-aminophenyldisufide for the adsorption of Cu^{2+}, Pb^{2+}, and Cd^{2+} ions. The results showed that the adsorption process of all the ions investigated was better fitted by the Langmuir isotherm model, whose maximum adsorption capacity reached 208.44 mg·g^{-1} for the adsorption of Cu^{2+} ions. In the study of Ciesielczyk et al. (2017) [44], Lignin/MgO-SiO_2, a hybrid sorbent, was applied to the adsorption of galvanic waste solutions of copper (II) ions, and the adsorption process was well fitted by the Langmuir model, obtaining an R^2 of 0.982 and adsorption capacity equal to 83.98 mg·g^{-1}. Xu et al. (2019) [39] investigated electrospun SiO_2–MgO hybrid fibers as adsorbents of Pb^{2+} and Cu^{2+} ions. The results showed that the adsorption of Cu^{2+} ions onto the electrospun SiO_2–MgO hybrid fibers at 298 K obtained an adsorption capacity equal to 492.96 mg·g^{-1}. In addition, the Langmuir isotherm was well fitted in this adsorption process. Recently, Ismail et al. (2022) [12] investigated the impact of Cu^{2+} ion removal on MgO nanostructures. To manufacture the MgO nanoparticles, a thermal pyrolysis process was used. The regression coefficient for the Langmuir isotherm model for the adsorption of Cu^{2+} ions onto the MgO nanoparticles was 0.9960, and the highest adsorption capability of Cu^{2+} was 546.45 mg·g^{-1}.

4.2. Adsorption Mechanisms

4.2.1. Energy Dispersive Spectroscopy (EDS)

Heavy metal adsorption has four mechanisms: precipitation, ion exchange, complexation of functional groups, and interaction between metal ions and π electrons. The samples were analyzed by energy dispersive spectroscopy (EDS) after Cu^{2+} adsorption to understand the occurring mechanism of copper ion adsorption (Table 5). The analyses were performed before adsorption (MgO-St, MgO-CP, and MgO-Av) and after adsorption at the lowest (0.95 g·L^{-1}) and highest concentration (11.0 g·L^{-1}).

The EDS data were analyzed before and after the adsorption of copper ions (Table 4). Before the Cu^{2+} ion removal process, a higher percentage of Mg ions was observed. The Mg percentage decreased as the copper concentration increased. Therefore, higher concentrations of Mg^{2+} were released during the Cu^{2+} adsorption. From the EDS data, we can suggest the occurrence of the cation exchange mechanism. In which the Mg ions that are part of the MgO crystalline network are replaced by Cu^{2+} cations in the network. This releases into the water magnesium ions that are atoxic and not harmful to the environment [12]. In general, as a solid-liquid interfacial process, the adsorption of ions onto adsorbents is initiated by electrostatic attraction and can also be followed by ion exchange, which usually occurs between the hydroxyl groups on the surface and the anions [45] (such as copper ions). The possible reaction promoted by MgO can be explained by Equations (6)–(8) [45].

$$MgO_{(s)} + H_2O \rightarrow Mg(OH)_2 \quad (6)$$

$$MgO_{(s)} + CuSO_4 \rightarrow CuO + Mg^{2+} + SO_4^{2-} \quad (7)$$

$$Mg(OH)_2 + CuSO_4 \rightarrow Cu(OH)_2 + Mg^{2+} + SO_4^{2-} \quad (8)$$

Table 5 presents the percentage by mass of the elements in the samples. Among the precursors, the *Aloe Vera* sample contained the highest percentage of magnesium. Considering the lowest concentration tested, the best material in the adsorption of copper was MgO-Av, while at the highest Cu^{2+} concentration tested, the highest percentage was adsorbed by the MgO-St sample. This may have occurred due to the surface area, pore volume, and diameter being higher than the other samples, as seen in the BET. The presence of sulfur originated from the precursor salt used for the tests.

Table 5. Compositional chemical analysis data by EDS (% element) for the samples before and after adsorption at lowest (Mg-St1, Mg-CP1, and Mg-Av1) and highest concentrations (Mg-St2, Mg-CP2, and Mg-Av2).

Samples	Mg%	O%	C%	Cu%	S%
MgO-St	37.01	51.00	11.99	-	-
Mg-St1	11.66	47.97	19.75	18.24	2.38
Mg-St2	0.44	41.66	8.34	42.91	6.66
MgO-Av	50.06	44.86	5.08	-	-
Mg-Av1	16.43	44.78	11.66	23.99	3.14
Mg-Av2	1.04	41.11	11.14	39.62	7.09
MgO-CP	41.55	48.41	10.04	-	-
Mg-CP1	12.75	48.50	19.11	17.60	2.04
Mg-Cp2	0.20	40.65	13.47	39.44	6.25

4.2.2. Fourier Transform Infrared Spectroscopy (FTIR)

To further explore the changes in the functional groups on the surface of the adsorbents before and after the copper adsorption in the two different concentrations of adsorption (0.95 and 11 $g{\cdot}L^{-1}$), we performed FTIR spectroscopy as shown in Figure 7. Figure 7a–c shows a narrow and intense band at 3700 and 3400 cm^{-1}, which was attributed to the vibrational stretching of the free -OH ions of the $Mg(OH)_2$, which were generated by the hydration of MgO [46]. These hydroxyl groups would be active adsorptive sites, and the hydrated Cu^{2+} ions can react with the hydroxyl groups electrostatically [47]. In addition, the adsorbed hydrated Cu^{2+} ions could partially hydrolyze and form Cu–OH, leading to the formation of Cu–O–Cu on the pore walls. Therefore, multilayer adsorption could take place on the porous MgO, exhibiting strong adsorptive performance and Freundlich-type adsorptive behavior [47], corroborating the results obtained for the synthesized oxides. The band at 3700 cm^{-1} was exhibited after adsorption for the lower concentration of copper by the MgO-St (Mg-St1, Figure 7a) and MgO-CP (Mg-CP1, Figure 7b) oxides. This OH stretching vibration band at 3700 cm^{-1} was changed in intensity after adsorption at the lower concentration, showing that the copper ions, Cu^{2+}, interacted with the OH in $Mg(OH)_2$ to form a Mg–O–Cu bond [12]. This result was confirmed by the shifting of the MgO vibration band located at 870 cm^{-1} for the lower and higher concentration of Cu^{2+}, and the intensity of it was increased with the higher concentration of Cu^{2+} [12,48]. However, for Mg-Av1 and Mg-Av2 (Figure 7c), Mg-St2 (Figure 7a), and Mg-CP2 (Figure 7b), no surface hydroxyl groups were detected around 3700 cm^{-1}, suggesting that the surface hydroxyl groups were too active to react with CO_2 in air and to subsequently form surface carbonate-like species [48]. The band at around 1632–1636 cm^{-1} for all the samples corresponded to the O–H bending mode, which is characterized by the bending vibration of the -OH group of the physiosorbed water molecules [46,49]. The bands at 1444 and 1114 cm^{-1} were attributed to the asymmetric and symmetric stretching of a unidentate carbonate on the surface, respectively [12,48]. The existence of carbonate ions could be estimated as the absorption of carbon dioxide gas from the atmosphere [50]. These bands explained the percentage of carbon identified by EDS in each oxide before and after adsorption. In addition, the intensities of the FTIR bands of the surface carbonate-like species were also changed after the Cu^{2+} sorption. That is to say that the surface unidentate carbonate species may have contributed to the Cu^{2+} removal mechanism [48]. Rhatingan and Nolan (2018) [51] presented a DFT study of TiO2 anatase (101) modified with SnO and MgO nanoclusters, with specific Sn_4O_4 and Mg_4O_4 compositions. The authors explained that point defects, such as oxygen vacancies, aid the chemistry on the metal oxide surface by providing sites for the adsorption and activation of molecules such as H_2O and CO_2. The results revealed that Cu^{2+} had a strong absorption band after being removed at 602–614 cm^{-1}, which was identified as the metal–oxygen (Cu–O) bond bending mode [50], suggesting that ionic exchange has occurred between Mg^{2+} and Cu^{2+}, forming this bond, associating with Equation (6) and corroborating with the EDS data. In comparing the three oxides synthesized before the copper removal (Figure 7d), a few differences were identified in the adsorption bands. The MgO-St oxide showed less intensity in the band at 3700 cm^{-1} and a small band at 1494 cm^{-1}, corresponding to the O–H bending mode [46,49].

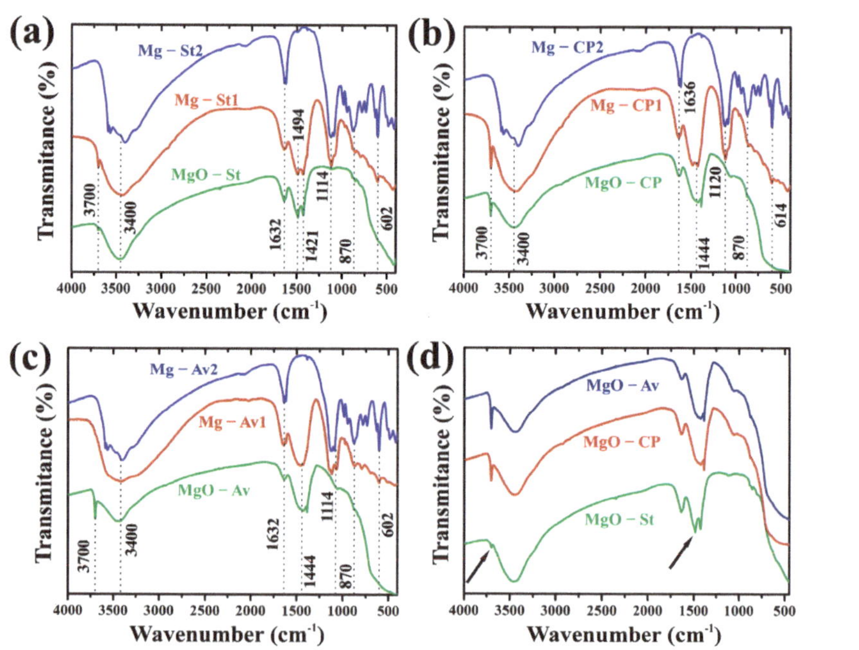

Figure 7. FTIR spectra of MgO before and after Cu^{2+} adsorption for MgO-St (**a**), MgO-CP (**b**), and MgO-Av (**c**) with an increased Cu^{2+} concentration, and comparison between the three synthesized oxides (**d**).

5. Conclusions

In summary, the MgO samples synthesized using polysaccharides as the reaction fuel showed a superior porosity and surface area compatible with porous materials that can remove metal ions, such as Cu^{2+}, from wastewater. The pores evidenced by SEM were from the gases released by the fuels used in the syntheses. The synthesized oxides showed the same crystalline structure, although they were synthesized from different polysaccharides. According to the Langmuir, Freundlich, and Temkin adsorption models, the adsorption patterns were fitted to the isotherms. For the Langmuir model, the MgO-CP material showed a maximum adsorption capacity of 6726.623 mg·g^{-1}, and the other samples also showed excellent results, with values of q_{max} of 6331.117 and 5831.244 mg·g^{-1} for MgO-St and MgO-Av, respectively. Before and after the adsorption tests, the oxides and the solid samples after the adsorption at concentrations of 0.95 and 11 g·L^{-1} were subjected to EDS analysis, which showed that as the insertion of Cu ions into the MgO structure occurred, Mg^{2+} ions were reduced. The same samples were also analyzed by FTIR, which showed bands referring to Mg–O–Cu bonds (870 cm^{-1}) and a Cu–O bond band at 602–614 cm^{-1}, confirming the adsorption of copper by the MgO-St, MgO-CP and MgO-Av oxides. Due to their high adsorption efficiency, the synthesized oxides with different polysaccharides are potential adsorbents for heavy metal removal.

Author Contributions: Conceptualization, N.B., F.J.A. and J.d.O.P.; methodology, N.B., J.d.S.C. and J.d.O.P.; validation, N.B., D.F.L.H., J.d.S.C. and S.J.; formal analysis, N.B., S.J. and J.d.O.P.; investigation, N.B. and D.F.L.H.; resources, C.B., H.J.A. and F.J.A.; data curation, N.B., J.d.O.P., D.F.L.H., J.d.S.C. and H.J.A.; writing—original draft preparation, N.B. and J.d.O.P.; writing—review and editing, N.B., C.B. and F.J.A.; visualization, N.B.; supervision, F.J.A. and C.B.; project administration, F.J.A.; funding acquisition, F.J.A. and C.B. All authors have read and agreed to the published version of the manuscript.

Funding: N.B. appreciates the CAPES (grant number 88887.628497/2021-00) for a graduate scholarship. F.J.A. is thankful for a CNPq Productivity grant (308625/2019-6) and the funding from CNPq (427127/2018-1). C.B. is a research associate of FR-FNRS, Belgium. C.B. and N.B. thank the collaborative grant Confap-WBI.

Institutional Review Board Statement: Not applicable.

Informed Consent Statement: Not applicable.

Data Availability Statement: The data presented in this study are available on request from the corresponding author.

Acknowledgments: The authors are grateful to the funding agencies CNPq, Capes, Finep, Fundação Araucária, WBI, and FRS-FNRS.

Conflicts of Interest: The authors declare no conflict of interest.

References

1. Adriano, D.C. *Trace Elements in the Terrestrial Environment*; Springer: Berlin/Heidelberg, Germany, 1986; ISBN 3540961585.
2. Abraham, M.R.; Susan, T.B. Water Contamination with Heavy Metals and Trace Elements from Kilembe Copper Mine and Tailing Sites in Western Uganda; Implications for Domestic Water Quality. *Chemosphere* **2017**, *169*, 281–287. [CrossRef]
3. Dias, C.A.; Da Costa, A.S.V.; Guedes, G.R.; Umbelino, G.J.d.M.; de Sousa, L.G.; Alves, J.H.; Silva, T.G.M. Impactos Do Rompimento Da Barragem de Mariana Na Qualidade Da Água Do Rio Doce/Samarco'sdam Failure and Its Consequences for the Quality of the Doce River Water. *Rev. Espinhaço* **2018**, *7*, 21–35.
4. Gabriel, F.Â.; Hauser-Davis, R.A.; Soares, L.; Mazzuco, A.C.A.; Rocha, R.C.C.; Pierre, T.D.S.; Saggioro, E.; Correia, F.V.; Ferreira, T.O.; Bernardino, A.F. Contamination and Oxidative Stress Biomarkers in Estuarine Fish Following a Mine Tailing Disaster. *PeerJ* **2020**, *8*, e10266. [CrossRef]
5. Ali, R.M.; Hamad, H.A.; Hussein, M.M.; Malash, G.F. Potential of Using Green Adsorbent of Heavy Metal Removal from Aqueous Solutions: Adsorption Kinetics, Isotherm, Thermodynamic, Mechanism and Economic Analysis. *Ecol. Eng.* **2016**, *91*, 317–332. [CrossRef]
6. Bhagat, B.; Jadeja, V.; Sharma, P.; Bandyopadhyay, R.; Mukherjee, K. Glycine Mediated Colorimetric Method along with Image Analyses Algorithms for Detection of Cu (II) in Water. *Mater. Sci. Eng. B Solid-State Mater. Adv. Technol.* **2022**, *286*, 116027. [CrossRef]
7. Dermentzis, K.; Christoforidis, A.; Valsamidou, E. Removal of Nickel, Copper, Zinc and Chromium from Synthetic and Industrial Wastewater by Electrocoagulation. *Int. J. Environ. Sci.* **2011**, *1*, 697–710.
8. Virolainen, S.; Wesselborg, T.; Kaukinen, A.; Sainio, T. Removal of Iron, Aluminium, Manganese and Copper from Leach Solutions of Lithium-Ion Battery Waste Using Ion Exchange. *Hydrometallurgy* **2021**, *202*, 105602. [CrossRef]

9. Duarte-Nass, C.; Rebolledo, K.; Valenzuela, T.; Kopp, M.; Jeison, D.; Rivas, M.; Azócar, L.; Torres-Aravena, Á.; Ciudad, G. Application of Microbe-Induced Carbonate Precipitation for Copper Removal from Copper-Enriched Waters: Challenges to Future Industrial Application. *J. Environ. Manag.* **2020**, *256*, 109938. [CrossRef]
10. Wei, Y.; Chen, T.; Qiu, Z.; Liu, H.; Xia, Y.; Wang, Z.; Zou, R.; Liu, C. Enhanced Lead and Copper Removal in Wastewater by Adsorption onto Magnesium Oxide Homogeneously Embedded Hierarchical Porous Biochar. *Bioresour. Technol.* **2022**, *365*, 128146. [CrossRef] [PubMed]
11. Qi, X.; Yin, H.; Zhu, M.; Yu, X.; Shao, P.; Dang, Z. MgO-Loaded Nitrogen and Phosphorus Self-Doped Biochar: High-Efficient Adsorption of Aquatic Cu^{2+}, Cd^{2+}, and Pb^{2+} and Its Remediation Efficiency on Heavy Metal Contaminated Soil. *Chemosphere* **2022**, *294*, 133733. [CrossRef] [PubMed]
12. Ismail, M.; Jobara, A.; Bekouche, H.; Abd Allateef, M.; Ben Aissa, M.A.; Modwi, A. Impact of Cu Ions Removal onto MgO Nanostructures: Adsorption Capacity and Mechanism. *J. Mater. Sci. Mater. Electron.* **2022**, *33*, 12500–12512. [CrossRef]
13. Dhal, J.P.; Sethi, M.; Mishra, B.G.; Hota, G. MgO Nanomaterials with Different Morphologies and Their Sorption Capacity for Removal of Toxic Dyes. *Mater. Lett.* **2015**, *141*, 267–271. [CrossRef]
14. Ge, L.; Wang, W.; Peng, Z.; Tan, F.; Wang, X.; Chen, J.; Qiao, X. Facile Fabrication of Fe@MgO Magnetic Nanocomposites for Efficient Removal of Heavy Metal Ion and Dye from Water. *Powder Technol.* **2018**, *326*, 393–401. [CrossRef]
15. Lv, G.; Zhu, C.; Zhang, H.; Su, Y.; Qian, P. Mechanism of CO_2 Adsorption on Point-Defective MgO Surfaces: First-Principles Study. *Appl. Surf. Sci.* **2022**, *604*, 154647. [CrossRef]
16. Luo, H.; Wang, Y.; Wen, X.; Cheng, S.; Li, J.; Lin, Q. Key Roles of the Crystal Structures of MgO-Biochar Nanocomposites for Enhancing Phosphate Adsorption. *Sci. Total Environ.* **2021**, *766*, 142618. [CrossRef] [PubMed]
17. Bassioni, G.; Farid, R.; Mohamed, M.; Hammouda, R.M.; Kühn, F.E. Effect of Different Parameters on Caustic Magnesia Hydration and Magnesium Hydroxide Rheology: A Review. *Mater. Adv.* **2021**, *2*, 6519–6531. [CrossRef]
18. Karthik, K.; Dhanuskodi, S.; Gobinath, C.; Prabukumar, S.; Sivaramakrishnan, S. Fabrication of MgO Nanostructures and Its Efficient Photocatalytic, Antibacterial and Anticancer Performance. *J. Photochem. Photobiol. B Biol.* **2019**, *190*, 8–20. [CrossRef]
19. Primo, J.d.O.; Bittencourt, C.; Acosta, S.; Sierra-Castillo, A.; Colomer, J.F.; Jaerger, S.; Teixeira, V.C.; Anaissi, F.J. Synthesis of Zinc Oxide Nanoparticles by Ecofriendly Routes: Adsorbent for Copper Removal From Wastewater. *Front. Chem.* **2020**, *8*, 571790. [CrossRef]
20. Mastuli, M.S.; Kamarulzaman, N.; Nawawi, M.A.; Mahat, A.M.; Rusdi, R.; Kamarudin, N. Growth Mechanisms of MgO Nanocrystals via a Sol-Gel Synthesis Using Different Complexing Agents. *Nanoscale Res. Lett.* **2014**, *9*, 134. [CrossRef]
21. Guo, L.; Lei, R.; Zhang, T.C.; Du, D.; Zhan, W. Insight into the Role and Mechanism of Polysaccharide in Polymorphous Magnesium Oxide Nanoparticle Synthesis for Arsenate Removal. *Chemosphere* **2022**, *296*, 133878. [CrossRef]
22. Brito, G.F.; Agrawal, P.; Araujo, E.M.; Melo, J.J.A. Biopolímeros, Polímeros Biodegradáveis e Polímeros Verdes, Revista Eletrônica de Materiais e Processos. *Rev. Eletrôn. Mater. Process.* **2011**, *6*, 127–139.
23. Dalpasquale, M.; Mariani, F.Q.; Müller, M.; Anaissi, F.J. Citrus Pectin as a Template for Synthesis of Colorful Aluminates. *Dyes Pigments* **2016**, *125*, 124–131. [CrossRef]
24. Choi, S.; Chung, M.H. A Review on the Relationship between Aloe Vera Components and Their Biologic Effects. *Semin. Integr. Med.* **2003**, *1*, 53–62. [CrossRef]
25. Primo, J.d.O.; Borth, K.W.; Peron, D.C.; Teixeira, V.d.C.; Galante, D.; Bittencourt, C.; Anaissi, F.J. Synthesis of Green Cool Pigments (CoxZn1-XO) for Application in NIR Radiation Reflectance. *J. Alloys Compd.* **2019**, *780*, 17–24. [CrossRef]
26. O'Brien, R.W.; Hunter, R.J. The Electrophoretic Mobility of Large Colloidal Particles. *Can. J. Chem.* **1981**, *59*, 1878–1887. [CrossRef]
27. Moulavi, M.H.; Kale, B.B.; Bankar, D.; Amalnerkar, D.P.; Vinu, A.; Kanade, K.G. Green Synthetic Methodology: An Evaluative Study for Impact of Surface Basicity of MnO_2 Doped MgO Nanocomposites in Wittig Reaction. *J. Solid State Chem.* **2019**, *269*, 167–174. [CrossRef]
28. Prado, D.C.; Fernández, I.; Rodríguez-Páez, J.E. MgO Nanostructures: Synthesis, Characterization and Tentative Mechanisms of Nanoparticles Formation. *Nano-Struct. Nano-Objects* **2020**, *23*, 100482. [CrossRef]
29. Wang, J.; Zhang, J.; Chen, R.; Yang, C.; Xiang, L.; Yi, M. A Vacuum Calcination Route to High-Surface-Area MgO Nanoplates for Superior Arsenate Adsorption and Catalytic Properties. *Vacuum* **2018**, *158*, 231–235. [CrossRef]
30. Ananda, A.; Ramakrishnappa, T.; Archana, S.; Yadav, L.S.R.; Shilpa, B.M.; Nagaraju, G.; Jayanna, B.K. Materials Today: Proceedings Green Synthesis of MgO Nanoparticles Using Phyllanthus Emblica for Evans Blue Degradation and Antibacterial Activity. *Mater. Today Proc.* **2022**, *49*, 801–810. [CrossRef]
31. Thommes, M.; Kaneko, K.; Neimark, A.V.; Olivier, J.P.; Rodriguez-Reinoso, F.; Rouquerol, J.; Sing, K.S.W. Physisorption of Gases, with Special Reference to the Evaluation of Surface Area and Pore Size Distribution (IUPAC Technical Report). *Pure Appl. Chem.* **2015**, *87*, 1051–1069. [CrossRef]
32. Isabel Alvarez Acevedo, N.; Cristina Guimarães Rocha, M.; Carlos Bertolino Professor, L. Determinação Da Área Superficial Específica e Da Porosidade de Duas Amostras de Argilas Provenientes Da Bacia de Taubaté- São Paulo/Determination of Specific Surface Area and Porosity of Two Clays Originated from Taubaté- Basin—São Paulo. *Braz. Appl. Sci. Rev.* **2021**, *5*, 39–57. [CrossRef]
33. Phillips, V.A.; Kolbe, J.L.; Opperhauser, H. Effect of PH on the Growth of Mg(OH)2 Crystals in an Aqueous Environment at 60 °C. *J. Cryst. Growth* **1977**, *41*, 228–234. [CrossRef]

34. Rodrigues, I.A.; Villalba, J.C.; Santos, M.J.; Melquiades, F.L.; Anaissi, F.J. Smectitic Clays Enriched with Ferric Ions for the Rapid Removal of Anionic Dyes in Aqueous Media. *Clay Miner.* **2020**, *55*, 12–23. [CrossRef]
35. Seif, S.; Marofi, S.; Mahdavi, S. Removal of Cr^{3+} Ion from Aqueous Solutions Using MgO and Montmorillonite Nanoparticles. *Environ. Earth Sci.* **2019**, *78*, 377. [CrossRef]
36. Feng, N.; Guo, X.; Liang, S. Adsorption Study of Copper (II) by Chemically Modified Orange Peel. *J. Hazard. Mater.* **2009**, *164*, 1286–1292. [CrossRef]
37. Baes, C.F.; Mesmer, R.S. The Hydrolysis of Cations. John Wiley & Sons, New York, London, Sydney, Toronto 1976. 489 Seiten, Preis: £ 18.60. In *Berichte der Bunsengesellschaft für Physikalische Chemie*; John Wiley & Sons: Hoboken, NJ, USA, 1977; Volume 81, pp. 245–246. [CrossRef]
38. Yan, H.; Bai, J.; Chen, X.; Wang, J.; Zhang, H.; Liu, Q.; Zhang, M.; Liu, L. High U(vi) Adsorption Capacity by Mesoporous $Mg(OH)_2$ Deriving from MgO Hydrolysis. *RSC Adv.* **2013**, *3*, 23278–23289. [CrossRef]
39. Xu, C.; Shi, S.; Wang, X.; Zhou, H.; Wang, L.; Zhu, L.; Zhang, G.; Xu, D. Electrospun SiO_2-MgO Hybrid Fibers for Heavy Metal Removal: Characterization and Adsorption Study of Pb(II) and Cu(II). *J. Hazard. Mater.* **2019**, *381*, 120974. [CrossRef]
40. Inyinbor, A.A.; Adekola, F.A.; Olatunji, G.A. Kinetics, Isotherms and Thermodynamic Modeling of Liquid Phase Adsorption of Rhodamine B Dye onto Raphia Hookerie Fruit Epicarp. *Water Resour. Ind.* **2016**, *15*, 14–27. [CrossRef]
41. Huang, L.; Yang, Z.H.; Yan, L.Ü.J.; Alhassan, S.I.; Gang, H.Y.; Wang, T.; Wang, H.Y. Preparation of 2D Carbon Ribbon/Al2O3 and Nitrogen-Doped Carbon Ribbon/Al2O3 by Using MOFs as Precursors for Removing High-Fluoride Water. *Trans. Nonferrous Met. Soc. China* **2021**, *31*, 2174–2188. [CrossRef]
42. Said, K.A.M.; Ismail, N.Z.; Jama'in, R.L.; Alipah, N.A.M.; Sutan, N.M.; Gadung, G.G.; Baini, R.; Zauzi, N.S.A. Application of Freundlich and Temkin Isotherm to Study the Removal of Pb(II) via Adsorption on Activated Carbon Equipped Polysulfone Membrane. *Int. J. Eng. Technol.* **2018**, *7*, 91–93. [CrossRef]
43. Moscofian, A.S.O.; Airoldi, C. Synthesized Layered Inorganic-Organic Magnesium Organosilicate Containing a Disulfide Moiety as a Promising Sorbent for Cations Removal. *J. Hazard. Mater.* **2008**, *160*, 63–69. [CrossRef]
44. Ciesielczyk, F.; Bartczak, P.; Klapiszewski, Ł.; Jesionowski, T. Treatment of Model and Galvanic Waste Solutions of Copper(II) Ions Using a Lignin/Inorganic Oxide Hybrid as an Effective Sorbent. *J. Hazard. Mater.* **2017**, *328*, 150–159. [CrossRef]
45. Cao, C.Y.; Qu, J.; Wei, F.; Liu, H.; Song, W.G. Superb Adsorption Capacity and Mechanism of Flowerlike Magnesium Oxide Nanostructures for Lead and Cadmium Ions. *ACS Appl. Mater. Interfaces* **2012**, *4*, 4283–4287. [CrossRef]
46. Li, N.; Dang, H.; Chang, Z.; Zhao, X.; Zhang, M.; Li, W.; Zhou, H.; Sun, C. Synthesis of Uniformly Distributed Magnesium Oxide Micro-/Nanostructured Materials with Deep Eutectic Solvent for Dye Adsorption. *J. Alloys Compd.* **2019**, *808*, 151571. [CrossRef]
47. Wang, X.; Cai, W.; Lin, Y.; Wang, G.; Liang, C. Mass production of micro/nanostructured porous ZnO plates and their strong structurally enhanced and selective adsorption performance for environmental remediation. *J. Mater. Chem.* **2010**, *20*, 8582–8590. [CrossRef]
48. Wu, P.Y.; Jiang, Y.P.; Zhang, Q.Y.; Jia, Y.; Peng, D.Y.; Xu, W. Comparative Study on Arsenate Removal Mechanism of MgO and MgO/TiO2 Composites: FTIR and XPS Analysis. *New J. Chem.* **2016**, *40*, 2878–2885. [CrossRef]
49. Siddqui, N.; Sarkar, B.; Pendem, C.; Khatun, R.; Sivakumar Konthala, L.N.; Sasaki, T.; Bordoloi, A.; Bal, R. Highly Selective Transfer Hydrogenation of α,β-Unsaturated Carbonyl Compounds Using Cu-Based Nanocatalysts. *Catal. Sci. Technol.* **2017**, *7*, 2828–2837. [CrossRef]
50. Shakir, I.; Almutairi, Z.; Shar, S.S.; Nafady, A. Fabrication of a Flower-like $Cu(OH)_2$ Nanoarchitecture and Its Composite with CNTs for Use as a Supercapacitor Electrode. *Ceram. Int.* **2022**, *48*, 11278–11285. [CrossRef]
51. Rhatigan, S.; Nolan, M. Impact of Surface Hydroxylation in MgO-/SnO-Nanocluster Modified TiO2 Anatase (101) Composites on Visible Light Absorption, Charge Separation and Reducibility. *Chin. Chem. Lett.* **2018**, *29*, 757–764. [CrossRef]

materials MDPI

Article

Utilization of Bioflocculants from Flaxseed Gum and Fenugreek Gum for the Removal of Arsenicals from Water

Deysi J. Venegas-García and Lee D. Wilson *

Department of Chemistry, University of Saskatchewan, 110 Science Place, Thorvaldson Building (Room 165), Saskatoon, SK S7N 5C9, Canada
* Correspondence: lee.wilson@usask.ca

Abstract: Mucilage-based flocculants are an alternative to synthetic flocculants and their use in sustainable water treatment relates to their non-toxic and biodegradable nature. Mucilage extracted from flaxseed (FSG) and fenugreek seed (FGG) was evaluated as natural flocculants in a coagulation–flocculation (CF) process for arsenic removal, and were compared against a commercial xanthan gum (XG). Mucilage materials were characterized by spectroscopy (FT-IR, ^{13}C NMR), point-of-zero charge (pH_{pzc}) and thermogravimetric analysis (TGA). Box–Behnken design (BBD) with response surface methodology (RSM) was used to determine optimal conditions for arsenic removal for the CF process for three independent variables: coagulant dosage, flocculant dosage and settling time. Two anionic systems were tested: **S1**, roxarsone (organic arsenate 50 mg L^{-1}) at pH 7 and **S2** inorganic arsenate (inorganic arsenate 50 mg L^{-1}) at pH 7.5. Variable arsenic removal (RE, %) was achieved: 92.0 (**S1**-FSG), 92.3 (**S1**-FGG), 92.8 (**S1**-XG), 77.0 (**S2**-FSG), 69.6 (**S2**-FGG) and 70.6 (**S2**-XG) based on the BBD optimization. An in situ kinetic method was used to investigate arsenic removal, where the pseudo-first-order model accounts for the kinetic process. The FSG and FGG materials offer a sustainable alternative for the controlled removal of arsenic in water using a facile CF treatment process with good efficiency, as compared with a commercial xanthan gum.

Keywords: mucilage-based flocculants; flaxseed gum; fenugreek gum; xanthan gum; coagulation–flocculation; roxarsone; arsenate; Box–Behnken Design

Citation: Venegas-García, D.J.; Wilson, L.D. Utilization of Bioflocculants from Flaxseed Gum and Fenugreek Gum for the Removal of Arsenicals from Water. *Materials* **2022**, *15*, 8691. https://doi.org/10.3390/ma15238691

Academic Editors: Andreas Taubert, Roberta G. Toro, Daniela Caschera and Abeer M. Adel

Received: 16 October 2022
Accepted: 25 November 2022
Published: 6 December 2022

Publisher's Note: MDPI stays neutral with regard to jurisdictional claims in published maps and institutional affiliations.

1. Introduction

Conventional treatment of water generally involves coagulation, flocculation, filtration, and disinfection methods [1]. The coagulation–flocculation (CF) process has been employed as a simple and effective way to destabilize, agglomerate and remove suspended particles from water and wastewater due to the effectiveness and colloidal properties of CF systems [2]. Currently, CF processes are facilitated with the use of inorganic coagulants [3]. Conventional flocculants are particularly important because they are efficient at low dosages and may form diverse types of colloidal nanostructured systems, which yield favourable flocs that result in efficient phase separation of pollutants through the CF process [4]. Among synthetic polymers, polyacrylamide as a well-known example of a flocculant that biodegrades poorly, whereas some of its degradation by-products and acrylamide residues have known toxicity [5]. Over the decades, the accessibility of synthetic polymers derived from non-renewable carbon sources has shifted to more sustainable alternatives as researchers explore more readily available natural biopolymer-derived materials from renewable biomass resources [6]. Natural biopolymers are generally acquired from renewable biomass (algae, plants, microbial and animal sources) that are either comprised of carbohydrate, lipids, or proteins [7]. Bioflocculants have gained increasing attention for water treatment due to their biodegradability, non-toxic properties, and effective flocculation performance, which are sometimes comparable with synthetic flocculants [8]. Flaxseed (*Linum usitatissimum* L.) is an ancient crop cultivated globally for its fiber and oil content [9]. The mucilage extracted from the outermost layer of flaxseed hulls has great potential utility as a biopolymer flocculant [10]. The flaxseed mucilage contains two major fractions: (*i*) rhamnogalacturonan-I (acidic fraction) and (*ii*) arabinoxylans (neutral fraction) [11]. Although flax has been used as a phytoremediation tool for the remediation of different heavy metals, there are limited reports on the use of flaxseed mucilage for wastewater treatment [12]. *Trigonella foenum-graceum* (fenugreek) is primarily cultivated in Asia, Northern Africa and the Middle East. Fenugreek is widely used as a flavoring agent and in folk medicine, where its seeds contain 23–26% protein, 6–7% fat and 58% carbohydrates (25% of which is dietary fiber), saponins, flavonoids and a gum that can be extracted from the endosperm of fenugreek seed [13]. Fenugreek gum (FGG) is mainly polysaccharide in nature, comprised of a typical type of galactomannan having a linear chain of β 1,4-linked D-mannose as the backbone, where a single unit of D-galactose is joined by α 1,6-linkage, with a 1:1 ratio D-mannose: D-galactose [14]. Limited studies are available that employ FGG in wastewater treatment, where available research indicates that FGG may serve as a promising material for CF-based processes [15,16].

Arsenic is a highly toxic element to animals and plants and it is widely distributed in the environment [17]. Long-term exposure to arsenic contaminated drinking water, even at low levels of exposure, will contribute risk to human health [18], including skin cancer, stomach cancer, respiratory tract cancer, and extensive liver damage [19]. Roxarsone (4-hydroxy-3-nitrobenzene arsonic acid) is an organoarsenical that was widely used as an antimicrobial feed additive to prevent parasitic diseases of poultry and greater livestock production through greater poultry weight gain [20]. A large fraction of roxarsone does not undergo metabolism that can be transferred to soil and water, whereas the inorganic by-products are the species that possess greater toxicity [21].

The main objective of this study was to evaluate the CF removal properties of flaxseed gum (FSG) and fenugreek gum (FGG) as natural flocculants for removal of organic and inorganic arsenic species, as compared with a commercial xanthan gum (XG) bioflocculant. The pH conditions were chosen herein to favour the formation of the diprotic As species, based on the pK_a values of arsenate and roxarsone, where maximum arsenic removal has been reported [22,23]. The Box–Behnken design (BBD) was employed to investigate the role of coagulant dosage, flocculant dosage and settling time on the removal efficiency (RE; %) of organic and inorganic arsenic species. The BBD method also affords determination of desirable operating conditions for achieving the maximum arsenic removal. Moreover, the adequacy of the model and the reliability of statistical analysis with various experimental parameters were determined by comparing the experimental and predicted RE (%) response for arsenic species in aqueous media.

2. Materials and Methods

2.1. Materials

All chemicals were of analytical reagent (AR) grade. Ferric chloride hexahydrate (97%), sodium hydroxide (99%), hydrochloric acid (35%), xanthan gum (from *Xantomonas campestris*), potassium antimony (III) tartrate hydrate (99%), antimony molybdate tetrahydrate (99.9%) and spectroscopic grade potassium bromide (99%) were purchased from Sigma-Aldrich (Oakville, ON, CA, USA). Sodium hydrogen arsenate heptahydrate (98%) was obtained from Alfa Aesar (Tewksbury, MA, USA). Potassium phosphate dibasic (99%) and potassium phosphate monobasic (99%) were obtained from Fisher Scientific (New York, NY, USA). L-ascorbic acid (99%) was obtained from BDH Chemicals Canada (Mississauga, ON, CA, USA). Flaxseed gum was obtained from the College of Agriculture and BioResources at the University of Saskatchewan (Saskatoon, SK, Canada). Fenugreek gum was obtained from Emerald seed products Ltd. (Avonlea, SK, Canada). Roxarsone was acquired from Haohua Industry Co. Ltd. (Changzhou, China) and was purified via recrystallization [23]. All materials were used as received unless specified otherwise. All stock solutions were prepared using 18 MΩ cm Millipore water.

2.2. Fourier-Transform Infrared (FT-IR) Spectroscopy

The FTIR spectra of the flocculants were obtained using a Bio-RAD FTS-40 IR spectrophotometer (Bio-Rad Laboratories, Inc., Philadelphia, PA, USA). Dried powder samples were mixed with pure spectroscopic grade KBr in a 1:10 weight ratio with grinding in a small mortar and pestle. The Diffuse Reflectance Infrared Fourier Transform (DRIFT) spectra were obtained in reflectance mode at 295 K with a resolution of 4 cm^{-1} over a spectral range of 500–4000 cm^{-1}. Multiple scans were recorded and corrected relative to a KBr background.

2.3. Thermogravimetric Analysis (TGA)

Thermal profiles of the FSG, FGG and XG were obtained using a TA Instruments Q50 TGA system (New Castle, DE, USA) with a heating rate of 5 °C min^{-1} up to 500 °C and nitrogen as the carrier gas. The results reported herein are shown as first derivative (DTG) plots of weight loss (%) with temperature (wt.%/°C) against temperature (°C).

2.4. pH at Point-of-Zero-Charge (pH_{pzc})

The point-of-zero-charge (pH_{pzc}) of materials was determined according to Kong & Wilson [24]. A stock solution of NaCl (0.01 M) was prepared and 20 mL portions were added into five vials (8-dram). The solution pH conditions of the samples were adjusted between pH 1 and 8 using aqueous NaOH or HCl solutions. The flocculants (65 mg) were added to each solution, and allowed to equilibrate for 48 h, before the final pH was recorded. The pH_{pzc} was estimated plotting final pH against initial pH, and the point of intersection of the resulting null pH refers to the point zero charge, pH_{pzc}.

2.5. Coagulation–Flocculation Process

The coagulation–flocculation parameters were based on the experimental design matrix obtained from the BBD and the RSM, using a program-controlled conventional jar test Phipps & Bird PB-900 apparatus (Richmond, VA, USA) with six 2 L jars and stirrers. Two types of CF systems were carried out independently: (**S1**) organic arsenic (roxarsone) and (**S2**) inorganic arsenic (arsenate). Both systems ensured an initial concentration of 50 mg L^{-1} arsenic. Approximately 1 L of simulated arsenic-containing sample was added to the jar tester vessel and the pH was adjusted using 0.1 M NaOH or 0.1 M HCl to pH 7 and pH 7.5, for **S1** and **S2**, respectively. An aliquot of the arsenic solution was sampled to measure initial arsenic concentration. The CF process was carried out by adapting a reported method [25]. A predetermined amount of Fe (III) salt (25, 37.5 or 50 mg L^{-1}) was added to the solution, followed by rapid stirring for 3 min at 295 rpm. Thereafter, the stirring rate was reduced to 25 rpm for 20 min. During this period, the bioflocculant (1, 60.5 or 120 mg L^{-1}) was added within the first 5 min. Then, the stirring was stopped and different settling times (10–90 min) were tested. For **S1**, after dilution of the sample, 5 mL was used for UV−vis spectral analysis by adding 5 mL of phosphate buffer. A calibration curve of roxarsone was obtained (λ = 244 nm) with standard solution buffered using a Varian Cary-6000 Scan UV-vis spectrophotometer (Palo Alto, CA, USA). For **S2**, after dilution, 3 mL of sample was used for UV−vis spectral analysis by adding 0.5 mL of molybdate reagent [26]. After addition of the reagent to the arsenic sample, a blue colored complex formed after 20 min, prior to collection of the UV−vis absorbance values. A calibration curve of arsenate was obtained using the molybdate colorimetric method (λ = 900 nm) using a SPECTRONIC 200 visible spectrophotometer (Waltham, MA, USA). Experiments were repeated in duplicate, where the average value was reported. The removal efficiency (RE; %) of arsenic and the adsorption capacity, q_e (arsenic; mg·g^{-1}) were calculated by Equations (1) and (2), respectively.

$$RE\ (\%) = \frac{C_o - C_e}{C_o} \times 100 \tag{1}$$

$$q_e = \frac{(C_o - C_e) \times V}{m} \tag{2}$$

Here, C_o and C_e are the initial and equilibrium concentrations (mg L^{-1}) of arsenic, V is the volume (L), and m is the total weight (g) of the system (coagulant + flocculant).

2.6. Box–Behnken Experimental Design

The BBD method was used to determine the effects of key experimental operating variables on the arsenic removal and to determine the combined effect of the variables that yield the maximum arsenic removal (cf. Equation (1)). BBD is a response surface methodology, which is a collection of mathematical and statistical techniques that are useful for the modelling and analysis of systems wherein a response of interest is influenced by several variables, where the goal is to optimize this response [27]. Preliminary experiments indicate that three important operating parameters in the CF process that affect arsenic removal are the coagulant ($FeCl_3$) dose, flocculant (FSG, FGG or XG) dose and settling time. These variables were chosen as the independent variables and designated as A, B and C, respectively. Coagulant dose (A) was changed between 25 and 50 mg L^{-1}, flocculant dose (B) was varied between 1 and 120 mg L^{-1} and settling time (C) varied from 10 to 90 min. In Table 1, the experimental design involved the three parameters (A, B and C), each at three levels, coded −1, 0, and +1 for low, middle, and high concentrations, respectively.

Table 1. Levels of each factor for the Box–Behnken design method.

Independent Factors	Units	Symbol	Coded Levels		
			−1	0	1
$FeCl_3$	mg L^{-1}	A	25	37.5	50
Gum	mg L^{-1}	B	1	60.5	120
Settling time	min	C	10	50	90

To account for variability of the independent variables on RE (%), 12 points with 3 central points and no orthogonal blocking were carried out according to the statistical matrices developed by the RSM. Experiments were repeated in a duplicate manner, and the average value was reported. In correlating the arsenic removal efficiency ($\hat{y}$) with other independent variables (A, B and C), a response surface function was utilized (cf. Equation (3)). Where $\hat{y}$ serves as the predicted response surface function, b_0 is the model constant, b_1, b_2 and b_3 linear coefficients, whereas b_{12}, b_{13}, and b_{23} are the cross-product coefficients, and b_{11}, b_{22}, and b_{33} are the quadratic coefficients in Equation (3). The response function coefficients were determined by regression analysis of the experimental data and the Minitab 19 DOE regression program.

$$\hat{y} = b_0 + b_1A + b_2B + b_3C + b_{12}AB + b_{13}AC + b_{23}BC + b_{11}A^2 + b_{22}B^2 + b_{33}C^2 \tag{3}$$

2.7. Kinetic Studies

The adsorption of arsenate species (inorganic and organic) on the surface of adsorbent (coagulant + flocculant) is a dynamic process, where kinetic experiments reflect the change in the adsorption rate and time with the environmental conditions.

Kinetic models of adsorption such as the pseudo-first order (PFO) and pseudo-second order (PSO) models were used to examine the arsenic adsorption data. The term "pseudo" can be taken to imply that a rate law for adsorption is being expressed in terms of an adsorbed amount q (i.e., occupied sites of adsorption) onto a solid phase, in contrast to the concentration of the adsorbing species in the solution phase [28].

The non-linear forms of pseudo-first order (PFO) and pseudo-second order (PSO) kinetic models are defined by Equations (4) and (5), respectively:

$$q_t = q_e\left(1 - exp^{-k_1 t}\right) \tag{4}$$

$$q_t = \frac{q_e{}^2\, k_2 t}{1 + q_e k_2 t} \tag{5}$$

q_t (mg g^{-1}) and q_e (mg g^{-1}) indicate that the adsorption capacity of Fe(III)-flocculant towards arsenic removal at time (t), and at dynamic equilibrium. k_1 (min^{-1}), and k_2 (mg g^{-1} min^{-1}) are the respective rate constants of PFO and PSO parameters.

Kinetic studies were performed in situ using a one-pot method described by Mohamed & Wilson [29]. Briefly, a 600 mL beaker containing 500 mL of 50 mg L^{-1} arsenic solution that was mixed by magnetic stirring. A filter paper (Whatman no. 40) was folded into a cone and attached to the beaker that was immersed in the solution at a depth of 2 cm whilst stirring at 25 rpm (cf. Figure 1). An initial 2.5 mL aliquot was sampled from within the filter cone interior and added to a 25 mL volumetric flask. Sampling within the filter cone began at time (t) = 0, when the CF system was added, which continued at 1 min intervals for 10 min, then for a further 10 min at 2 min intervals, and finally for 40 min at 5 min intervals. At this point, stirring was stopped (t = 60 min) and sampling continued for a further 40 min at 10 min intervals. Sample aliquots were prepared for UV-vis spectral analysis, as discussed above for each system. The adsorption capacity at different times was determined using Equation (2).

Figure 1. Illustration of the one-pot method for in situ kinetic studies.

3. Results and Discussion

3.1. FT-IR Spectroscopy

FT-IR spectra of FSG, FGG and XG are presented in Figure 2A. The FTIR spectra of FSG presented a broad and strong band observed at 3312 cm^{-1} revealed the presence of vibrational bands of the hydroxyl groups (–OH) that collectively form hydrogen bonds [30]. The existence of a weak vibrational band at 2932 cm^{-1} was ascribed to the presence of alkyl bonds of C–H (symmetric and asymmetric) [31]. The C=O asymmetric stretching vibration at 1604 cm^{-1} was attributed to the carboxyl group indicating the presence of uronic acids or bound water in FSG. The carboxyl group acts as a binding site for ions, which has a great influence on gelling and rheological properties of the flocculant. The absorption at 1420 cm^{-1} was referred to C–O stretching vibration and revealed the existence of uronic acids [32]. The area between 800 and 1200 cm^{-1} highlights the fingerprint region for carbohydrates, which was used as a good indicator to assess the structural differences of various gums. Furthermore, the peak at 1145 cm^{-1} supports the existence of the glycosidic linkages C–O–C and C–OH [33]. The IR spectrum of FGG shows a broad band at 3343 cm^{-1}

associated with stretching vibrations of primary –OH group and a band at 2928 cm^{-1} due to the C–H stretching vibrations [15]. Three bands at 1452 cm^{-1} and 1377 cm^{-1} and 983 cm^{-1} correspond to C–H bond bending and at 1034 cm^{-1} is due to vibration of C–O–C stretching, respectively. The peak at 1144 cm^{-1} and 873 cm^{-1} can be assigned to the C–O and C–C bond stretching mode [14]. The band centered at 1230 cm^{-1} is indicative of the stretching vibration of C–O, which supports the presence of polysaccharides, due to their unique IR absorption properties [33].

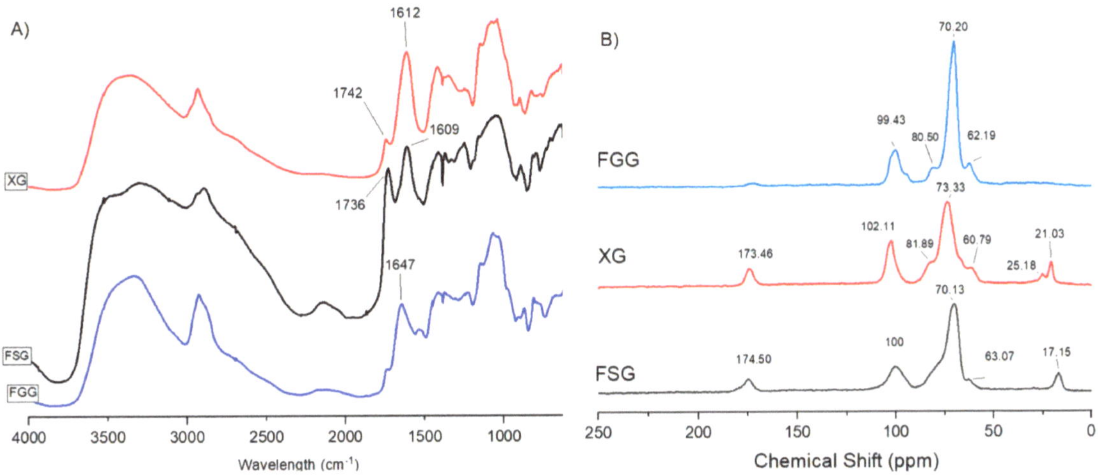

Figure 2. IR and ^{13}C solids NMR spectra for flaxseed gum (FSG), fenugreek gum (FGG) and xanthan gum (XG): (**A**) FT-IR spectra, (**B**) ^{13}C solids NMR spectra.

The spectrum for XG shows a broad signal at 3360 cm^{-1}, which indicates –OH bond stretching, and a sharp band at 2929 cm^{-1} relates to absorption of symmetrical and asymmetrical stretching of CH_3 or CH_2 groups [34]. The adsorption band at 1740 cm^{-1} describes –C=O stretching of carboxylic acids and esters, while the band at 1610 cm^{-1} denotes –C=O axial deformation in enols, such as β-diketones [35]. The band at 1412 cm^{-1} is attributed to –COO– groups and the absorption band at 1031 cm^{-1} indicates acetal groups [36]. The results described above display similar spectral profiles that are typical for IR spectra of such polysaccharide-based gums.

3.2. Solids ^{13}C-NMR Spectroscopy

Figure 2B shows the ^{13}C NMR spectra of FSG, XG and FGG. The ^{13}C NMR spectrum of FSG presents a signature at 174.5 ppm, which relates to a carboxylate anion (–COO–) from a carboxymethyl substituent typical of FSG polysaccharides. The signature at 100.0 ppm relates to the anomeric ^{13}C atoms of the polysaccharide, while C-2 and C-5 NMR lines of the polysaccharide overlap at 70.1 ppm. The primary carbon (C-6) resonance of FSG appears at 63.0 ppm. FSG contains two main polysaccharides that are acidic peptic-like with a neutral arabinoxylan polysaccharide, where the neutral carbohydrate moiety may contain uronic acid, depending on the overall molecular weight of FSG [37]. The resonances at 17.0 ppm represent the C-6 of the rhamnose. The ^{13}C NMR spectrum of FSG results concur with those from a previous study [30,38,39]. The spectrum of the FGG presented a characteristic signature at 99.4 ppm, which is assigned to C-1 of galactose and mannose units. The peaks at 70.2 ppm and 62.1 ppm represent the C-2 and C-5 and C-6 of the galactose and mannose units, respectively. In addition, the signal at 80.5 ppm relate to C-1 of the β-D-mannose reducing chain end of FGG. The results obtained herein concur with other reported NMR spectra. The strong signal at 173.4 ppm in the XG spectrum is due to C=O from proteins, fatty acids and a pyruvated exopolysaccharide of the gum structure. The spectral lines at 25.1 ppm and 102.1 ppm represent the methyl carbon at C-1 of the glucose of the xanthan gum and the anomeric carbon (C-2) of the pyruvic acid ketal group, respectively. The broad and strong signal at 73.3 ppm relates to C-2, C-3 and C-5 of the glucose polymer chain. The peak located at 21.0 ppm shows the ^{13}C ($-CH_3$) group of acetate from proteins, fatty acids and exopolysaccharides [40].

3.3. Thermogravimetric Analysis (TGA)

Thermogravimetric analysis (TGA) is a simple method for studying the decomposition pattern and the thermal stability of the biopolymers. The TGA results herein are shown (Figure 3a) as a DTG profile (first derivative of mass with temperature) and weight loss (%) versus temperature (Figure 3b), to characterize the trends in thermal stability for FSG, FGG and XG. The weight loss events for the thermal decomposition of FSG, FGG and XG occurs in three steps: firstly at 25 °C to 150 °C due to desorbed water and volatiles, secondly at 150–250 °C due to the loss of functional groups (–OH, –COOH) and thirdly above 250 °C due to breakdown of the biopolymer backbone.

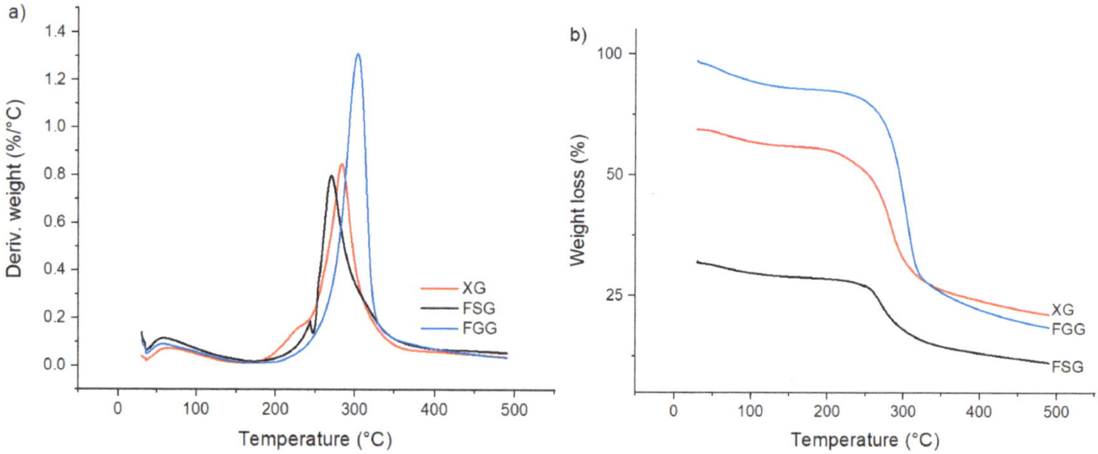

Figure 3. TGA results presented as a DTG profile (**a**) and weight loss versus temperature (**b**) for flaxseed gum (FSG), fenugreek gum (FGG) and xanthan gum (XG).

According to the TGA results in Figure 3a, the onset of degradation occurs at 275 °C, 300 °C and 290 °C for FSG, FGG and XG, respectively, in agreement with other reported TGA results [14,41,42].

3.4. pH at Point-of-Zero-Charge (pH_{pzc})

The pH_{pzc} parameter indicates the pH where the net surface charge of a material is zero, and it provides insight on the role of electrostatic interactions of a material surface with charged species. The pH_{pzc} of the FSG, FGG and XG were estimated according to the method reported by Kong & Wilson [24]. Figure 4 shows that the initial pH values of the XG, FSG and FGG are approximately 4.8, 4.9 and 5.9, respectively in agreement with other reports [43,44]. The surface charge of the adsorbents is positive at a pH below the pH_{pzc}, where adsorption of OH^- ions and other anionic species occur because the ζ-potential of the sorbent is positive overall. Meanwhile, it is negative when the solution pH is above pH_{pzc}, where adsorption of cation species occur due to deprotonation of the sorbent surface [20].

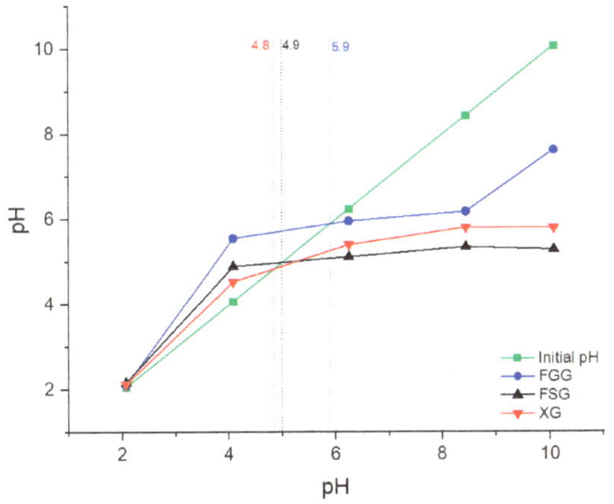

Figure 4. Zero-point-charge (pH_{pzc}) as a function of variable initial pH for FSG, FGG and XG.

3.5. Box–Behnken Experimental Design

3.5.1. Main Effects of the Independent Variables on the Response Functions

The removal of arsenic (RE; %) from water was investigated by a CF process. Three key operating parameters (coagulant dose, flocculant dose and settling time) were investigated using the BBD method. Plots depicting the main effects for the arsenic removal efficiency for system **S1** and system **S2** are presented in the Supplementary Materials for the coagulant and flocculant dosage and settling time (cf. Figures S1–S3). The corresponding plots illustrate important factors that are set at three levels for the designed experiment, which enables an in-depth analysis of the effects of factors on the CF process. These plots provide a preliminary estimate on the effects of independent variables: coagulant dose, flocculant dose and settling time on the response (RE; %). The statistical significance of the response function generated was verified by the F-test and ANOVA results (cf. Tables S1–S6, Supplementary Material). Factors having a p-value < 0.05 that indicate statistical significance. The effect of increasing coagulant dosage from 25 to 50 mg L^{-1} (cf. Figure S1, Supplementary Material) shows that the arsenic removal increase with increasing coagulant dosage, as expected from other reports [45–47]. The maximum arsenic removal (%) achieved was 88, 87, 88.5, 76.1, 67.5 and 66.7 for **S1** arsenic removal with FSG (a), FGG (b) and XG (c), and **S2** arsenic removal with FSG (d), FGG (e) and XG (f), considering $FeCl_3$ as the main factor. These results corroborate that the formation of the As-Fe oxides complex as the rate determining step in the process [48,49]. The role of flocculant dosage from 1 to 120 mg L^{-1} (cf. Figure S2, Supplementary Material) has a similar trend on arsenic removal (%) for **S1** and **S2** systems. An increased flocculant dosage results in greater arsenic removal (%) up to a maximum RE (%) value. Thereafter, the arsenic removal (%) decreases at higher flocculant concentration (120 mg L^{-1}). This is probably due to the higher gum concentrations causing a vertical accumulation of the gum in the water column, which prevents efficient settling [14,50]. The effect on settling time for **S1** and **S2** was set from 10 to 90 min (cf. Figure S3, Supplementary Material). Each of the flocculants (FSG, FGG and XG) had a similar effect on the arsenic removal for **S1**. An increase in the settling times from 10 to 90 min led to greater arsenic removal (%), where small changes in arsenic removal and negligible changes in sludge volume was observed. For **S2**, the settling times from 10 to 90 min showed that greater settling time increased the arsenic removal (%) to a maximum value, where the arsenic removal (%) decreased at 90 min and beyond. According to these results, settling was achieved within 40–50 min. The rapid sedimentation indicated that the process is controlled by charge neutralization and bridging effects [51]. Maximum removal occurs when no more surface sites are available to the arsenate complex. Moreover, the adsorption time is limited by the settling of the gum, which sinks more rapidly as it gets heavier with adsorbed As-Fe oxide particles [52].

In order to explore the interactions of various major operating factors on arsenic removal by the flocculation process, the Response Surface Methodology (RSM) was employed. Three-dimensional (3D) response surface plots and 2D contour plots (Supplementary Material, Figures S4–S9) were obtained from the model-predicted response by changing two independent variables within the experimental conditions at the same time as the third variable was fixed at a constant optimal level. These plots are useful to assess the interactive relationship between the independent variables and the response variable.

3.5.2. Box−Behnken Analysis

The coefficients of the response function for arsenic RE (%) are presented in Table 2. Predicted values of RE (%) for arsenic were determined by the response functions with the obtained coefficients. A comparison of experimental and predicted values for coagulant dosage, flocculant dosage and settling time parameters for arsenic removal (%) are presented in the Supplementary Material (Table S7 for **S1**, and Table S8 for **S2**). Predicted and experimental values of arsenic RE (%) for **S1** and the various bioflocculants are in good general agreement.

Table 2. Coefficients of the response function for arsenic removal (RE; %) at 295 K.

Coefficients	S1			S2		
	FSG	FGG	XG	FSG	FGG	XG
Const	88.0	87.0	88.5	76.167	67.5	66.667
$FeCl_3$ (A)	−10.25	−30.938	−18.25	−4.3125	−10.313	−9.1875
GUM (B)	2.125	−1.3125	−2.6875	−0.0625	3.25	−12.375
ST (C)	6.875	3.0	5.9375	6.0	1.8125	2.3125
AB	7.375	1.25	13.0	−7.0	4.625	4.0
AC	−4.375	−1.375	−1.0	−2.125	0.5	−0.125
BC	−1.375	1.125	3.375	−5.625	−0.625	2.75
AA	−28.938	−47.875	−34.563	−16.708	−12.625	−17.396
BB	−4.9375	−4.375	−18.188	−19.958	−27.5	−17.021
CC	−4.9375	−7.5	−6.4375	−15.583	−24.875	−6.8958

S1 (organic arsenic), **S2** (inorganic arsenic), FSG (Flaxseed gum), FGG (Fenugreek gum), XG (Xanthan gum).

The experimental data points in Box–Behnken experimental design were employed independently for each system, **S1** and **S2**, and three different flocculants (FSG, FGG and XG). The coefficients of determination (R^2) values for system **S1** were 0.97, 0.99 and 0.95 for FSG, FGG and XG, respectively.

It is evident from Tables S7 and S8 that the predicted responses are in good agreement based on the % error, which is less than the experimentally obtained responses. The coefficients of determination (R^2) values for **S2** were 0.94, 0.99 and 0.98, which show that the response models are highly reliable.

3.5.3. Confirmation and Validation

Optimal conditions for maximum arsenic removal were determined by the response models for the experimental data. The corresponding predicted arsenic removal (RE; %) for **S1**-FSG, **S1**-FGG, **S1**-XG, **S2**-FSG, **S2**-FGG, and **S2**-XG at these optimal conditions were 92.0 ± 0.6%, 92.3 ± 0.1%, 92.8 ± 0.1%, 77.0 ± 0.1%, 69.6 ± 0.6% and 70.6 ± 0.9%. The optimized conditions indicate a range between 32 and 36 mg L^{-1} of $FeCl_3$ was required to achieve an optimal arsenic removal for the six different flocculant systems, showing that the As/Fe ratio is the main effect that governs the arsenic removal. For **S1**, similar concentrations of FGG (52 mg L^{-1}) and of XG (50 mg L^{-1}), were required to achieve optimal arsenic removal; meanwhile, a slightly higher concentration (64 mg L^{-1}) was required for FSG. For **S2** the concentrations for FSG (60 mg L^{-1}) and FGG (62 mg L^{-1}) were similar, where the concentration required to maximize the arsenic removal efficiency for XG was lower (37 mg L^{-1}). The settling time for **S1** and **S2** requires at least 50 min to achieve the highest arsenic removal, presenting the lowest settling time for the system **S2**-FGG with 51 min versus the longer **S1**-FSG (81 min). A comparison of experimental and predicted values at optimized conditions for coagulant dosage, flocculant dosage and settling time for arsenic removal (%) are presented in Table 3. The predicted and experimental values of arsenic removal (%) are in good agreement.

Table 3. Results of confirmation experiments under optimized conditions at 295 K.

GUM	GUM Dose (mg L^{-1})	$FeCl_3$ (mg L^{-1})	Settling Time (min)	As Removal (%) (50 mg L^{-1})	
				S1 (Roxarsone)	
				Predicted	Experimental
FSG	64	35	81	91.8	92.0 ± 0.6
FGG	52	33	58	92.5	92.3 ± 0.1
XG	50	34	68	92.6	92.8 ± 0.1
				S2 (Arsenate)	
				Predicted	Experimental
FSG	60	36	58	77.0	77.0 ± 0.1
FGG	62	32	51	69.4	69.6 ± 0.6
XG	37	34	54	70.6	70.6 ± 0.9

The validity of the statistical methodology and the experimental results was confirmed by performing additional experiments in duplicate for different conditions within the experimental design for each individual gum system. The selected conditions for the coagulant and flocculant dosage, and settling time are listed in Tables S9 and S10 (cf. Supplementary Material) for **S1** and **S2**, respectively, including the predicted and the experimental results for RE (%) of arsenic. The experimental results for arsenic removal for each system are in good agreement with the predicted arsenic removal values.

Table 4 shows different mucilage-based and chitosan systems that serve as natural bioflocculants, where the conditions for arsenic removal, such as flocculant nature and dosage, coagulant nature and dosage, initial arsenic concentration and pH are listed. The RE (%) for FSG, FGG and XG presented herein show greater effectiveness based on the initial concentration.

Table 4. Comparison of maximum RE (%) of arsenic in aqueous medium using different biopolymers.

Flocculant	Concentration (mg L^{-1})	Coagulant	Concentration (mg L^{-1})	As (V) (mg L^{-1})	pH	RE (%)	Reference
Aloe vera gum	2	PAC	3	0.2–1	5	92.6	[53]
Chitosan	0.5	$FeCl_3$		0.2–2	7	~100	[45]
Opuntia ficus-indica gum	350	-	-	0.002–0.01	5.9	70	[54]
-	-	$FeCl_3$	27.029	1	5	98	[55]
-	-	$Fe_2(SO_4)_3$	100	5	6	99	[56]
-	-	$Al_2(SO_4)_3$	25–50	0.065–0.216	7–8	81	[57]
Flaxseed gum	64	$FeCl_3$	35	50	7	90	This work
Fenugreek gum	52	$FeCl_3$	33	50	7	90	This work
Xanthan gum	50	$FeCl_3$	34	50	7	93	This work
Flaxseed gum	60	$FeCl_3$	36	50	7.5	77	This work
Fenugreek gum	62	$FeCl_3$	32	50	7.5	69.6	This work
Xanthan gum	37	$FeCl_3$	34	50	7.5	70.6	This work

3.6. Flocculation Kinetics

To study the effect of Fe (III)-flocculant for **S1** and **S2**, the one-pot kinetic experiment was carried out. Table 5 lists the kinetic parameters, where high correlation with the PFO model provides insight on the dependence of one of the components of the CF process. Based on reports in the literature, the initial concentration of arsenate is the main factor that governs the CF process [58,59]. Based on R^2 values, the arsenic uptake by Fe (III)-FSG, Fe (III)-FGG and Fe (III)-XG systems obey the PFO model because of higher R^2 values over the PSO model. By comparing the correlation coefficient (R^2), the adsorption for arsenic onto the systems Fe(III)-gum is well described by the PFO model. Similar kinetic results for arsenic removal by biopolymers were previously reported [60].

Table 5. Adsorption kinetics parameters for **S1** and **S2** adsorption over FSG, FGG and XG at 295 K.

GUM	Pseudo-First-Order (PFO) Model			Pseudo-Second-Order (PSO) Model		
	k_1	q_e	R^2	k_2	q_e	R^2
	(min^{-1})	$(mg\ g^{-1})$		(min^{-1})	$(mg\ g^{-1})$	
			S1 (Roxarsone)			
FSG	0.168	518.6 ± 10.8	0.952	3.6×10^{-4}	580.3 ± 23.6	0.897
FGG	0.147	607.5 ± 16.7	0.924	2.7×10^{-4}	682.3 ± 32.9	0.865
XG	0.163	602.6 ± 11.5	0.956	3.2×10^{-4}	669.5 ± 25.1	0.901
			S2 (Arsenate)			
FSG	0.138	398.9 ± 13.7	0.896	3.7×10^{-4}	451.2 ± 25.6	0.842
FGG	0.112	296.7 ± 9.8	0.915	3.7×10^{-4}	343.2 ± 19.6	0.869
XG	0.168	537.5 ± 8.6	0.967	3.9×10^{-4}	595.1 ± 16.8	0.937

The kinetics profile of Fe (III)-flocculant system used for the removal of arsenic is shown in Figure 5, where the two kinetic models (PFO and PSO) were used to fit the adsorption profiles for arsenic removal versus time. The adsorption process for arsenic occurs in two stages: a fast-initial adsorption stage, where ~90% of equilibrium adsorption capacity was achieved, and the subsequent slow adsorption phase, where the adsorption capacity gradually reached its maximum value for theequilibrium capacity.

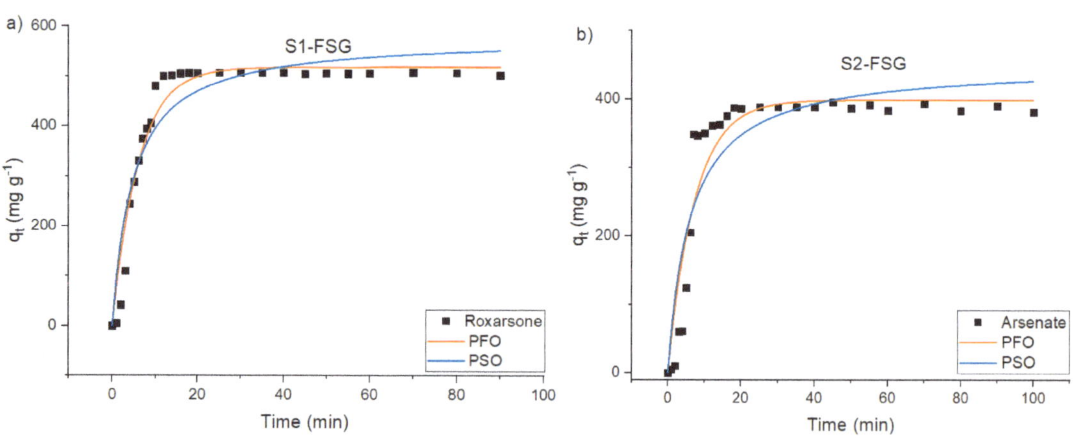

Figure 5. *Cont.*

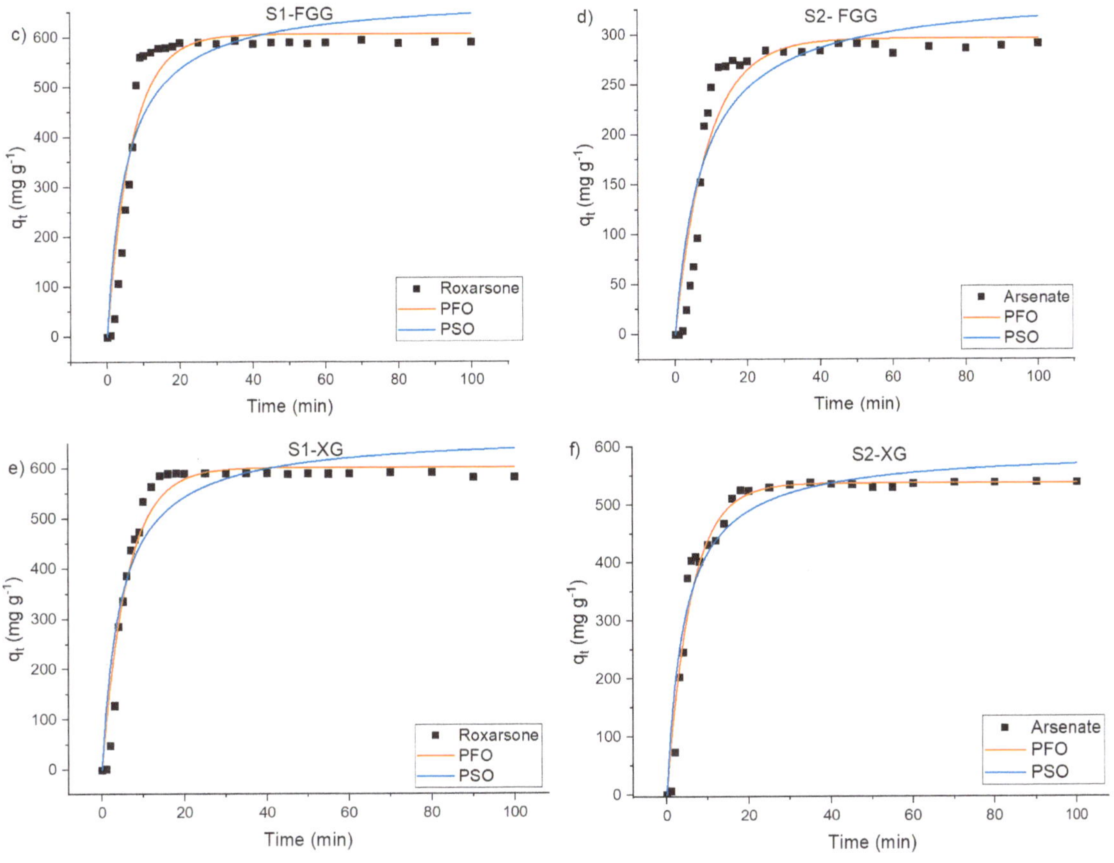

Figure 5. Kinetic profiles for the adsorption at 295 K: **S1**-FSG (**a**), **S2**-FSG (**b**), **S1**-FGG (**c**), **S2**-FGG (**d**), **S1**-XG (**e**), **S2**-XG (**f**), PFO and PSO kinetic-model best-fit results.

4. Conclusions

This study focused on the use of flaxseed gum and fenugreek gum as flocculants, which were compared against commercial xanthan gum, for the removal of organic and inorganic arsenic species in water using a coagulation–flocculation (CF) process. Structural and physicochemical characterization of the gum flocculants employed spectroscopic (FT-IR, ^{13}C NMR) and TGA to determine their relative composition and functional properties to guide the selection of materials for the flocculation process. A Box–Behnken statistical experimental design was used to determine the effects of coagulant dosage, flocculant dosage and settling time on the arsenic removal efficiency. The flocculation ability of the gum materials was investigated using an arsenic concentration of 50 mg L^{-1} and a fixed initial pH of 7 and 7.5, for the organic and inorganic arsenic species (roxarsone, **S1**; arsenate, **S2**), respectively. The Box–Behnken statistical experimental design and Response Surface Methodology were effective in determining the optimum conditions for arsenic removal by a CF process for several bioflocculants (FSG, FGG and XG). Experimental results in Table 3 demonstrated that the optimum coagulant dosage for maximum arsenic removal with ferric chloride was between 32 and 36 mg L^{-1} for **S1** and **S2,** which further validates the regression equations employed herein. Optimum flocculant dosages for maximum arsenic removal were found as follows: 64 mg L^{-1} (**S1**-FSG), 60 mg L^{-1} (**S2**-FSG), 52 mg L^{-1} (**S1**-FGG), 62 mg L^{-1} (**S2**-FGG), 50 mg L^{-1} (**S1**-XG) and 37 mg L^{-1} (**S2**-XG). The optimal settling time was estimated between 51 and 81 min. The kinetic results were studied in situ by using a one-pot kinetic method, where the PFO model provided a better fit relative to the kinetic profiles over the PSO model. FSG and FGG materials offer a sustainable alternative for the controlled removal of arsenic in water that employs the CF process. CF-based water treatment is a cost effective, environmentally friendly, facile and efficient treatment technology, as compared with other arsenic removal techniques. Therefore, the arsenic removal efficiency plus the low cost of FSG and FGG highlight the utility of these systems as promising bioflocculants for implementation in continuous water treatment systems.

Supplementary Materials: The following supporting information can be downloaded at: https://www.mdpi.com/article/10.3390/ma15238691/s1, Figures S1–S3: Arsenic removal efficiency (%)—Main factors effect for roxarsone and arsenate as a function of coagulant dose, flocculant dose, and settling time for the various bioflocculants; Figures S4–S9: 3D Box–Behnken response surface and 2D contour plots of arsenic removal (arsenate and roxarsone) with bioflocculants as a function of flocculant type and settling time, coagulant and settling time, and coagulant and flocculant; Tables S1–S6: ANOVA tables for arsenic removal (%) arsenic-bioflocculants systems at 295 K; Tables S7 and S8: Removal efficiency of the response function for roxarsone and arsenic removal at 295 K; Tables S9 and S10: Observed and predicted inorganic arsenic removal efficiency for arsenate and roxarsone (validation) at 295 K.

Author Contributions: Conceptualization, L.D.W.; methodology, D.J.V.-G.; software, D.J.V.-G.; validation, D.J.V.-G. and L.D.W.; formal analysis, D.J.V.-G.; investigation, D.J.V.-G.; resources, L.D.W.; data curation, D.J.V.-G.; writing—original draft preparation, D.J.V.-G.; writing—review and editing, D.J.V.-G. and L.D.W.; visualization, D.J.V.-G.; supervision, L.D.W.; project administration, L.D.W.; funding acquisition, L.D.W. All authors have read and agreed to the published version of the manuscript.

Funding: L.D.W. acknowledges the support provided by the Government of Canada through the Natural Sciences and Engineering Research Council of Canada (NSERC Discovery Grant Number: RGPIN 04315-2021) and the APC was funded by the Editorial Office of the journal Materials.

Institutional Review Board Statement: Not applicable.

Informed Consent Statement: Not applicable.

Data Availability Statement: The data presented in this study are available in this article and the accompanying Supplementary Materials herein.

Acknowledgments: The Saskatchewan Structural Sciences Centre (SSSC) is acknowledged for providing facilities to conduct this research. Funding from Canada Foundation for Innovation, Natural Sciences and Engineering Research Council of Canada and the University of Saskatchewan support research at the SSSC. Martin Reaney from the University of Saskatchewan is gratefully acknowledged for provision of Flax Seed Gum and Fenugreek Gum for this research.

Conflicts of Interest: The authors declare no conflict of interest.

References

1. Dayarathne, H.N.; Angove, M.J.; Aryal, R.; Abuel-Naga, H.; Mainali, B. Removal of Natural Organic Matter from Source Water: Review on Coagulants, Dual Coagulation, Alternative Coagulants, and Mechanisms. *J. Water Process Eng.* **2021**, *40*, 101820. [CrossRef]
2. Ge, J.; Guha, B.; Lippincott, L.; Cach, S.; Wei, J.; Su, T.L.; Meng, X. Challenges of Arsenic Removal from Municipal Wastewater by Coagulation with Ferric Chloride and Alum. *Sci. Total Environ.* **2020**, *725*, 138351. [CrossRef] [PubMed]
3. Bruno, P.; Campo, R.; Giustra, M.G.; De Marchis, M.; Di Bella, G. Bench Scale Continuous Coagulation-Flocculation of Saline Industrial Wastewater Contaminated by Hydrocarbons. *J. Water Process Eng.* **2020**, *34*, 101156. [CrossRef]
4. Wei, H.; Gao, B.; Ren, J.; Li, A.; Yang, H. Coagulation/Flocculation in Dewatering of Sludge: A Review. *Water Res.* **2018**, *143*, 608–631. [CrossRef]
5. Nasrollahzadeh, M.; Sajjadi, M.; Iravani, S.; Varma, R.S. Starch, Cellulose, Pectin, Gum, Alginate, Chitin and Chitosan Derived (Nano)Materials for Sustainable Water Treatment: A Review. *Carbohydr. Polym.* **2021**, *251*, 116986. [CrossRef] [PubMed]
6. Prabhakar, S.A.; Ojha, N.; Das, N. Application of Aloe Vera Mucilage as Bioflocculant for the Treatment of Textile Wastewater: Process Optimization. *Water Sci. Technol.* **2020**, *82*, 2446–2459. [CrossRef]

7. Kamel, R.; Afifi, S.M.; Kassem, I.A.A.; Elkasabgy, N.A.; Farag, M.A. Arabinoxylan and Rhamnogalacturonan Mucilage: Outgoing and Potential Trends of Pharmaceutical, Environmental, and Medicinal Merits. *Int. J. Biol. Macromol.* **2020**, *165*, 2550–2564. [CrossRef]
8. Ajao, V.; Fokkink, R.; Leermakers, F.; Bruning, H.; Rijnaarts, H.; Temmink, H. Bioflocculants from Wastewater: Insights into Adsorption Affinity, Flocculation Mechanisms and Mixed Particle Flocculation Based on Biopolymer Size-Fractionation. *J. Colloid Interface Sci.* **2021**, *581*, 533–544. [CrossRef]
9. Hu, Y.; Shim, Y.Y.; Reaney, M.J.T. Flaxseed Gum Solution Functional Properties. *Foods* **2020**, *9*, 681. [CrossRef]
10. Karami, N.; Kamkar, A.; Shahbazi, Y.; Misaghi, A. Electrospinning of Double-Layer Chitosan-Flaxseed Mucilage Nanofibers for Sustained Release of Ziziphora Clinopodioides Essential Oil and Sesame Oil. *Lwt* **2021**, *140*, 110812. [CrossRef]
11. Ding, H.H.; Qian, K.; Goff, H.D.; Wang, Q.; Cui, S.W. Structural and Conformational Characterization of Arabinoxylans from Flaxseed Mucilage. *Food Chem.* **2018**, *254*, 266–271. [CrossRef] [PubMed]
12. Saleem, M.; Bachmann, R.T. A Contemporary Review on Plant-Based Coagulants for Applications in Water Treatment. *J. Ind. Eng. Chem.* **2019**, *72*, 281–297. [CrossRef]
13. Kim, I.T.S.; Sethu, V.; Arumugasamy, S.K.; Selvarajoo, A. Fenugreek Seeds and Okra for the Treatment of Palm Oil Mill Effluent (POME)—Characterization Studies and Modeling with Backpropagation Feedforward Neural Network (BFNN). *J. Water Process Eng.* **2020**, *37*, 101500. [CrossRef]
14. Mishra, S.; Kundu, K. Synthesis, Characterization and Applications of Polyacrylamide Grafted Fenugreek Gum (FG-g-PAM) as Flocculant: Microwave vs. Thermal Synthesis Approach. *Int. J. Biol. Macromol.* **2019**, *141*, 792–808. [CrossRef] [PubMed]
15. Gu, G.; Mo, C.; Zhao, K.; Chen, Z.; Wang, X. Influence of Fenugreek-Gum and Particle Size on Performance of Talc Flotation. *Physicochem. Probl. Miner. Process.* **2018**, *54*, 1026–1033. [CrossRef]
16. Tangri, A. Trigonella Foenum-Graecum Mucilage: An Adsorbent for Removal of Sulphate Ions. *Int. J. Adv. Res. Chem. Sci.* **2014**, *1*, 34–42.
17. Sousa, G.V.; Teles, V.L.G.; Pereira, E.G.; Modolo, L.V.; Costa, L.M. Interactions between As and Se upon Long Exposure Time and Effects on Nutrients Translocation in Golden Flaxseed Seedlings. *J. Hazard. Mater.* **2021**, *402*, 123565. [CrossRef]
18. Cui, J.; Jing, C. A Review of Arsenic Interfacial Geochemistry in Groundwater and the Role of Organic Matter. *Ecotoxicol. Environ. Saf.* **2019**, *183*, 109550. [CrossRef]
19. Zhang, Z.; Li, J.; Sun, C.; Marhaba, T.F.; Zhang, W.; Zhang, Y. Arsenic Speciation by Sequential Extraction from As-Fe Precipitates Formed Under Different Coagulation Conditions. *Water. Air. Soil Pollut.* **2016**, *227*, 309. [CrossRef]
20. Zang, S.; Zuo, Y.; Wang, J.; Liu, X.; Gomez, M.A.; Wei, L. Adsorption Removal of Roxarsone, Arsenite(III), and Arsenate(V) Using Iron-Modified Sorghum Straw Biochar and Its Kinetics. *Acta Geochim.* **2021**, *40*, 409–418. [CrossRef]
21. Mahaninia, M.H.; Wilson, L.D. A Kinetic Uptake Study of Roxarsone Using Cross-Linked Chitosan Beads. *Ind. Eng. Chem. Res.* **2017**, *56*, 1704–1712. [CrossRef]
22. Kobya, M.; Soltani, R.D.C.; Omwene, P.I.; Khataee, A. A Review on Decontamination of Arsenic-Contained Water by Electrocoagulation: Reactor Configurations and Operating Cost along with Removal Mechanisms. *Environ. Technol. Innov.* **2020**, *17*, 100519. [CrossRef]
23. Kong, D.; Wilson, L.D. Synthesis and Characterization of Cellulose-Goethite Composites and Their Adsorption Properties with Roxarsone. *Carbohydr. Polym.* **2017**, *169*, 282–294. [CrossRef]
24. Kong, D.; Wilson, L.D. Uptake of Methylene Blue from Aqueous Solution by Pectin–Chitosan Binary Composites. *J. Compos. Sci.* **2020**, *4*, 95. [CrossRef]
25. Agbovi, H.K.; Wilson, L.D.; Tabil, L.G. Biopolymer Flocculants and Oat Hull Biomass To Aid the Removal of Orthophosphate in Wastewater Treatment. *Ind. Eng. Chem. Res.* **2017**, *56*, 37–46. [CrossRef]
26. Ibnul, N.K.; Tripp, C.P. A Solventless Method for Detecting Trace Level Phosphate and Arsenate in Water Using a Transparent Membrane and Visible Spectroscopy. *Talanta* **2021**, *225*, 122023. [CrossRef] [PubMed]
27. Agbovi, H.K.; Wilson, L.D. Optimisation of Orthophosphate and Turbidity Removal Using an Amphoteric Chitosan-Based Flocculant–Ferric Chloride Coagulant System. *Environ. Chem.* **2019**, *16*, 599. [CrossRef]
28. Bekhoukh, A.; Moulefera, I.; Zeggai, F.Z.; Benyoucef, A.; Bachari, K. Anionic Methyl Orange Removal from Aqueous Solutions by Activated Carbon Reinforced Conducting Polyaniline as Adsorbent: Synthesis, Characterization, Adsorption Behavior, Regeneration and Kinetics Study. *J. Polym. Environ.* **2022**, *30*, 886–895. [CrossRef]
29. Mohamed, M.H.; Wilson, L.D. Kinetic Uptake Studies of Powdered Materials in Solution. *Nanomaterials* **2015**, *5*, 969–980. [CrossRef] [PubMed]
30. Jiang, Y.; Reddy, C.K.; Huang, K.; Chen, L.; Xu, B. Hydrocolloidal Properties of Flaxseed Gum/Konjac Glucomannan Compound Gel. *Int. J. Biol. Macromol.* **2019**, *133*, 1156–1163. [CrossRef]
31. Liang, S.; Li, X.; Ma, X.; Li, A.; Wang, Y.; Reaney, M.J.T.; Shim, Y.Y. A Flaxseed Heteropolysaccharide Stimulates Immune Responses and Inhibits Hepatitis B Virus. *Int. J. Biol. Macromol.* **2019**, *136*, 230–240. [CrossRef] [PubMed]
32. Minevich, I.E.; Osipova, L.L.; Nechiporenko, A.P.; Melnikova, M.I.; Tsyganova, T.B. IR-Spectroscopy of Polysaccharide Flaxseed (*Linum usitatissimum* L.) Products. *Foods Raw Mater.* **2019**, *7*, 274–282. [CrossRef]
33. Safdar, B.; Pang, Z.; Liu, X.; Jatoi, M.A.; Mehmood, A.; Rashid, M.T.; Ali, N.; Naveed, M. Flaxseed Gum: Extraction, Bioactive Composition, Structural Characterization, and Its Potential Antioxidant Activity. *J. Food Biochem.* **2019**, *43*, 1–11. [CrossRef] [PubMed]

34. Faria, S.; De Oliveira Petkowicz, C.L.; De Morais, S.A.L.; Terrones, M.G.H.; De Resende, M.M.; De Frana, F.P.; Cardoso, V.L. Characterization of Xanthan Gum Produced from Sugar Cane Broth. *Carbohydr. Polym.* **2011**, *86*, 469–476. [CrossRef]
35. Ćirić, A.; Medarević, Đ.; Čalija, B.; Dobričić, V.; Mitrić, M.; Djekic, L. Study of Chitosan/Xanthan Gum Polyelectrolyte Complexes Formation, Solid State and Influence on Ibuprofen Release Kinetics. *Int. J. Biol. Macromol.* **2020**, *148*, 942–955. [CrossRef] [PubMed]
36. Inphonlek, S.; Niamsiri, N.; Sunintaboon, P.; Sirisinha, C. Chitosan/Xanthan Gum Porous Scaffolds Incorporated with in-Situ-Formed Poly(Lactic Acid) Particles: Their Fabrication and Ability to Adsorb Anionic Compounds. *Colloids Surf. A Physicochem. Eng. Asp.* **2020**, *603*, 125263. [CrossRef]
37. Qian, K.Y.; Cui, S.W.; Wu, Y.; Goff, H.D. Flaxseed Gum from Flaxseed Hulls: Extraction, Fractionation, and Characterization. *Food Hydrocoll.* **2012**, *28*, 275–283. [CrossRef]
38. Liu, J.; Shen, J.; Shim, Y.Y.; Reaney, M.J.T. Carboxymethyl Derivatives of Flaxseed (*Linum usitatissimum* L.) Gum: Characterisation and Solution Rheology. *Int. J. Food Sci. Technol.* **2016**, *51*, 530–541. [CrossRef]
39. Kaushik, P.; Dowling, K.; Adhikari, R.; Barrow, C.J.; Adhikari, B. Effect of Extraction Temperature on Composition, Structure and Functional Properties of Flaxseed Gum. *Food Chem.* **2017**, *215*, 333–340. [CrossRef]
40. Kang, Y.; Li, P.; Zeng, X.; Chen, X.; Xie, Y.; Zeng, Y.; Zhang, Y.; Xie, T. Biosynthesis, Structure and Antioxidant Activities of Xanthan Gum from Xanthomonas Campestris with Additional Furfural. *Carbohydr. Polym.* **2019**, *216*, 369–375. [CrossRef]
41. Rashid, F.; Ahmed, Z.; Hussain, S.; Huang, J.Y.; Ahmad, A. *Linum usitatissimum* L. Seeds: Flax Gum Extraction, Physicochemical and Functional Characterization. *Carbohydr. Polym.* **2019**, *215*, 29–38. [CrossRef]
42. Makhado, E.; Pandey, S.; Ramontja, J. Microwave-Assisted Green Synthesis of Xanthan Gum Grafted Diethylamino Ethyl Methacrylate: An Efficient Adsorption of Hexavalent Chromium. *Carbohydr. Polym.* **2019**, *222*, 114989. [CrossRef]
43. Salarbashi, D.; Bazeli, J.; Fahmideh-Rad, E. Fenugreek Seed Gum: Biological Properties, Chemical Modifications, and Structural Analysis—A Review. *Int. J. Biol. Macromol.* **2019**, *138*, 386–393. [CrossRef]
44. Lan, Y.; Ohm, J.-B.; Chen, B.; Rao, J. Physicochemical Properties and Aroma Profiles of Flaxseed Proteins Extracted from Whole Flaxseed and Flaxseed Meal. *Food Hydrocoll.* **2020**, *104*, 105731. [CrossRef]
45. Hesami, F.; Bina, B.; Ebrahimi, A.; Amin, M. Arsenic Removal by Coagulation Using Ferric Chloride and Chitosan from Water. *Int. J. Environ. Health Eng.* **2013**, *2*, 17. [CrossRef]
46. Qiao, J.; Jiang, Z.; Sun, B.; Sun, Y.; Wang, Q.; Guan, X. Arsenate and Arsenite Removal by FeCl3: Effects of PH, As/Fe Ratio, Initial As Concentration and Co-Existing Solutes. *Sep. Purif. Technol.* **2012**, *92*, 106–114. [CrossRef]
47. Fox, D.I.; Stebbins, D.M.; Alcantar, N.A. Combining Ferric Salt and Cactus Mucilage for Arsenic Removal from Water. *Environ. Sci. Technol.* **2016**, *50*, 2507–2513. [CrossRef]
48. Chiavola, A.; D'Amato, E.; Sirini, P.; Caretti, C.; Gori, R. Arsenic Removal from a Highly Contaminated Groundwater by a Combined Coagulation-Filtration-Adsorption Process. *Water Air Soil Pollut.* **2019**, *230*, 87. [CrossRef]
49. Liu, B.; Liu, Z.; Wu, H.; Pan, S.; Cheng, X.; Sun, Y.; Xu, Y. Effective and Simultaneous Removal of Organic/Inorganic Arsenic Using Polymer-Based Hydrated Iron Oxide Adsorbent: Capacity Evaluation and Mechanism. *Sci. Total Environ.* **2020**, *742*, 140508. [CrossRef]
50. Zhang, Z.; Wang, H.; Chen, T.; Zhang, H.; Liang, J.; Kong, W.; Yao, J.; Zhang, J.; Wang, J. Synthesis and Structure Characterization of Sulfated Galactomannan from Fenugreek Gum. *Int. J. Biol. Macromol.* **2019**, *125*, 1184–1191. [CrossRef] [PubMed]
51. Jiang, X.; Li, Y.; Tang, X.; Jiang, J.; He, Q.; Xiong, Z.; Zheng, H. Biopolymer-Based Flocculants: A Review of Recent Technologies. *Environ. Sci. Pollut. Res.* **2021**, *28*, 46934–46963. [CrossRef] [PubMed]
52. Le, O.T.H.; Tran, L.N.; Doan, V.T.; Van Pham, Q.; Van Ngo, A.; Nguyen, H.H. Mucilage Extracted from Dragon Fruit Peel (*Hylocereus Undatus*) as Flocculant for Treatment of Dye Wastewater by Coagulation and Flocculation Process. *Int. J. Polym. Sci.* **2020**, *2020*, 1–9. [CrossRef]
53. Katubi, K.M.; Amari, A.; Harharah, H.N.; Eldirderi, M.M.; Tahoon, M.A.; Ben Rebah, F. Aloe Vera as Promising Material for Water Treatment: A Review. *Processes* **2021**, *9*, 782. [CrossRef]
54. Vargas-Solano, S.V.; Rodríguez-González, F.; Martínez-Velarde, R.; Morales-García, S.S.; Jonathan, M.P. Removal of Heavy Metals Present in Water from the Yautepec River Morelos México, Using Opuntia Ficus-Indica Mucilage. *Environ. Adv.* **2022**, *7*, 100160. [CrossRef]
55. Inam, M.A.; Khan, R.; Park, D.R.; Ali, B.A.; Uddin, A.; Yeom, I.T. Influence of PH and Contaminant Redox Form on the Competitive Removal of Arsenic and Antimony from Aqueous Media by Coagulation. *Minerals* **2018**, *8*, 574. [CrossRef]
56. Song, S.; Lopez-Valdivieso, A.; Hernandez-Campos, D.J.; Peng, C.; Monroy-Fernandez, M.G.; Razo-Soto, I. Arsenic Removal from High-Arsenic Water by Enhanced Coagulation with Ferric Ions and Coarse Calcite. *Water Res.* **2006**, *40*, 364–372. [CrossRef] [PubMed]
57. Khan, M.M.T.; Yamamoto, K.; Ahmed, M.F. A Low Cost Technique of Arsenic Removal from Drinking Water by Coagulation Using Ferric Chloride Salt and Alum. *Water Supply* **2002**, *2*, 281–288. [CrossRef]
58. Asmel, N.K.; Yusoff, A.R.M.; Sivarama Krishna, L.; Majid, Z.A.; Salmiati, S. High Concentration Arsenic Removal from Aqueous Solution Using Nano-Iron Ion Enrich Material (NIIEM) Super Adsorbent. *Chem. Eng. J.* **2017**, *317*, 343–355. [CrossRef]
59. Baskan, M.B.; Pala, A. Determination of Arsenic Removal Efficiency by Ferric Ions Using Response Surface Methodology. *J. Hazard. Mater.* **2009**, *166*, 796–801. [CrossRef] [PubMed]
60. Perez Mora, B.; Bellú, S.; Mangiameli, M.F.; Frascaroli, M.I.; González, J.C. Response Surface Methodology and Optimization of Arsenic Continuous Sorption Process from Contaminated Water Using Chitosan. *J. Water Process Eng.* **2019**, *32*, 100913. [CrossRef]

materials

MDPI

Article

Effects of Functionalized Materials and Bacterial Metabolites on Quality Indicators in Composts

Krzysztof Gondek [1], Piotr Micek [2], Monika Mierzwa-Hersztek [1,*], Jerzy Kowal [3], Krzysztof Andres [4], Katarzyna Szczurowska [1], Marcin Lis [3] and Krzysztof Smoroń [5]

1 Department of Agricultural and Environmental Chemistry, University of Agriculture in Krakow, Al. Mickiewicza 21, 31-120 Krakow, Poland
2 Department of Nutrition, Animal Biotechnology and Fisheries, University of Agriculture in Krakow, Al. Mickiewicza 24/28, 30-059 Krakow, Poland
3 Department of Zoology and Animal Welfare, University of Agriculture in Krakow, Al. Mickiewicza 24/28, 30-059 Krakow, Poland
4 Department of Animal Reproduction, Anatomy and Genomics, University of Agriculture in Krakow, Al. Mickiewicza 24/28, 30-059 Krakow, Poland
5 Specialized Mining Company "Górtech" sp. z o. o., ul. Wielicka 50, 30-552 Krakow, Poland
* Correspondence: monika6_mierzwa@wp.pl

Abstract: The addition of functionalized materials (biochar, zeolite, and diatomite) and lyophilized metabolic products of *Pseudomonas* sp. and *Bacillus subtilis* to composted biomass may bring many technological and environmental benefits. In this study, we verify the effects of biochar, zeolite Na-P1 ($Na_6Si_{10}Al_6O_{32}\cdot 12\ H_2O$), diatomite ($SiO_2\_nH_2O$), and bacterial metabolites on the composting of biomass prepared from poultry litter, corn straw, grass, leonardite, and brown coal. The experimental design included the following treatments: C—biomass without the addition of functionalized materials and bacterial metabolites, CB—biomass with the addition of biochar, CBM—biomass with the addition of biochar and bacterial metabolites, CZ—biomass with the addition of zeolite, CZM—biomass with the addition of zeolite and bacterial metabolites, CD—biomass with the addition of diatomite, and CDM—biomass with the addition of diatomite and bacterial metabolites. Composts were analyzed for enzymatic and respiratory activities, mobility of heavy metals, and the presence of parasites. The results of this study revealed that, among the analyzed functionalized materials, the addition of diatomite to the composted biomass (CD and CDM) resulted in the most effective immobilization of Cd, Zn, Pb, and Cu. Zinc immobilization factors (IFHM) for diatomite-amended composts averaged 30%. For copper, each functionalized material was found to enhance mobilization of the element in bioavailable forms; similar observations were made for lead, except for the compost to which biochar and bacterial metabolites were added (CBM). The determined values of biochemical indicators proved the different effects of the applied functionalized materials and bacterial metabolites on the microbial communities colonizing individual composts. The dehydrogenase activity (DhA) was lower in all combinations as compared with the control, indicating an intensification of the rate of processes in the studied composts. The highest basal respiration (BR) and substrate-induced respiration (SIR) activities were determined in composts with the addition of bacterial metabolites (CBM, CZM, and CDM). The addition of functionalized materials completely inactivated *Eimeria* sp. in all combinations. In the case of *Capillaria* sp., complete inactivation was recorded for the combination with zeolite as well as biochar and diatomite without bacterial metabolites (CB, CZ, and CD).

Keywords: compost; poultry litter; functionalized materials; bacterial metabolites

Citation: Gondek, K.; Micek, P.; Mierzwa-Hersztek, M.; Kowal, J.; Andres, K.; Szczurowska, K.; Lis, M.; Smoroń, K. Effects of Functionalized Materials and Bacterial Metabolites on Quality Indicators in Composts. *Materials* **2022**, *15*, 8564. https://doi.org/10.3390/ma15238564

Academic Editors: Andrei V. Petukhov, Roberta G. Toro, Daniela Caschera and Abeer M. Adel

Received: 24 September 2022
Accepted: 28 November 2022
Published: 1 December 2022

1. Introduction

Composting is an aerobic process in which organic materials are degraded and transformed by microorganisms into organic and inorganic products [1,2]. Composting is a way to transform organic materials, including waste from crop and livestock production, into products that can be safely and profitably used in the environment for soil fertilization, reclamation of degraded land, or revitalization of brownfields [2,3]. Composts can also play a fungistatic role in the soil environment for the benefit of crop health [4].

The quantitative and qualitative composition of microbial populations during composting characterizes their ability to degrade composted biomass [5]. Changes in the quantitative and qualitative composition of microbial populations during composting depend on the qualitative composition of biomass, the availability of nutrients, and the environmental conditions. The dominant groups of microorganisms in the composting process are bacteria and fungi. The type of organic materials used and the quantitative and qualitative composition of the microbial population involved in composting strongly influence compost quality [6,7]. Compost quality considered from the chemical, biochemical, and sanitation point of view is a key element that indicates process efficiency, and also determines possible directions of compost management. The amount of waste in the world is enormous, and it should be noted that nearly 50% of this waste contains organic matter which mainly includes food waste, waste from animal and agricultural production, waste from the agro-food industry, and wood waste [8]. Given such a rich resource of organic materials to use, designing the composition of biomass for composting should not be problematic. However, it should be stressed that the size and type of organic material resources are not the only elements that determine a successful composting process. According to Sokač et al. [9], the operational efficiency of the composting process depends on numerous variables such as temperature, pH, moisture content, and the C/N ratio. The cited authors have stated that in order to achieve optimal composting efficiency, all variables and their relationships must be considered at the same time.

Currently, many researchers are working to develop integral composting models capable of describing and predicting all physicochemical and biochemical changes that occur during the biological transformation of waste [9]. At the same time, there is research being conducted to identify alternative additives to use in composted biomass to optimize the biochemical processes taking place during waste conversion. The research on alternative additives has mainly focused on materials that would improve the environmental conditions for microbial growth. These materials should be characterized by: the ability to optimize process efficiency; favorable influences on the physical properties of composted biomass (porosity, specific surface area, and cation exchange capacity); the ability to effectively reduce ammonia (NH_3), nitrous oxide (N_2O), and methane (CH_4) emissions; the bioavailability of toxic trace elements; as well as health risks.

The use of various types of additives, even in small amounts, can significantly modify the composting process [10]. As shown by studies to date, mineral, organic, and biological additives stimulate microbial activity, thus, altering the duration of individual composting phases. According to Himanen and Hänninen [11], the duration of the thermophilic phase increased from 2 to 3 weeks after adding commercial products containing zeolite, kaolinite, chalk, ash, and sulfates to composted biomass. Many studies have reported rapid temperature increases in composted animal manure, food waste, or plant waste following the addition of mineral and organic materials [12–14]. The faster temperature increase when biomass is enriched with various materials may result from an increase in the microbial population, and thus, a higher intensity of the biochemical processes taking place. As stated by Waqas et al. [14], as compared with the control, composted materials supplemented with biochar reached the thermophilic temperature faster and increased organic matter degradation by 14.4–15.3%, the concentration of NH_4^+ by 37.8–45.6%, and the concentration of NO_3^- by 50–62%. However, on the one hand, it should not be assumed that the addition of each material will cause changes in the temperature profile of the composting process [15,16]. On the other hand, the parameters that may change are the chemical composition and the value of sanitation indicators in composts [6,17,18].

Supplementation of composted biomass with functionalized materials (before setting up the composting process) should be an indispensable element of the design, and also the management of composting in terms of the environmental impact of the process and product in terms of climate neutrality. Therefore, it is important to search for functionalized materials with a broader spectrum of action, going beyond the process itself, i.e., acting after the compost is introduced into the soil. The effects of adding biochar, zeolite, and diatomite on composting, as well as some chemical and biological indicators of the product have already been described in the literature [18–20]. However, there is still no information on the synergistic effect of functionalized materials such as biochar, zeolite Na-P1 ($Na_6Si_{10}Al_6O_{32}\cdot 12\ H_2O$), diatomite ($SiO_2\_nH_2O$), and biologically active substances, i.e., surfactants and siderophores, which are products of microbial metabolism. Designing composted biomass to contain materials of very different origins and functions may bring many measurable technological (e.g., facilitating the proper structure of composted biomass, maintaining adequate humidity and better air exchange, and appropriate pH) and environmental (e.g., using a wide range of waste materials with different organic matter decomposition rates, better conditions for carbon sequestration, immobilization of heavy metals, and hygienization of treated waste) benefits. Additionally, the addition of bacterial metabolites such as siderophores and surfactants to the composted biomass is of great significance.

Surfactants or surface active agents are a group of substances with specific properties. Such compounds consist of two parts: one portion is hydrophilic and the other portion is hydrophobic. The hydrophilic, polar portion shows high affinity to water and polar liquids, and also has lyophobic properties. Siderophores are organic compounds with low molecular masses that are produced by microorganisms and plants growing under low iron conditions. The primary function of these compounds is to chelate the ferric iron [Fe(III)] from different terrestrial and aquatic habitats, and thereby, make it available for microbial and plant cells. Siderophores have received much attention in recent years because of their potential roles and applications in various areas of environmental research. Their significance in these applications is because siderophores have the ability to bind a variety of metals in addition to iron, and they have a wide range of chemical structures and specific properties. For instance, siderophores function as biocontrols, biosensors, and bioremediation and chelation agents, in addition to their important role in weathering soil minerals and enhancing plant growth [21]. Xi et al. [22] argued that the addition of biosurfactant to solid waste containing Hg and sodium pentachlorophenolate could effectively improve composting by removing both contaminants faster. According to Gong et al. [23], the addition of rhamnolipid biosurfactant and/or microorganisms alone or in all combinations significantly increased the growth rate of *E. fetida* as well as the number of juveniles and cocoons. It also resulted in denser populations of cellulolytic fungi and Azotobacter and increased cellulase and urease activities in vermicomposts [23]. O'Brien et al. [24] and O'Brien and Buckling [25] reported that siderophores could act as public foods by inactivating toxic heavy metals, in which case, they bind to metals, preventing their uptake by cells and rendering the metal (and thus the environment) non-toxic [24,26]. Hesse et al. [27] also reported on the detoxifying role of siderophores. However, the available literature still lacks information on whether the addition of functionalized materials and bacterial metabolites to composted biomass can reduce animal endoparasites or their developmental forms.

Given the above, it was hypothesized that the combined use of functionalized materials (biochar, zeolite, and diatomite) and microbial metabolic products (siderophores and surfactants) has synergistic effects and (i) reduces heavy metal mobility, (ii) optimizes biochemical processes, and (iii) improves the sanitation condition of poultry litter composts.

2. Materials and Methods

2.1. Materials Used for Composting and Process Conditions

This study was carried out under laboratory conditions (air temperature in the room 20 ± 2 °C) on composts prepared from poultry litter, corn straw, fresh grass, leonardite, and brown coal. The proportions of components are given in Table 1.

Table 1. The share of individual components in biomass for composting.

Material	Composts						
	C	CB	CBM	CZ	CZM	CD	CDM
	Share in Dry Matter (%)						
Poultry litter	50	50	50	50	50	50	50
Corn straw	23	23	23	23	23	23	23
Fresh grass	20	20	20	20	20	20	20
Brown coal	1	1	1	1	1	1	1
Leonardite	1	1	1	1	1	1	1
Biochar	-	5	5	-	-	-	-
Zeolite	-	-	-	5	5	-	-
Diatomite	-	-	-	-	-	5	5
Pseudomonas sp. metabolites	-	-	0.20 *	-	0.20 *	-	0.20 *
Bacillus subtilis metabolities	-	-	0.20 *	-	0.20 *	-	0.20 *

* g/6.5 kg of biomass; C—biomass without the addition of functionalized materials and bacterial metabolites; CB—biomass with the addition of biochar; CBM—biomass with the addition of biochar and bacterial metabolites; CZ—biomass with the addition of zeolite; CZM—biomass with the addition of zeolite and bacterial metabolites; CD—biomass with the addition of diatomite; CDM—biomass with the addition of diatomite and bacterial metabolites.

The composted biomass was supplemented with biochar, synthetic zeolite Na-P1, and diatomite after calcination. The addition of functionalized materials was 5% (relative to the dry matter of the mixture). Lyophilized metabolic products of *Pseudomonas* sp. and *Bacillus subtilis* (mixed in a 1:1 ratio) were also added to the respective combinations at 0.400 g (Σ lyophilized metabolic products of *Pseudomonas* sp. and *Bacillus subtilis*). The lyophilisate contained the transformation products of microorganisms in the amount of at least 10^8 cfu/g.

Poultry litter came from a domestic guinea fowl farm. The birds were kept in a floor system on peat bedding at a stocking rate of 6 birds/m^2. Peat was spread on the floor at 6 L per 1 m^2 once before introducing the guinea fowl. The animals were fed an industrial compound feed containing 14.9% crude protein, 5.1% crude fat, and 4.6% crude fiber. The compound feed consisted of wheat, corn, wheat bran, soybean oil, and post-extraction soybean and sunflower meal. Corn straw came from the field cultivation of corn for grain, while the fresh grass was from green space maintenance. Brown coal used in the study came from the output of Kopalnia Węgla Brunatnego Sieniawa Sp. z o.o. (Brown Coal Mine Sieniawa, Poland). Biochar was produced by Carbon Team (Poland) from coniferous tree waste at 450 °C. Diatomite after calcination (750–800 °C) was produced by Specjalistyczne Przedsiębiorstwo Górnicze GÓRTECH Sp. z o.o. (Poland). In addition, synthetic zeolite Na-P1 was produced by hydrothermal method from fly ash generated from conventional coal combustion at the Kozienice Power Plant. Bacterial metabolites were obtained from RDLS Sp. z o.o., and their production process is described in the publication and patent application [28,29]. Selected chemical parameters of the materials used for composting are given in Table 2.

Table 2. Selected chemical parameters of the materials used for composting (mean ± SD).

Material	Dry Matter	Ash	N	C	pH	EC *	Cd	Cu	Pb	Zn
	g kg^{-1}		g kg^{-1} dm			mS cm^{-1}		mg kg^{-1} dm		
Poultry litter	795 ± 72	345 ± 30	34.66 ± 1.77	307 ± 7	6.93 ± 0.00	3.99 ± 0.19	0.485 ± 0.031	13.14 ± 0.75	4.86 ± 0.05	96.85 ± 1.08
Corn straw	925 ± 46	85 ± 1	13.62 ± 0.11	438 ± 22	6.13 ± 0.02	3.02 ± 0.080	0.237 ± 0.047	2.93 ± 0.07	1.96 ± 0.53	59.27 ± 1.26
Fresh grass	313 ± 30	94 ± 2	14.75 ± 0.46	442 ± 4	5.83 ± 0.05	3.98 ± 0.16	0.391 ± 0.001	2.91 ± 0.01	2.04 ± 0.33	25.11 ± 0.14
Brown coal	682 ± 47	82 ± 2	8.53 ± 0.02	647 ± 1	4.58 ± 0.28	0.58 ± 0.04	0.133 ± 0.025	1.05 ± 0.01	1.78 ± 0.07	5.37 ± 0.19
Leonardite	598 ± 36	421 ± 14	4.77 ± 0.08	426 ± 9	5.77 ± 0.25	0.32 ± 0.02	0.264 ± 0.007	4.32 ± 0.22	2.57 ± 0.14	69.86 ± 0.97
Biochar	661 ± 72	94 ± 6	6.48 ± 0.36	420 ± 3	7.47 ± 0.16	0.44 ± 0.09	0.247 ± 0.032	5.48 ± 0.14	1.63 ± 0.01	29.06 ± 0.73
Zeolite	970 ± 29	943 ± 1	0.81 ± 0.09	32 ± 1	11.80 ± 0.07	1.86 ± 0.05	0.346 ± 0.028	21.70 ± 0.86	11.71 ± 0.89	35.13 ± 1.89
Diatomite	999 ± 15	999 ± 0	0.39 ± 0.06	<0.001 ± 0	10.02 ± 0.68	0.61 ± 0.03	0.353 ± 0.022	24.38 ± 0.68	12.38 ± 0.97	37.78 ± 2.07
Pseudomonas sp. metabolites	-	197 ± 1	68.09 ± 0.11	453 ± 0	-	-	0.051 ± 0.015	0.76 ± 0.21	1.00 ± 0.48	5.49 ± 1.52
Bacillus subtilis metabolites	-	344 ± 2	100.21 ± 0.67	336 ± 3	-	-	0.005 ± 0.000	0.64 ± 0.06	0.61 ± 0.42	20.41 ± 3.03

* EC—electrical conductivity.

Oocysts and parasite eggs were found by the McMaster method in the feces of the litter used for the study. Contamination with developmental forms of parasites in the feces of the litter used for the study was as follows (the sum of oocysts and eggs per gram): *Eimeria* sp., 20; *Heterakis* sp., 40; and *Capillaria* sp., 40.

Composting was carried out for 9 weeks (from mid-July to the first week of September 2021). The process was conducted in a 0.8 × 0.9 × 0.6 m bioreactor; perforated-wall containers holding 6.50 kg of fresh composted material were placed into it. The bioreactor chamber was isolated from external conditions and aerated at 0.015 m^3 min^{-1}, 6 times a day. For better aeration and homogenization, the composted biomass was taken out of the bioreactor once a week and mixed by hand.

The experimental design included the following treatments: C—biomass without the addition of functionalized materials and bacterial metabolites, CB—biomass with the addition of biochar, CBM—biomass with the addition of biochar and bacterial metabolites, CZ—biomass with the addition of zeolite, CZM—biomass with the addition of zeolite and bacterial metabolites, CD—biomass with the addition of diatomite, CDM—biomass with the addition of diatomite and bacterial metabolites.

2.2. Chemical Analyses in Materials and Composts

In the materials used for composting and in composts, the pH was determined potentiometrically in a suspension of compost and water (compost/water = 1:5) with a pH-meter (CP-505), and electrical conductivity (EC) (compost/water = 1:5) with a conductivity meter (conductivity/oxygen meter CCO-501). The C and N contents in composts were determined using a CNS analyzer (Vario MAX Cube, Elementar) [30]. Total contents of the studied elements were determined in composts after digesting the samples in concentrated HCl and HNO_3 (3:1 *v/v*) in a microwave oven. The total digestion time was 30 min, the maximum digestion temperature was 240 °C, and the maintained pressure was 40 bar. Mobile forms of Cu, Cd, Pb, and Zn were extracted from composts with redistilled water (compost/water = 1:10) for 2 h. The elemental contents in the filtrates and extracts were determined using an argon inductively coupled plasma optical emission spectrometer (ICP-OES Optima 7300DV, Perkin Elmer). Heavy metal immobilization factors (IFHM) in the composted biomass were calculated from the following equation [31]:

$$\text{Immobilized metal } (\%) = \frac{(\text{water metal for the control} - \text{water metal for the treated sample}) \times 100}{\text{water metal for the control}} \quad (1)$$

2.3. Biochemical Analyses in Composts

The basal respiration (BR) was determined using the method according to ISO 16072 [32]. Moist compost subsamples (20 g) were incubated at 20 ± 2 °C for 24 h. The released CO_2 was absorbed in 0.05 M NaOH solution and precipitated as barium carbonate by adding

0.5 M $BaCl_2$ solution. Non-consumed sodium hydroxide was titrated with 0.1 M HCl in the presence of phenolphthalein as an indicator and the amount of CO_2 was calculated [32]. The substrate-induced respiration (SIR) data were determined by CO_2 evolution 6 h after adding glucose (10 g kg^{-1} aqueous solution) [33]. The respiratory-activation quotient (QR) was calculated by dividing the BR rate by the SIR rate according to ISO 17155 [34].

The dehydrogenases activity (DhA) activity was determined by converting colorless, water-soluble 2,3,5-triphenyltetrazolium chloride (TTC) into red-water-insoluble 1,3,5-triphenylformazan (TPF). Samples (5 g) were incubated at 37 $\pm$ 2 °C for 24 h. The enzymatic activity of dehydrogenases was determined by colorimetric analysis using a Backman DU 640 spectrophotometer at a wavelength of 485 nm.

2.4. *Parasitological Analyses in Composts*

Parasitological analyses were performed in composts with natural water content. The presence of *Eimeria* sp. oocysts and *Capillaria* sp. eggs was tested in 100 g samples using the method developed by Dada (1979). The determination involved mixing 100 g of compost with 200 mL of 5% NaOH solution. Then, the samples were allowed to stand for 1 h, shaken for 20 min, and transferred to 50 mL centrifuge tubes, straining the suspension through a 1 mm mesh sieve. After centrifugation (3 min, 1500 rpm), the solution from above the precipitate was decanted, and water was poured over the precipitate. The samples were centrifuged again, and the remaining precipitate was mixed and flooded with flotation fluid (saturated solution of NaCl and sucrose, d = 1.26 g mL^{-1}). After re-centrifugation and refilling the flotation fluid to the convex meniscus, a cover slip was placed over the test tube. After 20 min, the cover slips were transferred from the test tubes to glass slides with a drop of glycerol. The developmental forms of parasites were searched for in the resulting preparations under the microscope at 100× and 400× magnifications.

2.5. *Statistics*

The experiment was carried out in a completely randomized design with three replicates. One-way analysis of variance (ANOVA) was carried out using Statistica 12.0 (TIBCO Software Inc., Palo Alto, CA, USA). The significance of differences between the means was conducted using a Duncan test at $p \leq 0.05$. Variability of results within the treatments was discerned by calculating standard deviation (SD).

3. Results

3.1. *Temperature Changes during Composting*

The temperature of composted biomass increased most rapidly in treatments with the addition of biochar (CB) and biochar and bacterial metabolites (CBM) (Figure 1). The highest temperatures were recorded in the third week of the process in the same treatments and in the treatment with the addition of diatomite and bacterial metabolites (CDM). The lowest temperature was determined for the control biomass (C), where the maximum temperature was reached with a week delay (beginning of the 4th week of composting). Regardless of the treatment, the temperature range of the composted biomass was generally typical for this type of process. However, it should be noted that during the whole composting period, the highest temperature, as compared with other treatments, was recorded for CBM.

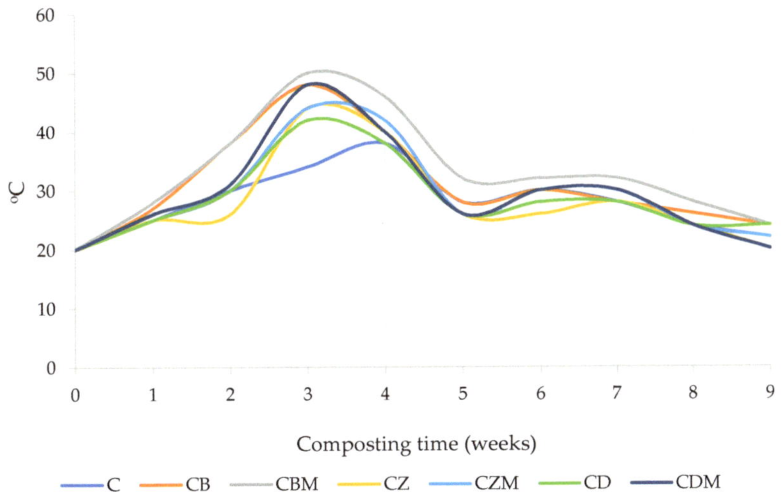

Figure 1. Temperature changes during composting. C—biomass without the addition of functionalized materials and bacterial metabolites, CB—biomass with the addition of biochar, CBM—biomass with the addition of biochar and bacterial metabolites, CZ—biomass with the addition of zeolite, CZM—biomass with the addition of zeolite and bacterial metabolites, CD—biomass with the addition of diatomite, CDM—biomass with the addition of diatomite and bacterial metabolites.

3.2. Effects of Functionalized Materials and Bacterial Metabolites on Selected Chemical Properties of Composts

The content of dry matter in composts did not differ significantly (Table 2). The greatest amount of dry matter was determined in composts with the additions of diatomite (CD) and diatomite and bacterial metabolites (CDM).

Composts supplemented with biochar (BC) and biochar and bacterial metabolites (CBM) contained the significantly highest amount of carbon (Table 3). Lower carbon contents were determined in other composts as compared with the control (C). The reduction in carbon content averaged 11.1% in composts with zeolite (CZ, CZM) and over 20.5% in composts with diatomite (CD, CDM) as compared with the control compost. In general, the nitrogen content was significantly lower in all composts as compared with C (Table 3). The lowest N loss was observed in CB and CBM composts (average 3.3%).

Table 3. Selected chemical properties of composts (mean ± SD).

Compost	Dry Matter	pH	EC	C	N	Cd *	Cu *	Pb *	Zn *
	g kg^{-1}		mS cm^{-1}	g kg^{-1} dm		mg kg^{-1} dm			
C	470.0 ** ± 44.7 a	7.54 ± 0.01 a	4.36 ± 0.07 b	245.0 ± 2.6 c	27.1 d ± 0.8 d	0.047 ± 0.002 b	2.50 ± 0.14 a	0.463 ± 0.065 a	11.19 ± 0.3 d
CB	475.6 ± 57.1 a	7.58 ± 0.01 b	4.19 ± 0.14 b	326.5 ± 5.5 d	26.7 ± 0.5 cd	0.084 ± 0.004 d	2.95 ± 0.01 a	0.479 ± 0.018 a	11.92 ± 0.2 e
CBM	468.7 ± 45.7 a	7.61 ± 0.02 b	4.60 ± 0.29 c	321.1 ± 15.4 d	25.7 ± 1.6 c	0.083 ± 0.005 d	3.01 ± 0.08 a	0.410 ± 0.054 a	12.07 ± 0.1 e
CZ	457.3 ± 54.9 a	7.87 ± 0.01 c	3.64 ± 0.03 a	217.8 ± 0.6 b	21.3 ± 0.1 a	0.042 ± 0.003 b	4.46 ± 0.17 b	0.736 ± 0.035 c	10.81 ± 0.4 c
CZM	493.0 ± 49.1 a	7.83 ± 0.04 c	3.77 ± 0.08 a	217.9 ± 3.6 b	23.4 ± 0.1 b	0.064 ± 0.000 c	5.29 ± 0.09 c	0.626 ± 0.041 b	11.52 ± 0.0 d
CD	518.0 ± 59.2 a	7.61 ± 0.02 b	3.61 ± 0.08 a	193.8 ± 1.0 a	21.6 ± 0.3 a	0.017 ± 0.008 a	6.47 ± 0.61 d	0.543 ± 0.082 b	8.37 ± 0.1 a
CDM	506.9 ± 51.9 a	7.60 ± 0.00 b	3.76 ± 0.03 a	195.6 ± 5.7 a	21.8 ± 0.7 a	0.023 ± 0.005 a	5.19 ± 0.78 d	0.542 ± 0.010 b	9.26 ± 0.0 b

C—biomass without the addition of functionalized materials and bacterial metabolites, CB—biomass with the addition of biochar, CBM—biomass with the addition of biochar and bacterial metabolites, CZ—biomass with the addition of zeolite, CZM—biomass with the addition of zeolite and bacterial metabolites, CD—biomass with the addition of diatomite, CDM—biomass with the addition of diatomite and bacterial metabolites, EC—electrolytical conductivity. * Contents of bioavailable forms extracted with H_2O. ** Each value represents the mean of three replicates ± SD. The different letters within a column indicate a significant difference at $p \leq 0.05$ according to Duncan's multiple range tests

The significantly lowest pH value was determined in compost C (Table 3). The addition of biochar and diatomite to the composted biomass had a comparable effect on the pH of composts, shaping it at a value of about 7.60. The pH values recorded for CZ and CZM composts were over 0.2 higher.

The electrical conductivity (EC) value in the C compost was 4.36 mS cm^{-1} (Table 3). The significantly highest EC value was determined in the CBM compost (4.60 mS cm^{-1}). The addition of both zeolite and diatomite rendered the EC values of composts significantly lower as compared with biochar-enriched composts, regardless of the supplementation of composted biomass with bacterial metabolites.

3.3. *Effects of Functionalized Materials and Bacterial Metabolites on the Heavy Metal Content in Composts*

The content of bioavailable forms of heavy metals differed not only by the type of the analyzed element, but also by the functionalized additive introduced into the composted biomass (Table 3). Significantly, most numerous bioavailable forms of cadmium were recorded in CB and CBM composts. The determined contents were also reflected in the highest share of these cadmium forms in the total element content (Figure 2). As compared with other composts, the contents of bioavailable cadmium in CD and CDM were significantly the lowest. However, it should be noted that a better effect in reducing cadmium bioavailability was observed in the combination without bacterial metabolites. The positive effect of diatomite addition on cadmium immobilization was also confirmed by the immobilization factor (IFHM) values determined for CD and CDM composts (Figure 3).

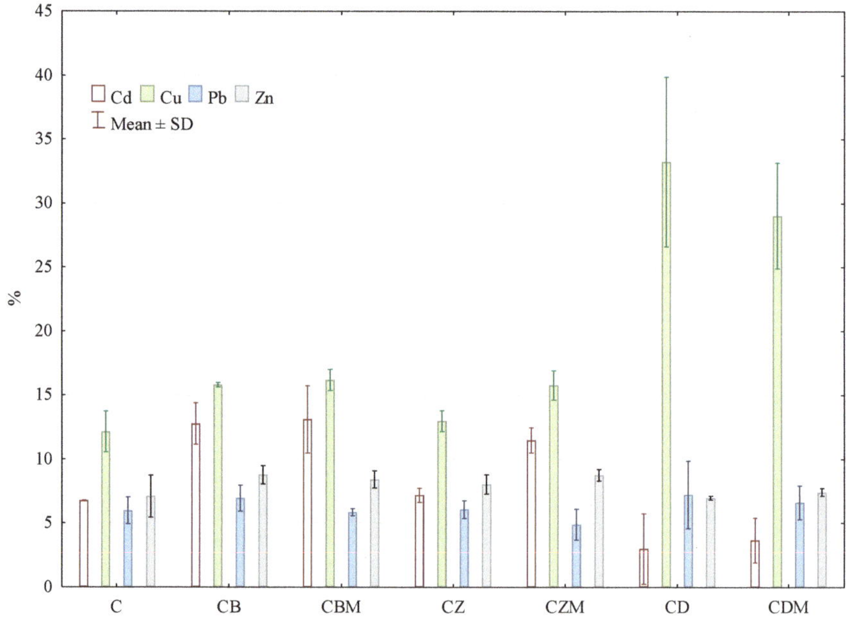

Figure 2. Share of water-extracted heavy metals in the total heavy metal content in composts. C—biomass without the addition of functionalized materials and bacterial metabolites, CB—biomass with the addition of biochar, CBM—biomass with the addition of biochar and bacterial metabolites, CZ—biomass with the addition of zeolite, CZM—biomass with the addition of zeolite and bacterial metabolites, CD—biomass with the addition of diatomite, CDM—biomass with the addition of diatomite and bacterial metabolites.

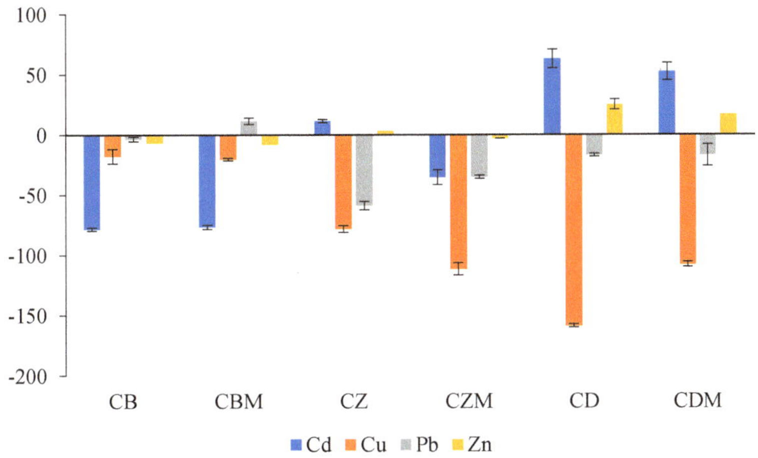

Figure 3. Value of the heavy metal immobilization factor (IFHM) in composts. C—biomass without the addition of functionalized materials and bacterial metabolites, CB—biomass with the addition of biochar, CBM—biomass with the addition of biochar and bacterial metabolites, CZ—biomass with the addition of zeolite, CZM—biomass with the addition of zeolite and bacterial metabolites, CD—biomass with the addition of diatomite, CDM—biomass with the addition of diatomite and bacterial metabolites.

The content of bioavailable copper was significantly higher in composts supplemented with zeolite and diatomite as compared with C, CB, and CBM composts (Table 3). A significant share of bioavailable copper in its total content was observed for CD and CDM combinations, totaling more than 33% and 29%, respectively (Figure 2). It is noteworthy that in the treatments enriched with biochar and zeolite, the share of bioavailable copper in their total contents increased in CBM and CZM as compared with CB and CZ. Taking the control treatment (C) as a reference, each functionalized material was found to deteriorate the copper immobilization conditions (Figure 3).

The content of bioavailable forms of lead was lowest in C, CB, and CBM composts (Table 3). The content of bioavailable lead increased the most in the CZ and CZM composts as compared with the control. There were no significant differences in the share of bioavailable lead in its total content for all functionalized materials added (Figure 2). The immobilization factor IFHM values confirmed that the addition of zeolite and diatomite to the composted biomass increased the potentially bioavailable forms of this element (Figure 3).

The contents of bioavailable forms of zinc were similar, except for CD and CDM composts (Table 3). The addition of diatomite to the composted biomass, especially in the combination without bacterial metabolites (CD), significantly reduced the content of bioavailable zinc. The share of bioavailable forms of lead in the total content did not show much variation and ranged from 6.97% to 8.79% (Figure 2). The lowest contents of bioavailable zinc were reflected in IFHM values, which indicated a clear element immobilization for CD and CDM composts (Figure 3).

3.4. *Effects of Functionalized Materials and Bacterial Metabolites on Selected Biochemical Properties of Composts*

Dehydrogenase activity (DhA) is related to the transformation of carbon and nitrogen compounds. The highest DhA was observed in the control compost (Table 4). Supplementation of the composted biomass decreased DhA, most in CBM and CD composts. Of all the functionalized materials, zeolite, when added to composted materials, had the best effect on DhA, with both CZ and CZM showing the least reduction in DhA as compared with C. As compared with the control, the decrease in DhA activity was 11.9% in CZ and 2.5% in CZM.

Table 4. Biochemical indicators of composts (mean ± SD).

Compost	DhA	BR	SIR	QR
	μg TPF g dm h		μg CO_2 g dm h	
C	508.4 * ± 15.1 d	220.5 ± 3.0 d	465.4 ± 19.8 c	0.475 ± 0.027 bc
CB	433.9 ± 15.5 ab	177.6 ± 5.0 b	465.2 ± 7.5 c	0.382 ± 0.005 a
CBM	416.6 ± 15.1 a	195.8 ± 10.6 c	568.3 ± 4.8 d	0.347 ± 0.048 a
CZ	448.1 ± 8.5 bc	202.5 ± 2.7 c	385.6 ± 4.0 a	0.525 ± 0.001 d
CZM	495.5 ± 6.9 d	222.7 ± 15.9 d	434.1 ± 9.3 bc	0.513 ± 0.026 cd
CD	410.4 ± 11.6 a	155.2 ± 3.4 a	400.3 ± 9.0 ab	0.388 ± 0.017 a
CDM	461.9 ± 21.9 c	196.5 ± 4.9 c	434.3 ± 2.2 bc	0.452 ± 0.009 b

C—biomass without the addition of functionalized materials and bacterial metabolites, CB—biomass with the addition of biochar, CBM—biomass with the addition of biochar and bacterial metabolites, CZ—biomass with the addition of zeolite, CZM—biomass with the addition of zeolite and bacterial metabolites, CD—biomass with the addition of diatomite, CDM—biomass with the addition of diatomite and bacterial metabolites, DhA—dehydrogenases activity, BR—basal respiration, SIR—induced respiration, QR—respiratory-activation quotient. * Each value represents the mean of three replicates ± SD. The different letters within a column indicate a significant difference at $p \leq 0.05$, according to Duncan's multiple range tests.

The basal respiration (BR) value determined in the study differed significantly among the composts. The highest BR value, as compared with the C, was found in the compost with the addition of zeolite and bacterial metabolites (CZM) (Table 4). Other composts showed lower BR values.

The addition of glucose to the analyzed compost samples significantly increased the basal respiration rate, which was caused by the appearance of a source of readily available carbon in the living environment of microorganisms. It was found that the greatest effect on the SIR value had the addition of biochar, especially biochar and bacterial metabolites (CBM), to the composted biomass. The greatest increase in the SIR value was determined in the CBM compost (as compared with the C compost, the value increased by over 22%).

The calculated respiratory-activation quotient (QR), reflecting the amount of microorganisms present in the dormant or active state, ranged from 0.347 for CBM to 0.525 for CZ (Table 4).

3.5. *Effects of Functionalized Materials and Bacterial Metabolites on Developmental Forms of Selected Parasites*

Eimeria sp. oocysts and *Capillaria* sp. eggs were found in the tested compost samples, but no *Heterakis* sp. eggs, which were earlier present in the feces of birds, were found. *Eimeria* sp. coccidia and the highest number of *Capillaria* sp. eggs were found in the C compost (Table 5). CBM and CDM were characterized by the lowest number of *Capillaria* sp. eggs. No *Capillaria* sp. eggs were found in the CZM compost.

Table 5. Developmental forms of parasites found in the tested composts.

Sample	*Eimeria* sp.	*Capillaria* sp.
C	+	++
CB	−	−
CBM	−	+
CZ	−	−
CZM	−	−
CD	−	−
CDM	−	+

The following were found: ++, numerous developmental forms; +, single developmental forms; −, no developmental forms. C—biomass without the addition of functionalized materials and bacterial metabolites, CB—biomass with the addition of biochar, CBM—biomass with the addition of biochar and bacterial metabolites, CZ—biomass with the addition of zeolite, CZM—biomass with the addition of zeolite and bacterial metabolites, CD—biomass with the addition of diatomite, CDM—biomass with the addition of diatomite and bacterial metabolites.

4. Discussion

The problem of proper waste management has recently become a global issue. It affects practically every environmental sphere and poses a threat to every form of life on earth. The numerous processes and technologies used for waste disposal and transformation are powerful tools to manage waste. However, changing trends and production and environmental conditions are forcing the search for new or the improvement of currently used waste processing technologies. Composting represents the group of processes termed as environmentally friendly. Currently, many researchers are working to develop integral composting models capable of describing and predicting various physicochemical and biochemical changes that occur during the biological transformation of waste [9]. However, this is not the only field where intensive research work is being carried out. At the same time, there is a search for alternative additives to composted biomass to optimize biochemical processes taking place during waste conversion. In this study, functionalized organic and mineral additives were introduced into a multi-material composition with the main share of poultry litter. Biochar, synthetic zeolite, and diatomite after calcination were used for composting. Metabolites of *Pseudomonas* sp. and *Bacillus subtilis* served as biologically active substances. The introduced materials affected the composting process in different ways, which was also demonstrated by Barthod et al. [10]. Considering the temperature criterion, the temperature reached and the rate of temperature change during the initial composting phase both indicate the greatest reduction of organic matter in biomass supplemented with biochar, especially with the addition of biochar and bacterial metabolites [35]. Other combinations showed a milder temperature profile as well as a lower maximum temperature of the composted biomass. As stated by Waqas et al. [14], composted materials enriched with biochar, irrespective of its production temperature, reached the thermophilic temperature faster than the control. The opposite thesis was presented by Malinowski et al. [36], who found that the addition of biochar to the organic fraction of municipal solid waste did not alter the composting process. The study by Guo et al. [37] on the composting of pig manure with rice straw showed that materials enriched with biochar tended to loosen during composting, and biochar increased the total porosity of compost aggregates by about 90% as compared with other treatments, which was mainly attributed to decomposition effects. Therefore, it can be concluded that the addition of biochar to composted biomass refines its physical structure, and the synergistic effect of biochar and bacterial metabolites can significantly improve the living conditions of microorganisms involved in the process, which is reflected in the temperature—a measurable effect of microorganisms' activity. The introduction of mineral functionalized materials (zeolite and diatomite) to the composted biomass, even enriched with bacterial metabolites, caused the maximum biomass temperature to take lower values during composting.

The pH values of the obtained composts ranged from 7.54 to 7.87 and were higher for the combination with synthetic zeolite, which had the highest pH value among the functionalized materials used in this study. A relatively high pH in all combinations used in this study may be due to the mineralization of nitrogen compounds, including those derived from litter, which mainly concerns the release of ammonia and ammonium ion [38].

Electrical conductivity (EC) is an important element that determines the natural use of compost. This property can have a significant impact on seed germination as well as plant growth and development. As a result of the high concentration of easily soluble salts, for example, in the soil solution, water uptake by the plant root system may be reduced after applying compost. The effect of microbial decomposition of organic materials is the release of, inter alia, calcium or magnesium ions which increase the electrical conductivity of the compost [38]. In the present study, the addition of both diatomite and zeolite, regardless of the enrichment of biomass with bacterial metabolites, significantly reduced the EC value of composts as compared with the control and biochar combinations, which resulted from the unique meso-microporous structure of both materials and, consequently, their high sorption capacity. Salinity reduction in zeolite-enriched composts was reported by Soudejani et al. [39].

The obtained C/N ratio values indicated that the composted waste was effectively degraded. These values, especially for the combination with functionalized mineral materials (zeolite and diatomite) may suggest that the composting process was carried out too long. It should be noted that the addition of bacterial metabolites was of little importance here. Compost amended with biochar proved to be more stable in terms of the C/N ratio. Biochar contains carbon structures that are difficult to decompose. According to Barthod et al. [10], biochar, as a highly aromatic pyrolysis product, is stable when added to composting, and therefore, can increase the carbon sequestration potential, contributing to climate change mitigation.

Heavy metals in compost, and especially their bioavailable forms, can have adverse effects on plant growth and development, and consequently, on animal and human health [38]. Therefore, determining both the content and bioavailability of heavy metals is important for assessing the quality of composts made from waste substances, including those of animal or municipal origin [10]. During composting, the transformation of organic matter (which receives most attention) takes place and also involves changes in the quantity and speciation of heavy metals, which very often lead to excessive mobilization of unavailable forms of heavy metals. The authors of the present study hypothesized that the use of functionalized materials as well as siderophores and surfactants in composting would reduce the content of water-soluble forms of cadmium, copper, lead, and zinc, a thesis that has not been fully confirmed. The addition of diatomite had a positive effect on the immobilization of cadmium and zinc; for copper, each functionalized material was found to deteriorate the conditions of element immobilization; similar observations were made for lead except for the compost supplemented with biochar and bacterial metabolites. These results proved that there are specific conditions that should be met to limit the bioavailability of heavy metals, where the chemical behavior of the element in such a complex matrix as compost is not without significance. On the one hand, when using biochar as an additive, differences in the binding of heavy metals may relate, for example, to the conditions in which the material was produced, which in turn determines the number of surface functional groups capable of binding heavy metals. On the other hand, when using mineral additives in composted biomass, the effect of heavy metal immobilization is mainly based on adsorption and ion exchange processes.

Modification of composting by introducing various organic or mineral additives affects the microbial communities inhabiting the composted biomass through the influence of these materials on the pH, moisture content, or aeration of the pile, as well as nutrient content [10]. The obtained value of biochemical indicators in the obtained composts proved different effects of the applied functionalized materials on microbial communities. The determined dehydrogenase activity (DhA) value in all combinations was lower than that determined in the control compost; in the case of basal respiration (BR), a slightly higher parameter value was recorded for the compost amended with zeolite and bacterial metabolites; for substrate-induced respiration (SIR), a higher value was determined in the compost with the addition of biochar and bacterial metabolites. The obtained results suggest that, despite many studies on the subject, the high variability in the composition of biomass used for composting, as well as the specific properties of functionalized materials make it very difficult to clearly predict the phases and effects of the process. Functionalized materials such as biochar, due to their properties (mainly porous structure), can increase microbial activity by affecting the moisture content and aeration of the composted biomass [10,14], which in turn impacts the temperature profile. Mineral functionalized additives such as zeolites or diatomite can fulfill their functions in terms of, for example, binding or neutralization of contaminants, while their beneficial effect on microbial activity in nutrient supply may be limited [40,41].

Barthod et al. [10] stated that mineral additives could indirectly alter microbial activity, while the addition of microbial consortium directly changed a microbial community and compost activity. As reported by Sasaki et al. [42], a commercial microbial additive changed the temperature profile of composting and ammonia emissions due to the growth of mesophilic and thermophilic bacteria as compared with normal composting without additives. In this study, no live cultures of microorganisms were used, but their metabolic products (surfactants and siderophores) instead, which did not significantly affect the biochemical parameters of composts.

Metabolic reactions during aerobic transformation of waste lead, among other things, to an increase in the temperature of the composted material. That has a significant effect on the type, number, and species of microorganisms involved in the biological process, but also inactivates most pathogens [43]. According to Hassen et al. [44] and Polprasert [45], a temperature of 55–60 °C maintained for at least 24 h at about 50% humidity in a compost pile is crucial for the removal of most pathogens. In our study, inactivation of *Eimeria* sp. was confirmed in all combinations as compared with the control compost. In the case of *Capillaria* sp., a decrease in the number of eggs was found in the composts with the addition of biochar and bacterial metabolites as well as diatomite and bacterial metabolites, and their complete devastation in the combinations with zeolite and biochar as well as diatomite without bacterial metabolites. Oocysts of *Eimeria* sp. and eggs of *Hetarakis* sp. and *Capillaria* sp. are listed as very and moderately resistant to external factors [46], so their partial or complete elimination in composting with the use of the mentioned additives may indicate efficient hygienization, which is very beneficial from a parasitological point of view. In addition, according to Malinowski et al. [36], the number of potentially pathogenic microorganisms decreased significantly when composting the organic fraction of municipal solid waste with the addition of biochar. Sobrinho et al. [47] determined that most of the composted waste from rabbits, guinea pigs, mice, and hamsters from animal care facilities eliminated more than 90% of Escherichia coli, Salmonella sp., as well as protozoa oocysts and helminth eggs.

5. Conclusions

The use of diatomite in composting significantly affected the immobilization of cadmium and zinc; for copper, each functionalized material was found to enhance mobilization of the element in bioavailable forms; similar observations were made for lead, except for the compost to which biochar and bacterial metabolites were added. The determined value of biochemical indicators proved different effects of the applied functionalized materials and bacterial metabolites on the microbial communities colonizing individual composts. The determined activity of dehydrogenases (DhA) was lower in all combinations as compared with the control, indicating a higher rate of processes taking place during composting. Basal respiration (BR) was slightly higher in the compost supplemented with zeolite and bacterial metabolites, while substrate-induced respiration (SIR) was higher in the compost amended with biochar and bacterial metabolites. An increased SIR may suggest that there were no inhibitory substances in that compost or they were in concentrations that did not inhibit the viability of dormant microorganisms. Our results prove a beneficial hygienizing effect of functionalized materials but without the addition of bacterial metabolites. The obtained results do not clearly indicate a synergistic effect of functionalized materials and bacterial metabolites. Appropriate selection of the type of functionalized material and bacterial metabolites, as well as the amount of their supplementation to the composted biomass, requires further studies on different waste materials.

Author Contributions: Conceptualization, K.G.; Methodology, K.G.; Software, K.G.; Validation: K.G.; Formal analysis: K.G.; Investigation, K.G.; Resources, K.G.; Data Curation, K.G., K.S. (Katarzyna Szczurowska), and J.K.; Writing—Original Draft, K.G. and J.K.; Writing—Review and Editing, K.G, P.M., M.L., M.M.-H. and K.A.; Visualization, K.G.; Supervision, K.G. and M.M.-H.; Project administration, P.M.; Funding acquisition, P.M. and K.S. (Krzysztof Smoroń). All authors have read and agreed to the published version of the manuscript.

Funding: The research work was financed by the Ministry of Science and Higher Education of the Republic of Poland.

Institutional Review Board Statement: Not applicable.

Informed Consent Statement: Not applicable.

Data Availability Statement: Not applicable.

Acknowledgments: This research was included in the grant entitled: Implementation of the concept of new products based on activated diatomite, intended for use in poultry production. Grant number: BZ/4240/WHiBZ/2022.

Conflicts of Interest: The authors declare no conflict of interest.

References

1. Toledo, M.; Siles, J.; Gutiérrez, M.; Martín, M. Monitoring of the composting process of different agroindustrial waste: Influence of the operational variables on the odorous impact. *Waste Manag.* **2018**, *76*, 266–274. [CrossRef] [PubMed]
2. Kacprzak, M.; Kupich, I.; Jasinska, A.; Fijalkowski, K. Bio-based waste' substrates for degraded soil improvement—Advantages and challenges in European context. *Energies* **2022**, *15*, 385. [CrossRef]
3. Yu, H.; Xie, B.; Khan, R.; Shen, G. The changes in carbon, nitrogen components and humic substances during organic-inorganic aerobic co-composting. *Bioresour. Technol.* **2018**, *271*, 228–235. [CrossRef] [PubMed]
4. Mierzwa-Hersztek, M.; Gleń-Karolczyk, K.; Gondek, K. Fungistatic activity of composts with the addition of polymers obtained from thermoplastic corn starch and polyethylene—An innovative cleaner production alternative. *Sci. Total Environ.* **2018**, *635*, 1063–1075. [CrossRef] [PubMed]
5. Ling, N.; Deng, K.; Song, Y.; Wu, Y.; Zhao, J.; Raza, W.; Huang, Q.; Shen, Q. Variation of rhizosphere bacterial community in watermelon continuous mono-cropping soil by long-term application of a novel bioorganic fertilizer. *Microbiol. Res.* **2014**, *169*, 570–578. [CrossRef]
6. Kopeć, M.; Gondek, K.; Mierzwa-Hersztek, M.; Antonkiewicz, J. Factors influencing chemical quality of composted poultry waste. *Saudi J. Biol. Sci.* **2018**, *25*, 1678–1686. [CrossRef]
7. Rastogi, M.; Nandal, M.; Khosla, B. Microbes as vital additives for solid waste composting. *Heliyon* **2020**, *6*, e03343. [CrossRef]
8. Walling, E.; Trémier, A.; Vaneeckhaute, C. A review of mathematical models for composting. *Waste Manag.* **2020**, *113*, 379–394. [CrossRef]
9. Sokač, T.; Valinger, D.; Benković, M.; Jurina, T.; Gajdoš Kljusurić, J.; Radojčić Redovniković, I.; Jurinjak Tušek, A. Application of optimization and modeling for the composting processes enhancement. *Processes* **2022**, *10*, 229. [CrossRef]
10. Barthod, J.; Rumpel, C.; Dignac, M.-F. Composting with additives to improve organic amendments. A review. *Agron. Sustain. Dev.* **2018**, *38*, 17. [CrossRef]
11. Himanen, M.; Hänninen, K. Effect of commercial mineral-based additives on composting and compost quality. *Waste Manag.* **2009**, *29*, 2265–2273. [CrossRef] [PubMed]
12. Venglovsky, J.; Sasakova, N.; Vargova, M.; Pacajova, Z.; Placha, I.; Petrovsky, M.; Harichova, D. Evolution of temperature and chemical parameters during composting of the pig slurry solid fraction amended with natural zeolite. *Bioresour. Technol.* **2005**, *96*, 181–189. [CrossRef] [PubMed]
13. Czekała, W.; Malińska, K.; Cáceres, R.; Janczak, D.; Dach, J.; Lewicki, A. Co-composting of poultry manure mixtures amended with biochar—The effect of biochar on temperature and C-CO2 emission. *Bioresour. Technol.* **2016**, *200*, 921–927. [CrossRef] [PubMed]

14. Waqas, M.; Nizami, A.; Aburiazaiza, A.; Barakat, M.; Ismail, I.; Rashid, M. Optimization of food waste compost with the use of biochar. *J. Environ. Manag.* **2017**, *216*, 70–81. [CrossRef]
15. Gabhane, J.; William, S.P.; Bidyadhar, R.; Bhilawe, P.; Anand, D.; Vaidya, A.N.; Wate, S.R. Additives aided composting of green waste: Effects on organic matter degradation, compost maturity, and quality of the finished compost. *Bioresour. Technol.* **2012**, *114*, 382–388. [CrossRef] [PubMed]
16. Li, H.; Xin, H.; Liang, Y.; Burns, R.T. Reduction of ammonia emissions from stored laying hen manure through topical appli-cation of zeolite, Al+Clear, Ferix-3, or poultry litter treatment. *J. App. Poultry. Res.* **2008**, *17*, 421–431. [CrossRef]
17. Gondek, K.; Tabak, M.; Koncewicz-Baran, M.; Kopeć, M. Effects of the Addition of Foils Produced from Polyethylene and Maize Starch to Composted Biomass on Quantitative and Qualitative Composition of Humic Compounds and Optic Parameters of Humic Acids. *Pol. J. Environ. Stud.* **2015**, *24*, 2397–2403. [CrossRef] [PubMed]
18. Ren, X.; Wang, Q.; Chen, X.; Zhang, Y.; Sun, Y.; Li, R.; Li, J.; Zhang, Z. Elucidating the optimum added dosage of Diatomite during co-composting of pig manure and sawdust: Carbon dynamics and microbial community. *Sci. Total Environ.* **2021**, *777*, 146058. [CrossRef]
19. Özogul, F.; Šimat, V.; Gokdogan, S.; Regenstein, J.M.; Özogul, Y. Effect of Natural Zeolite (Clinoptilolite) on in vitro Biogenic Amine Production by Gram Positive and Gram Negative Pathogens. *Front. Microbiol.* **2018**, *9*, 2585. [CrossRef] [PubMed]
20. Sanchez-Monedero, M.; Cayuela, M.; Roig, A.; Jindo, K.; Mondini, C.; Bolan, N. Role of biochar as an additive in organic waste composting. *Bioresour. Technol.* **2018**, *247*, 1155–1164. [CrossRef]
21. Ahmed, E.; Holmström, S.J.M. Siderophores in environmental research: Roles and applications. *Microb. Biotechnol.* **2014**, *7*, 196–208. [CrossRef]
22. Xi, B.-D.; Liu, H.-L.; Huang, G.H.; Zhang, B.; Qin, X. Effect of bio-surfactant on municipal solid waste composting process. *J. Environ. Sci.* **2005**, *17*, 409–413.
23. Gong, X.; Wei, L.; Yu, X.; Li, S.; Sun, X.; Wang, X. Effects of Rhamnolipid and Microbial Inoculants on the Vermicomposting of Green Waste with Eisenia fetida. *PLoS ONE* **2017**, *12*, e0170820. [CrossRef]
24. O'Brien, S.; Hodgson, D.J.; Buckling, A. Social evolution of toxic metal bioremediation in *Pseudomonas aeruginosa*. *Proc. R. Soc. B Lond.* **2014**, *281*, 20140858. [CrossRef]
25. O'Brien, S.; Buckling, A. The sociality of bioremediation: Hijacking the social lives of microbial populations to clean up heavy metal contamination. *EMBO Rep.* **2015**, *16*, 1241–1245. [CrossRef] [PubMed]
26. Braud, A.; Geoffroy, V.; Hoegy, F.; Mislin, G.L.A.; Schalk, I.J. Presence of the siderophores pyoverdine and pyochelin in the extracellular medium reduces toxic metal accumulation in *Pseudomonas aeruginosa* and increases bacterial metal tolerance. *Environ. Microbiol. Rep.* **2010**, *2*, 419–425. [CrossRef]
27. Hesse, E.; O'Brien, S.; Luján, A.M.; Sanders, D.; Bayer, F.; Veen, E.M.; Hodgson, D.J.; Buckling, A. Author response for Stress causes interspecific facilitation within a compost community. *Ecol. Lett.* **2021**, *10*, 2169–2177. [CrossRef] [PubMed]
28. Styczynski, M.; Dziewit, L.; Debiec-Andrzejewska, K. New Strains of Psychrotolerant Bacteria, Composition, Their Application and Method of Soil Bioaugmentation. Polish Patent Appl. P.437824, 5 November 2021.
29. Styczynski, M.; Biegniewski, G.; Decewicz, P.; Rewerski, B.; Debiec-Andrzejewska, K.; Dziewit, L. Application of psychrotol-erant Antarctic bacteria and their metabolites as efficient plant growth promoting agents. *Front. Bioengineer. Biotechnol.* **2022**, *10*, 772891. [CrossRef]
30. *Operating Instructions Vario MAX Cube*; Elementar Analysensysteme GmbH: Langenselbold, Germany, 2013; p. 407.
31. Gondek, K.; Mierzwa-Hersztek, M.; Kopeć, M. Mobility of heavy metals in sandy soil after application of composts produced from corn straw, sewage sludge and biochar. *J. Environ. Manag.* **2018**, *210*, 87–95. [CrossRef] [PubMed]
32. *ISO 16072:2002*; Soil Quality—Laboratory Methods for Determination of Microbial Soil Respiration. International Organization for Standardization: Geneva, Switzerland, 2002.
33. *ISO PN-EN ISO 14240-1:2011*; Soil Quality—Determination of Soil Microbial Biomass—Part 1: Substrate-Induced Respiration Method (ISO 14240-1:1997). International Organization for Standardization: Geneva, Switzerland, 2011.
34. *ISO 17155:2012*; Soil Quality—Determination of Abundance and Activity of Soil Microflora Using Respiration Curves. International Organization for Standardization: Geneva, Switzerland, 2012.
35. Xiao, Y.; Zeng, G.-M.; Yang, Z.-H.; Shi, W.-J.; Huang, C.; Fan, C.-Z.; Xu, Z.-Y. Continuous thermophilic composting (CTC) for rapid biodegradation and maturation of organic municipal solid waste. *Bioresour. Technol.* **2009**, *100*, 4807–4813. [CrossRef]
36. Malinowski, M.; Wolny-Koładka, K.; Vaverková, M.D. Effect of biochar addition on the OFMSW composting process under real conditions. *Waste Manag.* **2019**, *84*, 364–372. [CrossRef]
37. Guo, X.-X.; Wu, S.-B.; Wang, X.-Q.; Liu, H.-T. Impact of biochar addition on three-dimensional structural changes in aggregates associated with humus during swine manure composting. *J. Clean. Prod.* **2021**, *280*, 124380. [CrossRef]
38. Jalalipour, H.; Jaafarzadeh, N.; Morscheck, G.; Narra, S.; Nelles, M. Potential of Producing Compost from Source-Separated Municipal Organic Waste (A Case Study in Shiraz, Iran). *Sustainability* **2020**, *12*, 9704. [CrossRef]
39. Soudejani, H.T.; Kazemian, H.; Inglezakis, V.; Zorpas, A.A. Application of zeolites in organic waste composting: A review. *Biocatal. Agric. Biotechnol.* **2019**, *22*, 101396. [CrossRef]
40. Wong, J.C.; Fang, M.; Li, G.X.; Wong, M.H. Feasibility of Using Coal Ash Residues as CO-Composting Materials for Sewage Sludge. *Environ. Technol.* **1997**, *18*, 563–568. [CrossRef]

41. Fang, M.; Wong, J.W.C. Effects of lime amendment on availability of heavy metals and maturation in sewage sludge composting. *Environ. Pollut.* **1999**, *106*, 83–89. [CrossRef]
42. Sasaki, H.; Kitazume, O.; Nonaka, J.; Hikosaka, K.; Otawa, K.; Itoh, K.; Nakai, Y. Effect of a commercial microbiological additive on beef manure compost in the composting process. *Anim. Sci. J.* **2006**, *77*, 545–548. [CrossRef]
43. Gajalakshmi, S.; Abbasi, S.A. Solid Waste Management by Composting: State of the Art. *Crit. Rev. Environ. Sci. Technol.* **2008**, *38*, 311–400. [CrossRef]
44. Hassen, A.; Belguith, K.; Jedidi, N.; Cherif, A.; Cherif, M.; Boudabous, A. Microbial characterization during composting of municipal solid waste. *Bioresour. Technol.* **2001**, *80*, 217–225. [CrossRef] [PubMed]
45. Polprasert, C. *Organic Waste Recycling. Technology and Management*, 3rd ed.; IWA Publishing: Bangkok, Thailand, 2007; pp. 88–134.
46. Strauch, D. Survival of pathogenic micro-organisms and parasites in excreta, manure and sewage sludge. *OIE Rev. Sci. Tech.* **1991**, *10*, 813–846. [CrossRef]
47. Sobrinho, E.M.; De Almeida, A.C.; Colen, F.; De Souza, R.M.; Menezes, I.R.; Vieira, V.A.; Oliveira, L.N.; Da Fonseca, M.P.; Santos, H.O.; Brandi, I.V.; et al. Composting as alternative treatment of solid wastes from laboratory animal care facilities. *Acta Vet. Bras.* **2011**, *5*, 184–191.

materials

MDPI

Article

Effect of Oxygen Plasma Pre-Treatment on the Surface Properties of Si-Modified Cotton Membranes for Oil/Water Separations

Leila Ghorbani [1], Daniela Caschera [2] and Babak Shokri [1,3,*]

[1] Laser and Plasma Research Institute, Shahid Beheshti University, Tehran 19839, Iran
[2] Institute for the Study of Nanostructured Materials, Strada Provinciale 35 d, n. 9, Montelibretti, 00010 Rome, Italy
[3] Faculty of Physics, Shahid Beheshti University, Tehran 19839, Iran
* Correspondence: b-shokri@sbu.ac.ir

Abstract: Hydrophobic and oleophilic Si-based cotton fabrics have recently gained a lot of attention in oil/water separation due to their high efficiency. In this study, we present the effect of O_2 plasma pre-treatment on the final properties of two Si-based cotton membranes obtained from dip coating and plasma polymerization, using polydimethylsiloxane (PDMS) as starting polymeric precursor. The structural characterizations indicate the presence of Si bond on both the modified cotton surfaces, with an increase of the carbon bond, assuring the success in surface modification. On the other hand, employing O_2 plasma strongly changes the cotton morphology, inducing specific roughness and affecting the hydrophobicity durability and separation efficiency. In particular, the wettability has been retained after 20 laundry tests at 40 °C and 80 °C, and, for separation efficiency, even after 30 cycles, an improvement in the range of 10–15%, both at room temperature and at 90 °C can be observed. These results clearly demonstrate that O_2 plasma pre-treatment, an eco-friendly, non-toxic, solvent-free, and one-step method for inducing specific functionalities on surfaces, is very effective in enhancing the oil/water separation properties for Si-based cotton membranes, especially in combination with plasma polymerization procedure for Si-based deposition.

Keywords: oxygen plasma pre-treatment; dip coating; plasma polymerization; PDMS; oil/water separation; cotton textile

Citation: Ghorbani, L.; Caschera, D.; Shokri, B. Effect of Oxygen Plasma Pre-Treatment on the Surface Properties of Si-Modified Cotton Membranes for Oil/Water Separations. *Materials* **2022**, *15*, 8551. https://doi.org/10.3390/ma15238551

Academic Editor: Alenka Vesel

Received: 18 October 2022
Accepted: 28 November 2022
Published: 30 November 2022

1. Introduction

In the last years, the increasing production of oily wastewater from textile, food, leather, Metal processing, oil and gas industries, and frequent requests for effective treatment for oil spill removal have given a lot of importance to the scientific research of oil-water separation, with the aim of reducing the induced severe environmental and economic impacts [1–7]. In particular, the development of membranes for oil/water separation grasps more attention with respect to other separation methods due to their high efficiency and non-secondary pollution [1,2,4,7–10]. Several studies have been conducted on oil/water separation membranes by using different materials and physical and chemical methods for developing separation filters. In particular, Si-based materials have been largely studied and employed for membrane functionalization, and several works proposed the combination of materials and procedures to develop oil/water separation membranes. Yeom et al. fabricated a superhydrophobic mesh and sponge through a simple and fast dip-coating of silica nanoparticles (SiNP), which improved the hydrophobicity of the surface [11]. Ma et al. suggested polymerized reaction of silica nanoparticles on electrospun PI nanofibers to fabricate a separation membrane [12]. In another study, Woo et al. fabricated a nanostructured glass microfiber by using the vertical polymerization coating method [13]. Zhang et al. grew nanofilaments on the textile via chemical vapor deposition of trichloromethylsilane for obtaining oil-water separation membranes [14], while in a study by Song et al., the nano-SiO_2 solution was prepared by sol-gel method and deposited on the cotton textile by a dip-coating approach to obtain superhydrophobic/superoleophilic fabrics [15]. Si-modified membranes present very promising performances in oil/water separation, but these systems often undergo poor adhesion and low durability, especially under drastic work conditions (i.e., high temperatures and long work-cycles [16–20]. In this context, the research of more effective methods to enhance the wettability and oil separation performances of Si-based membranes using green technologies is particularly appealing.

One of the most used materials for the fabrication of Si-based membranes for the oil/water separation process is polydimethylsiloxane (PDMS) due to its intrinsic hydrophobicity [16–18,21]. Several methods could be considered for depositing PDMS, such as dip-coating [22], plasma polymerization [23,24], electrospinning [17], vapor deposition [20], and chemical vapor deposition [18].

In this study, we took into consideration two methods, dip-coating, and plasma polymerization, for the fabrication of Si-based coating on cotton, using PDMS as the precursor. We also investigated the influence of an O_2 plasma pre-treatment on the final properties of the as-produced systems.

Plasma pre-treatment is one of the most effective methods to modify a surface, introducing specific functional groups [25] and/or changing its roughness [26]. The surface, as modified, changes its intrinsic properties as adhesion capability, wettability, and morphology [27,28]. Furthermore, specific plasma pre-treatment can also influence the properties of the coating that can be deposited on the modified surface due to the combination of imposed roughness and the presence of chemical groups on the substrate surface [29].

In particular, it has been observed that an O_2 plasma treatment could be particularly effective in influencing the wettability properties of textile-based systems [30]. In this view, it could introduce the fabrication of oil/water membranes procedure [2,31].

This work aims to demonstrate that O_2 plasma pre-treatment has an effective influence in enhancing the performances of Si-modified cotton as the membrane for oil/water separation. In this view, structural and functional properties of the as-fabricated Si-cotton membranes have been evaluated, with and without a pre-treatment of O_2 plasma on the cotton substrate. It has already been demonstrated that plasma pre-treatment can significantly affect the hydrophobicity of the surface by modifying the roughness of the surface [32]. As a consequence, our results indicated that O_2 plasma pre-treatment has a significant impact on the final morphology of Si-coatings, more evident for samples fabricated by the plasma polymerization method. Thanks to the specific hierarchical structure imposed by this pre-treatment, an increase in the contact angle values and separation efficiency of both filters have been observed. In particular, Si-cotton membrane fabricated by using only plasma processes (O_2 pre-treatment + plasma polymerization) showed an improvement in efficiency up to 15%, and this ability has been retained even after 30 separation cycles. In this contest, a one-step and eco-friendly O_2 plasma pre-treatment can be considered an influencing factor for the fabrication of oil/water separation filters with very encouraging properties even for applications in extreme conditions and environments.

2. Experiment

2.1. Materials

Cotton (from a local store) was cut into 10 cm × 10 cm samples, washed in a 5 wt.% aqueous detergent solution (obtained by diluting in distilled water a proper amount of a commercial clothes soap), and finally dried in an oven at 50 °C for 40 min. To prepare a final homogenous Si precursor solution with low viscosity, PDMS (Dow Corning, Midland, MI, USA, Sylgard186 elastomer base-2g) and PDMS (Sigma Aldrich, St. Louis, MO, USA, Sylgard186 curing agent-0.2 g) were mixed into 50 mL of toluene. All other materials were purchased from Sinopharm Chemical Reagent Co., Ltd., Shanghai, China.

In Figure 1, the polymeric chemical structure of PDMS is reported.

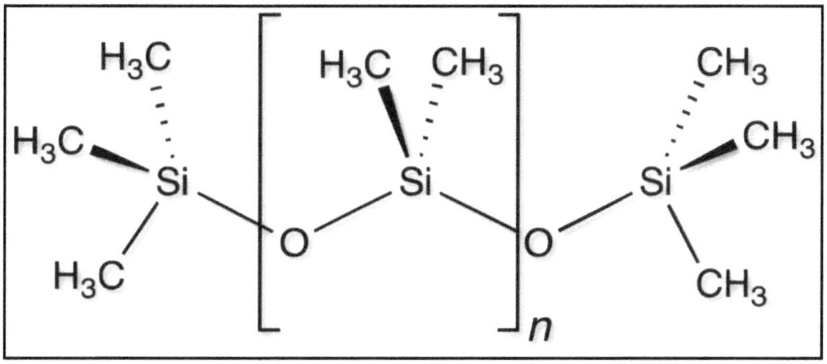

Figure 1. PDMS polymeric structure [33].

2.2. *Dip-Coating*

The cotton fabric was immersed in the PDMS toluene solution for one minute, squeezed to remove the excess, and then cured at about 125 °C in the oven for 20 min. The membrane obtained by this procedure has been labeled as Si-Cot_Dip.

2.3. *Plasma Polymerization*

A homemade low-pressure capacitively coupled plasma (PECVD) system was employed for plasma deposition. The cotton textile was placed on the power electrode (14 cm in diameter, RF power 13.56 MHz), connected to an automatic impedance matching network.

Argon was used as a carrier gas for introducing the PDMS-based solution vapor into the plasma chamber. The deposition process was carried out at 20 W RF power for 20 min at 18.6 Pa (working pressure). The PDMS/Ar mixture flow was fixed at 20 sccm. The volume of the plasma chamber was 15 L. This filter was labeled as Si-Cot_Plasma.

2.4. *O_2 Plasma Pre-Treatment*

O_2 plasma pre-treatment was performed on cotton substrates, using the same PECVD system, before PDMS deposition. The plasma conditions were set as follows: 20 W RF power, 20 sccm O_2 gas flow, 10 Pa working pressure, and 30 min treatment time. The process was carried out at 25 °C and checked by a thermocouple.

Then the as-treated cotton samples were modified with PDMS according to the procedures previously reported (2.2 and 2.3 paragraphs), and the as-fabricated Si-cotton membranes were labeled as follows:

Si-Cot_Dip/O_2, for the filter, pre-treated with O_2 plasma and then modified with PDMS by dip-coating.

Si-Cot_Plasma/O_2, for the filter, pre-treated with O_2 plasma and then deposited with PDMS by plasma polymerization.

2.5. *Structural and Morphological Characterization*

X-ray photoelectron spectroscopy (XPS) analysis was carried out using a Thermo Scientific KA1066 spectrometer with monochromatic Al Kα (with energy 1486.6 eV) X-ray radiation. The spectra were collected with 0.1 eV step at 50 eV constant pass energy. The calibration of the binding energy (BE) scale was achieved with respect to the Au 4f7/2 peak, set at BE = 84.0 eV. The accuracy of the binding energy (BE) scale was ±0.1 eV. Spectroscopic data were processed by CasaXPS software, subtracting the Shirley background and using a peak-fitting routine with a symmetrical product mix of Gaussian−Lorentzian functions L = 30% and G = 70%.

Fourier-transformed infrared (FTIR) spectra were performed using the Thermo Nicolet model nexus 470 in the range of 700–4000 cm^{-1}.

SEM analysis was collected by a Hitachi, with an accelerating voltage of 15 kV, pre-coating the samples with a gold layer to prevent any charges.

Wettability measurements were performed by the sessile drop method in standard conditions (at atmospheric pressure and room temperature). The cotton samples were fixed on a microscope slide with adhesive tape to ensure the planarity of the surface. A droplet of 4 µL of deionized water was dropped on the surface, and the photos of the drops were taken with a Canon camera and put on the side of the experimental set-up. The reported water contact angles (WCA) were averaged over three measurements in different spots. The contact angles were evaluated using the Image J software (1.52v).

2.6. *Laundry Test*

The durability of the surface properties of the systems, consequently the adhesion of the Si-coating, and the wettability performances of the membranes were evaluated by the laundry test. The Si-coated cotton (10 cm × 10 cm) was immersed in 200 mL distilled water at 40 °C and 80 °C, respectively, and two stirring bars of about 2 cm. It was left under magnetic stirring at an average speed of 600 rpm for 10 min, then dried in an oven at 50 °C. The test was performed at 40 °C and 80 °C. Twenty wash cycles of 10 min each have been performed for each sample, and five WCA measurements were performed for each cycle, reporting the average value [34]. The WCA measurements were repeated 20 times.

2.7. *Separation Efficiency*

The separation efficiency of the Si-coating textiles was tested at room temperature and 90 °C by using several oils: paraffin (viscosity 13–18 cSt), gasoline (viscosity 0.006 Poise), pump oil (viscosity > 500 cSt), and hexadecane (viscosity 0.0034 Poise), mixed with water (viscosity 1 cSt). In detail, a room temperature mixture and hot oil/hot water mixture with the 1:1 volume ratio were prepared. The hot oil/water systems were firstly mixed at room temperature and then heated at 100 °C, in a water bath, due to the high flammability.

These selected experimental conditions result in being closer to those normally detectable in practical situations, including the dispersion of oil in seawater or industrial wastewater.

The Si-modified cotton was placed on a funnel, and the oily solutions were poured on. Then, after the separation process, the efficiency was calculated by the following equation:

$$\text{separation efficiency } \% = (V_1/V_0) \times 100$$

where V_1 is the volume of the water collected on the filter after the separation test, and V_0 denotes the starting volume of the initial water [4,35,36]. The tests were repeated 30 times for different oil/water mixtures. The separation efficiency was calculated five times for each cycle, reporting the average values and calculating the corresponding errors, as recommended in [34].

3. Results and Discussion

The achievement of the cotton surface functionalization with Si-based coatings has been investigated by studying the chemical structure and composition of the Si-Cot_Dip/O_2 and Si-Cot_Plasma/O_2 systems, using XPS and FT-IR ATR measurements. The comparison of the XPS survey of samples (Figure 2a,e,i) shows several changes in the relative intensity of C1s and O1s signals. Furthermore, the presence of new peaks at 101.8 and 152.8 eV, related to Si2p and Si2s [37], on the Si-Cot_Dip/O_2 and Si-Cot_Plasma/O_2 surfaces (Figure 2e,i) is clearly observable. This evidence indicates the success of both methods (dip-coating and plasma polymerization) in adding Si-based coatings on the cotton surface.

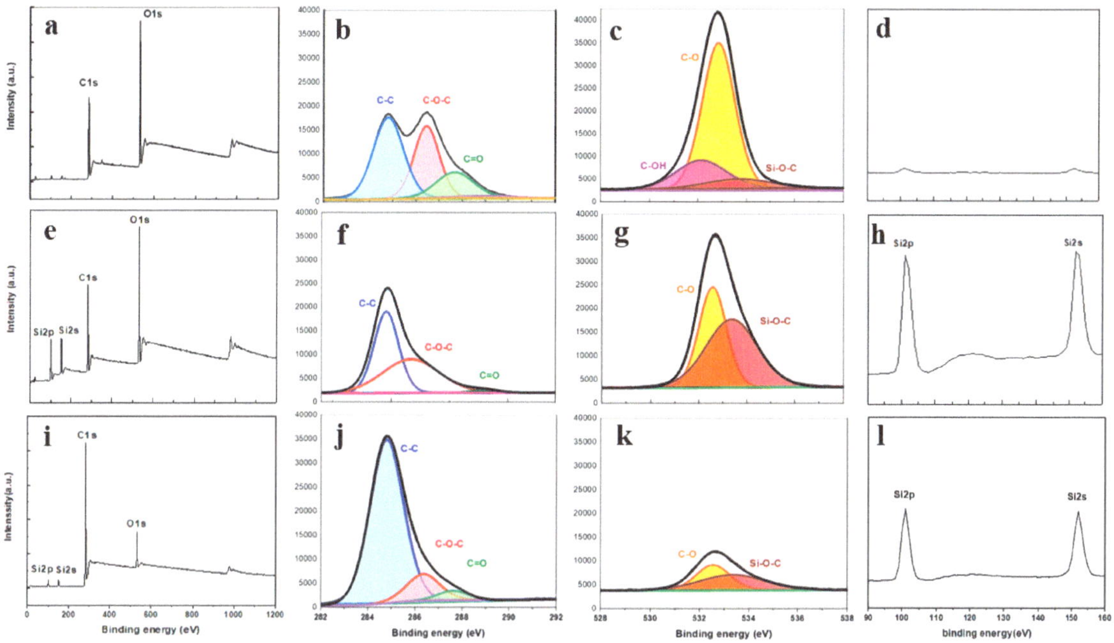

Figure 2. XPS measurements for pristine cotton (**a**) survey spectrum, (**b**) C1s spectra, (**c**) O1s spectra, (**d**) Si spectra; Si-Cot_Dip/O_2 (**e**) survey spectrum, (**f**) C1s spectra, (**g**) O1s spectra, (**h**) Si spectra; Si-Cot_Plasma/O_2 (**i**) survey spectrum, (**j**) C1s spectra (**k**) O1s spectra (**l**) Si spectra.

In Figure 2b, the C1s spectrum of the pristine sample presents the three typical bands for cotton, at 284.8, 286.5, and 287.7 eV, related to C-C, C-O-C, and C=O bonds, respectively [38]. The intensity of C-C, C-O-C, and C=O after dip-coating (Si-Cot_Dip/O_2) decreased in comparison with pristine cotton (Figure 2f). In contrast, after plasma polymerization (Si-Cot_Plasma/O_2) (Figure 2j), the C-C signal increases in intensity, while C-O-C and C=O signals are reduced with respect to cotton. This effect is due to the fact that the plasma polymerization process causes the incorporation of carbon species, with respect to oxygen, on the coating. In contrast, the dip-coating method is responsible for creating more Si species and decreasing the oxygen ones. On the other hand, the O1s spectrum demonstrates that the O_2 amount can be resolved into several components. In Figure 2c, the pristine cotton, the band around 532.6 eV is assigned to C-OH, related to unsaturated bonds on the surface [39]. This signal disappears in Si-Cot_Dip/O_2 and Si-Cot_Plasma/O_2 samples (Figure 2g,k), indicating that both methods can obtain unsaturated bonds on the surface passively. Moreover, the signals around 532.6 and 533.4 eV are due to C-O and Si-O-C, respectively [21]. The intensity of the Si-O-C bond increases more in Si-Cot_Dip/O_2 sample than that in the Si-Cot_Plasma/O_2 sample [40,41].

Furthermore, the evaluation of the atomic percentages of silicon, oxygen, and carbon (Figure S1) confirms the success of the two methods in depositing a Si-based coating.

Figure 3 shows the ATR-FTIR spectra of pristine cotton, O_2 plasma pretreated cotton, O_2 pre-treatment plasma, Si-Cot_Dip/O_2, and Si-Cot_Plasma/O_2 samples. In pristine cotton, the absorption peaks related to the O-H bond and the asymmetric stretching vibration C-H are present at 3335 and 2902 cm^{-1}, respectively [32]. Furthermore, the peaks at 1054–1026 cm^{-1} are attributed to the C-O-C stretching vibrations. The pristine cotton spectrum is comparable with the literature [32], confirming that no specific impurities are present. FT-IR spectrum of O_2 plasma pretreated (Figure 3b) cotton is very similar to that of the cotton pristine, but some differences are visible at those frequencies attributed to the

C=O vibration groups in the carbonyl structure and O–H stretching vibration, i.e., in the range around 1500–1750 and 2700–3100 cm^{-1}, respectively [32]. In particular, the effect of plasma surface oxidation can be related to the appearance of a small peak at 1540 cm^{-1} due to COO–stretching vibrations [42–45].

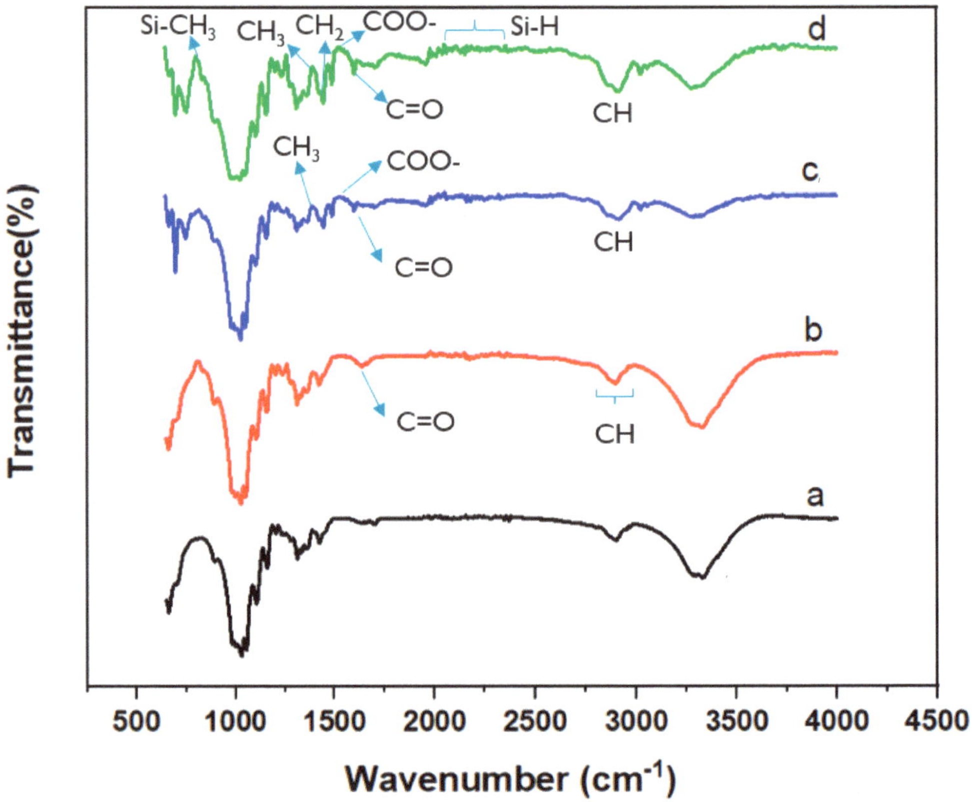

Figure 3. ATR-FTIR spectroscopy of (a) pristine cotton, (b) O_2 plasma pre-treatment, (c) Si-Cot_Dip/O_2, (d) Si-Cot_Plasma/O_2 samples.

The introduction of these specific C-O and O-H groups on the cotton surface, thanks to the O_2 plasma, leads be beneficial for the improvement of the Si-based coating adhesion since it can give the possibility to create more stable bonds between the coating and the cotton substrate [46,47].

The bands at 1007 and 1072 cm^{-1} in both the FT-IR spectra of Si-Cot_Dip/O_2 and Si-Cot_Plasma/O_2 samples (Figure 3c,d) are characteristic of Si–O–Si asymmetric and symmetric stretching vibrations [48]. However, the intensity of these bonds is higher in Si-Cot_Dip/O_2. Also, symmetric rocking at 864 cm^{-1} and stretching at 785 cm^{-1} vibrations in both coating samples can be attributed to the Si–CH_3 group. Related to Si-Cot_Dip/O_2 and Si-Cot_Plasma/O_2, the absorption band at 1410 cm^{-1} is assigned to CH_3 asymmetric bending in Si–CH_3 bonds, while the signals at 2965 and 2906 cm^{-1} are the asymmetric stretching and symmetric stretching of the CH_3 group, respectively [49,50]. Also, the low-intensity band at 1450 cm^{-1} is attributed to the asymmetric bending vibrations of the CH_2 group in the Si–CH_2–CH_2–Si link [49,50]. In addition, the Si–H stretching vibration presents signals at 2100–2300 cm^{-1}, while the band at 1629 cm^{-1} is due to the stretching of the C=O [49,50]. The presence of Si-related peaks in the FT-ATR spectra of the Si-Cot_Dip/O_2

and Si-Cot_Plasma/O_2 confirms the success of the deposition processes. Moreover, the intensities of the Si–H bonds at 1030, 1050, 2122, 2160, 2179, 2250, and 2333 cm^{-1} for Si-Cot_Dip/O_2 are more evident than those for Si-Cot_Plasma/O_2 sample, while C-H bonds at 2895, 3024, 3050, and 3337 cm^{-1} and C-C bonds at 1443 and 1495 cm^{-1} for Si-Cot_Plasma/O_2 are more intense than those for Si-Cot_Dip/O_2. These results confirm the observation of XPS measurements about the percentage of Si and C in the samples fabricated by two dip-coating and plasma polymerization techniques.

Although the plasma pre-treatments have almost no influence on the chemical composition of the coatings [51], it is well known that O_2 plasma pre-treatments can strongly modify the surface of the substrate, inducing, in the following deposited coatings, specific hierarchical growth and opportune morphologies, and contributing to the final surface properties [52,53]. To assess the influence of plasma pre-treatment, SEM analysis has been carried out, and the corresponding images, before and after O_2 plasma, for both dip-coating and plasma polymerization methods are shown in Figure 4. Figure 4a shows that pristine cotton fibers present a relatively smooth surface. When the cotton is covered with Si-based coatings by dip-coating and plasma polymerization methods, differences in morphologies could be visible. For the dip-coating method, the surface of the Si-Cot_Dip appears more monotonous and homogeneously covered, and no recognizable structures are visible (Figure 4b). On the contrary, the coating obtained from plasma polymerization, Si-Cot_Plasma, (Figure 4c), presents a rougher surface associated with visible nanostructures on the fiber. In Figure 4d, the effect of 30 min of O_2 plasma treatment on the cotton surface is presented: the roughness and the morphology of the fibers result in being significantly altered, and many holes are created on the surface due to O_2 plasma treatment. The influence of oxygen pre-treatment on the coating morphology and growth is vividly recognized in Figure 4e,f. According to Figure 4e, the Si-coating obtained by the dip method (Si-Cot_Dip/O_2) retains most of its morphological features, showing a compact and homogeneous coverage of about 1 μm in thickness, even if some effects of the O_2 plasma pre-treatment start to be visible on the Si growth. Furthermore, in Figure 4f, Si-Cot_Plasma/O_2 coating displays a more irregular structure, created on the surface fibers, with respect to the Si-Cot_Plasma coating obtained without a pre-treatment of oxygen (Figure 4c), permitting the growth of columnar nanostructures with a size of $\approx$1 μm.

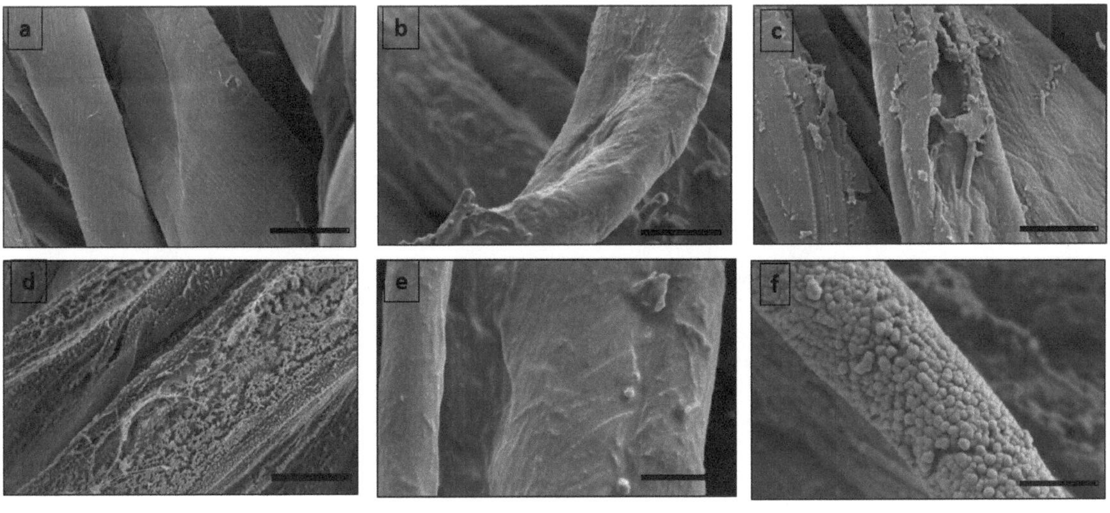

Figure 4. SEM images of (**a**) Pristine cotton, (**b**) Si-Cot_Dip, (**c**) Si-Cot_Plasma, (**d**) Cotton after oxygen plasma pre-treatment, (**e**) Si-Cot_Dip/O_2, (**f**) Si-Cot_Plasma/O_2 (Scale bar: 10 μm).

According to the Cassie–Baxter's theory [54], the generation of micro and nanostructures on the surface can increase the hydrophobicity thanks to the air pockets that can be trapped in these structures [53]. Previous research has also shown that O_2 plasma pre-treatment on textiles could strongly affect the adhesion surface [55], making the following Si-based coating more adhesive on cotton.

Figure 5 shows the macroscopic hydrophobicity and oleophilicity of both Si-Cot_Dip/O_2 and Si-Cot_Plasma/O_2. The contact angle (CA) results are summarized in Table 1, in which cotton pristine is added for comparison.

Figure 5. Water and oil on the surface of (**a**) Si-Cot_Dip/O_2, (**b**) Si-Cot_Plasma/O_2.

Table 1. WCA and Sliding angle measurements for pristine cotton and Si-coating samples.

Sample	Water Contact Angle (°)	Sliding Angle (°)
Pristine cotton	0	-
Si-Cot_Dip	137 ± 3	50 ± 2
Si-Cot_Dip/O_2	145 ± 3	30 ± 2
Si-Cot_Plasma	134 ± 3	25 ± 2
Si-Cot_Plasma/O_2	144 ± 3	15 ± 2

Si-coating samples without O_2 plasma pre-treatment present quite similar WCA values and great hydrophobic behavior. The presence of Si bonds and the changing of the surface morphology are conditions for final higher WCA values [56,57]. Additionally, increasing the amount of carbon on the cotton surface after coating, as previously observed by XPS, could be an important issue for predicting an increase in surface hydrophobicity [58]. More interestingly, O_2 plasma treatment has a positive influence by increasing the WCAs, and at the same time, by decreasing the sliding angle values for both Si-Cot_Dip/O_2 and Si-Cot_Plasma/O_2. In particular, Si-Cot_Plasma/O_2 shows the most appropriate combination of WCA and sliding angle for having a hydrophobic surface that should be used as a filter. The micro and nano hierarchical structures of Si-Cot_Plasma/O_2, similar to the surface of blue lotus [59], ensure the suitable wettability behavior. In addition, the low sliding angle of Si-Cot_Plasma/O_2 permits its use not only as a vertical filter but in a wide range of angles (15–90°) since it can repel water even at angles of 15°, very close to the horizontal position.

Laundering tests permit the evaluation of the durability of Si-based coatings on cotton. Figure 6 shows the variation of the contact angle of Si-Cot_Dip, Si-Cot_Dip/O_2, Si-Cot_Plasma, and Si-Cot_Plasma/O_2 after laundering at 40 °C and 80 °C. It is evident that after 20 laundry cycles, the WCA values for the Si-Cot_Dip and Si-Cot_Plasma (Figure 6a,c) decrease at about 125–128°, independently from the laundering temperature. It is noteworthy that the O_2 plasma pre-treatment results in an increase in the starting WCA values of about 10° and 15° for Si-Cot_Dip/O_2 and Si-Cot_Plasma/O_2, respectively. Furthermore, the contact angle remains quite constant in the range 135–140° for both Si-Cot_Dip/O_2 and Si-Cot_Plasma/O_2, even after twenty laundering cycles at 40 °C and 80 °C (Figure 6b,d). This behavior indicates that O_2 plasma pre-treatment not only can increase the hydrophobicity of the filter but can also have a great influence on the adhesion of the Si-based coatings. The effect is to obtain filters with stable and higher hydrophobicity durability. In addition, it is well known that the generation of specific hierarchical nanostructures on the cotton can also affect the oleophilicity of the fiber as an effect of capillary [60].

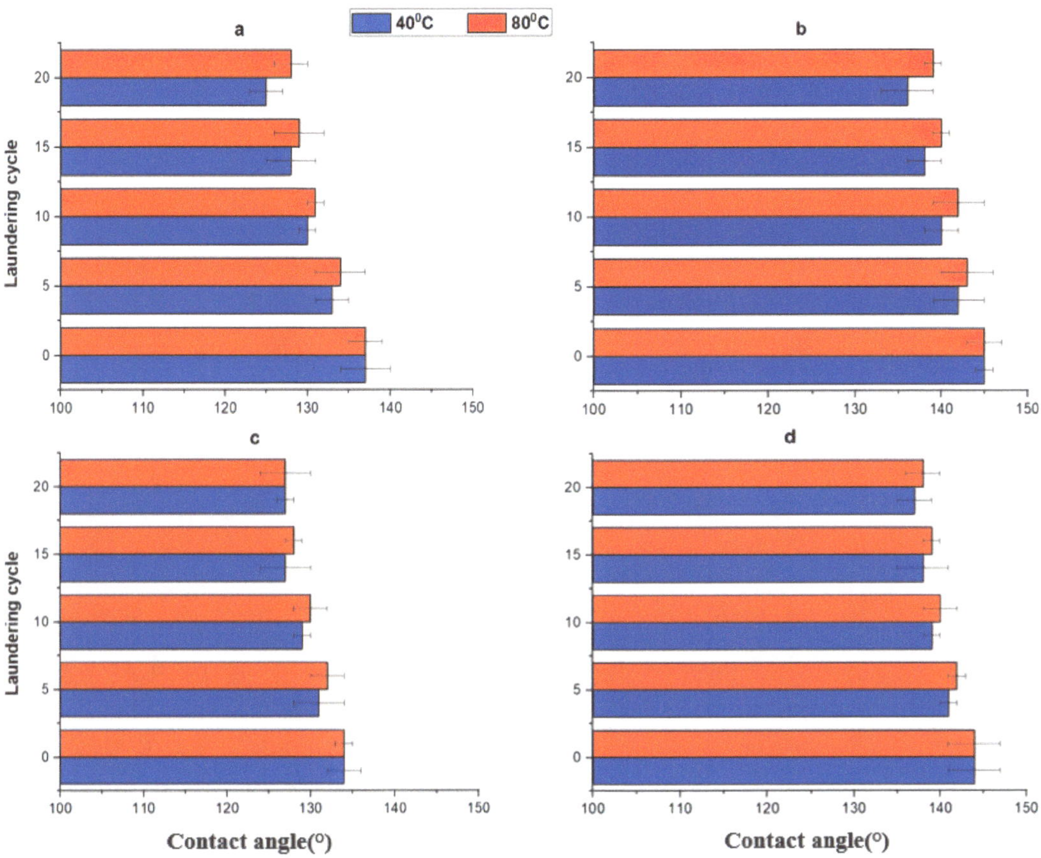

Figure 6. Variation of WCA of (**a**) Si-Cot_Dip and (**b**) Si-Cot_Dip/O_2 (**c**) Si-Cot_Plasma (**d**) Si-Cot_Plasma/O_2 after laundering tests at different temperatures.

The separation efficiency of the Si-Cot_Dip and Si-Cot_Dip/O_2, Si-Cot_Plasma and Si-Cot_Plasma/O_2 as oil/water filters is evaluated up to 30 separation cycles, at room temperature (25 °C) and 90 °C, for mixtures of oils/water (paraffin, gasoline, pump oil, and hexadecane) (Figures 7 and 8). It is evident that the separation efficiency of Si-Cot_Dip for mixture at 25 °C (Figure 7a) is retained even up to about 85% after 30 tests. This separation efficiency after using O_2 pre-treatment increases by about 10%, reaching approximately 95% at room temperature (Figures 7b and 8b).

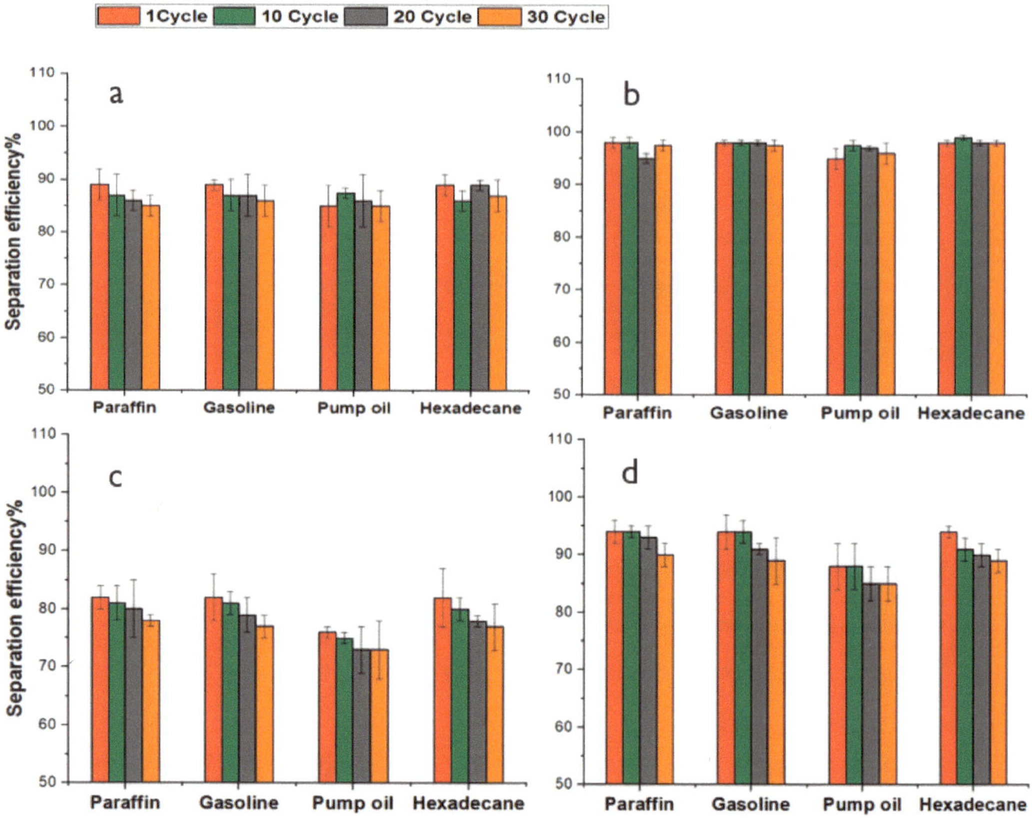

Figure 7. The separation efficiency of (**a**) Si-Cot_Dip (**b**) Si-Cot_Dip/O_2 (**c**) Si-Cot_Plasma (**d**) Si-Cot_Plasma/O_2 at room temperature for oil/water mixtures.

The separation efficiency for Si-Cot_Plasma is about 80–75% even after 30 cycles (Figures 7c and 8c). The O_2 pre-treatment, instead, permits an increase in separation efficiency of about 15%, efficiency reaching approximately 90% even at higher temperatures (Figures 7d and 8d). The slight efficiency increase for Si-Cot_Plasma/O_2 (15%) with respect to the Si-Cot_Dip/O_2 (10%) could be attributed to the different final morphologies, as observed from SEM analysis (Figure 4e, f): the combination of the two plasma processes (O_2 pre-treatment and plasma polymerization) induces a rougher and columnar growth of the Si-based coating on cotton. This particular structure has been demonstrated to be more effective in oil/water separation procedures [58]. Nevertheless, the observed soaring properties also for Si-Cot_Dip/O_2 confirm the positive effect of O_2 plasma pre-treatment on the development of a separation membrane system with higher separation efficiency, even in drastic conditions, without the need for any auxiliary force.

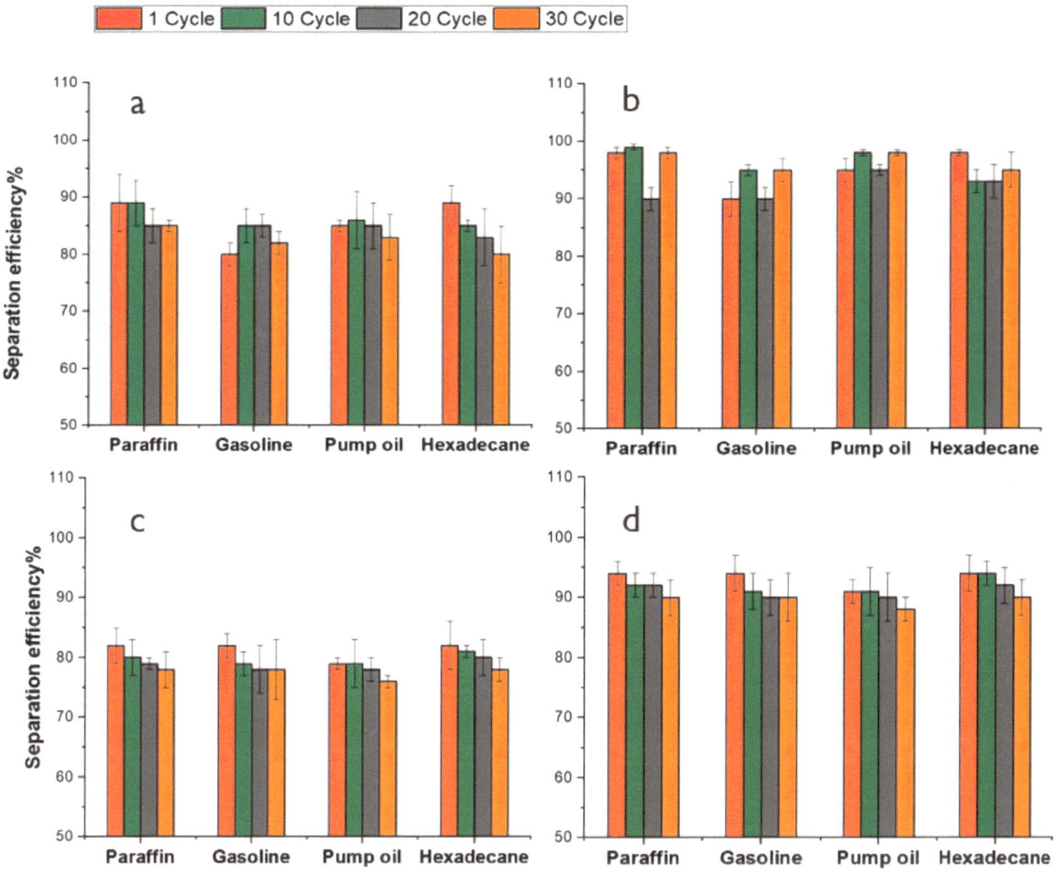

Figure 8. The separation efficiency of (**a**) Si-Cot_Dip (**b**) Si-Cot_Dip/O_2 (**c**) Si-Cot_Plasma (**d**) Si-Cot_Plasma/O_2 at 90 °C for several oil/water mixtures, after different cycles.

To prove the promising performances of the O_2 plasma pre-treatment as an effective way to increase separation efficiency even for the hot oil/water mixture, a comparison with other filtering cotton systems is provided in Table 2. It is evident that the Si-based cotton filter proposed in this work possesses identical or even superior properties at room temperature with respect to other filters. Furthermore, very few investigations have been carried out on the reusability and separation efficiency of filters in the presence of high-temperature mixtures. When higher filtration temperatures are considered, the analysis of Table 2 highlights that most of the materials proposed in the literature present low efficiency, or, if the performances are acceptable, usually the starting materials are not eco-friendly, or they are prepared using complex or expensive methods. On the contrary, the Si-based cotton filters show suitable performance in separation efficiency at high temperatures. Furthermore, the proposed synthetic methods are greener than traditional approaches. In addition, O_2 plasma pre-treatment that increases the efficiency and durability of the filter is a facile and eco-friendly method. These considerations make the O_2 plasma pre-treatment a very appealing and suitable method for increasing the efficiency of oil/water separation at different temperatures.

Table 2. Separation efficiency (SE) of different materials used as filters for hot oil/water mixture.

Material	Method	Type of Oils	SE of Room Temperature Mixture	Number of Cycles	SE of High Temperature Mixture	Number of Cycles	Reference
PDMS	Plasma polymerization	Gasoline Hexadecane	90	30	90	30 (Hot oil/Hot water)	This work
PDMS	Dip coating	Gasoline Hexadecane	95	30	95	30 (Hot oil/Hot water)	This work
DLC	Coating by plasma	Pump oil	100	1	-	-	[58]
Acrylamide and acrylonitrile	Graft copoly-merization	Crude Olive Diesel	95–99	1	-	-	[61]
HMDSO	Plasma polymer	Oil Hexane	99	1	-	-	[62]
HDTMS and SA	Dip coating	Hexadecane Diesel	100	1	-	-	[63]
HDTS	Dip coating	chloroform	98	10	-	-	[64]
PDMS	Vapor deposition	Hexane	99	1	-	-	[20]
PDA-Ca complex	Dip coating	Gasoline	-	-	97	80 (Hot water)	[65]
LiCl/DMAc	Dip coating	hexane	98.5	60	-	-	[66]
PEDOT-PSS hydrogel	Chemical polymerization	Diesel	-	-	99	50 (Hot water)	[67]
Mg $(NO_3)_2$	Dip coating	Diesel	10	96	-	-	[68]
Stearoyl	Dip coating	octan	10	100	-	-	[69]
PDMS-Fe3O4@MF	Co-deposition	toluene	30	99.9	-	-	[36]

4. Conclusions

In this paper, we explored the effect of O_2 plasma pre-treatment on the oil/water separation properties of Si-based cotton filters. The modified membranes have been realized using different methods, a dip-coating procedure, and plasma polymerization, using the eco-friendly PDMS as Si precursor. Structural characterizations have been carried out to assess the success of the Si functionalization by XPS and FT-IR spectroscopies. Furthermore, a deep analysis of the morphologies of the membranes has been evaluated, taking into account the effect of the plasma pre-treatment on the roughness and surface structures. SEM highlighted that O_2 plasma pre-treatment is very effective in inducing the growth of a specific hierarchical structure on the cotton surface. This feature is partially retained even in the membrane functionalized by the dip-coating method. As expected, both Si-Cot_Dip and Si-Cot_Plasma show very promising performances in oil/water separation tests, but their capability and durability result in an enhancement when the Si deposition is preceded by O_2 plasma pre-treatment. These results clearly point out the positive effect of O_2 plasma pre-treatment on the hydrophobic/oleophilic properties of the Si-based cotton membrane. The possibility of the plasma process to opportunely create and modulate a hierarchical structure on the fiber surface, combined with the Si-based coating properties, permits an increase in the separation efficiency of the membrane, also imparting great durability even after several cycles of usage and laundering. In particular, the combination of plasma processes (O_2 pre-treatment + plasma polymerization) results in the most effective way of developing oil/water separation filters, opportunely controlling the surface properties of the membranes.

Supplementary Materials: The following supporting information can be downloaded at: https://www.mdpi.com/article/10.3390/ma15238551/s1, Figure S1: Atomic percentage for Pristine, Si-Cot_Dip/O_2, Si-Cot_Plasma/O_2 samples, and Video S1: Oil/water separation process.

Author Contributions: Conceptualization, L.G.; Validation, D.C.; Investigation, L.G.; Writing—original draft, L.G.; Writing—review & editing, D.C.; Supervision, B.S.; project administration, B.S. All authors have read and agreed to the published version of the manuscript.

Funding: This research received no external funding.

Acknowledgments: The authors are very thankful to Ali Khatibi for tips in advancing the work and thanks to Saeed Javadi Anaghizi for his assistance in SEM images and Saravana B. Basuvalingam for his assistance in the XPS test. This work has been supported by the plasma laboratory of Shahid Beheshti University.

Conflicts of Interest: The authors declare no conflict of interest.

References

1. Sam, E.K.; Liu, J.; Lv, X. Surface Engineering Materials of Superhydrophobic Sponges for Oil/Water Separation: A Review. *Ind. Eng. Chem. Res.* **2021**, *60*, 2353–2364. [CrossRef]
2. Rasouli, S.; Rezaei, N.; Hamedi, H.; Zendehboudi, S.; Duan, X. Superhydrophobic and superoleophilic membranes for oil-water separation application: A comprehensive review. *Mater. Des.* **2021**, *204*, 109599. [CrossRef]
3. Vellopollath, K.; Gore, P.M.; Kandasubramanian, B. Foamed materials for oil-water separation. *Chem. Eng. J. Adv.* **2021**, *5*, 100076. [CrossRef]
4. Lee, T.L.D. Cotton fabrics modified for use in oil/water separation: A perspective review. *Cellulose* **2021**, *28*, 4575–4594. [CrossRef]
5. Zhang, N.; Qi, Y.; Zhang, Y.; Luo, J.; Cui, P.; Jiang, W. A review on oil/water mixture separation material. *Ind. Eng. Chem. Res.* **2020**, *59*, 14546–14568. [CrossRef]
6. Gupta, R.K.; Dunderdale, G.J.; England, M.W.; Hozumi, A. Oil/water separation techniques: A review of recent progresses and future directions. *J. Mater. Chem. A.* **2017**, *5*, 16025–16058. [CrossRef]
7. Baig, U.; Faizan, M.; Sajid, M. Multifunctional membranes with super-wetting characteristics for oil-water separation and removal of hazardous environmental pollutants from water: A review. *Adv. Colloid Interface Sci.* **2020**, *285*, 102276. [CrossRef] [PubMed]
8. Liu, H.; Gao, S.W.; Cai, J.S.; He, C.L.; Mao, J.J.; Zhu, T.X.; Chen, Z.; Huang, J.Y.; Meng, K.; Zhang, K.Q.; et al. Recent progress in fabrication and applications of superhydrophobic coating on cellulose-based substrates. *Materials* **2016**, *9*, 124. [CrossRef]
9. Nthunya, L.N.; Gutierrez, L.; Derese, S.; Nxumalo, E.N.; Verliefde, A.R.; Mamba, B.B.; Mhlanga, S.D. A review of nanoparticle-enhanced membrane distillation membranes: Membrane synthesis and applications in water treatment. *J. Chem. Technol. Biotechnol.* **2019**, *94*, 2757–2771. [CrossRef]
10. Padaki, M.; Murali, R.S.; Abdullah, M.S.; Misdan, N.; Moslehyani, A.; Kassim, M.A.; Hilal, N.; Ismail, A.F. Membrane technology enhancement in oil-water separation. A Review. *Desalination* **2015**, *357*, 197–207. [CrossRef]
11. Yeom, C.; Kim, Y. Purification of oily seawater/wastewater using superhydrophobic nano-silica coated mesh and sponge. *J. Ind. Eng. Chem.* **2016**, *40*, 47–53. [CrossRef]
12. Ma, W.; Zhang, M.; Liu, Z.; Kang, M.; Huang, C.; Fu, G. Fabrication of highly durable and robust superhydrophobic-superoleophilic nanofibrous membranes based on a fluorine-free system for efficient oil/water separation. *J. Memb. Sci.* **2019**, *570*, 303–313. [CrossRef]
13. Woo, S.; Park, H.R.; Park, J.; Yi, J.; Hwang, W. Robust and continuous oil/water separation with superhydrophobic glass microfiber membrane by vertical polymerization under harsh conditions. *Sci. Rep.* **2020**, *10*, 21413. [CrossRef] [PubMed]
14. Zhang, J.; Seeger, S. Polyester materials with superwetting silicone nanofilaments for oil/water separation and selective oil absorption. *Adv. Funct. Mater.* **2011**, *21*, 4699–4704. [CrossRef]
15. Song, H.; Zhao, Z.; Xu, G.; Wang, Y.; Hu, J. Facial preparation of superhydrophobic and superoleophilic textiles by depositing nano-SiO2 for oil–water separation. *J. Text. Inst.* **2017**, *108*, 1297–1301. [CrossRef]
16. Jin, Y.; Jiang, P.; Ke, Q.; Cheng, F.; Zhu, Y.; Zhang, Y. Superhydrophobic and superoleophilic polydimethylsiloxane-coated cotton for oil-water separation process: An evidence of the relationship between its loading capacity and oil absorption ability. *J. Hazard. Mater.* **2015**, *300*, 175–181. [CrossRef]
17. Shin, J.H.; Heo, J.H.; Jeon, S.; Park, J.H.; Kim, S.; Kang, H.W. Bio-inspired hollow PDMS sponge for enhanced oil–water separation. *J. Hazard. Mater.* **2019**, *365*, 494–501. [CrossRef]
18. Lee, J.H.; Kim, D.H.; Kim, Y.D. High-performance, recyclable and superhydrophobic oil absorbents consisting of cotton with a polydimethylsiloxane shell. *J. Ind. Eng. Chem.* **2016**, *35*, 140–145. [CrossRef]
19. Gao, S.; Dong, X.; Huang, J.; Dong, J.; Cheng, Y.; Chen, Z.; Lai, Y. Co-solvent induced self-roughness superhydrophobic coatings with self-healing property for versatile oil-water separation. *Appl. Surf. Sci.* **2018**, *459*, 512–519. [CrossRef]
20. Wen, N.; Miao, X.; Yang, X.; Long, M.; Deng, W.; Zhou, Q.; Deng, W. An alternative fabrication of underoil superhydrophobic or underwater superoleophobic stainless steel meshes for oil-water separation: Originating from one-step vapor deposition of polydimethylsiloxane. *Sep. Purif. Technol.* **2018**, *204*, 116–126. [CrossRef]

21. Wang, Y.; Wang, M.; Wang, J.; Wang, H.; Men, X.; Zhang, Z. A rapid, facile and practical fabrication of robust PDMS@starch coatings for oil-water separation. *J. Taiwan Inst. Chem. Eng.* **2019**, *99*, 215–223. [CrossRef]
22. Doan, H.N.; Vo, P.P.; Hayashi, K.; Kinashi, K.; Sakai, W.; Tsutsumi, N. Recycled PET as a PDMS-Functionalized electrospun fibrous membrane for oil-water separation. *J. Environ. Chem. Eng.* **2020**, *8*, 103921. [CrossRef]
23. Ghorbani, L.; Khatibi, A.; Basuvalingam, S.B.; Caschera, D.; Shokri, B. Fabrication of a Flexible Si-cotton Filter Membrane for Efficient Hot Oil/Hot Water Separation. *Fibers Polym.* **2021**, *23*, 843–851. [CrossRef]
24. Ghorbani, L.; Khatibi, A.; Shokri, B. Hydrophobic and oleophilic cotton fabrics for efficient oil-water separation through low-pressure plasma polymerization. *Iran. J. Phys. Res.* **2020**, *19*, 683–690. [CrossRef]
25. Cheng, S.Y.; Yuen, C.W.M.; Kan, C.W.; Cheuk, K.K.L.; Daoud, W.A.; Lam, P.L.; Tsoi, W.Y.I. Influence of atmospheric pressure plasma treatment on various fibrous materials: Performance properties and surface adhesion analysis. *Vacuum* **2010**, *84*, 1466–1470. [CrossRef]
26. Balu, B.; Breedveld, V.; Hess, D.W. Fabrication of "roll-off" and "sticky" superhydrophobic cellulose surfaces-via plasma processing. *Langmuir* **2008**, *24*, 4785–4790. [CrossRef]
27. Khalifa, I.B.; Ladhari, N. Hydrophobic behavior of cotton fabric activated with air atmospheric-pressure plasma. *J. Text. Inst.* **2020**, *111*, 1191–1197. [CrossRef]
28. Shanmugavelayutham, G.; Anupriyanka, T.; Bhagyashree, P.; Premasudha, P. Plasma Surface Modification of Cotton Fabric by Using Low Pressure Plasma. *IEEE Trans. Plasma Sci.* **2021**, *49*, 497–501. [CrossRef]
29. Montarsolo, A.; Periolatto, M.; Zerbola, M.; Mossotti, R.; Ferrero, F. Hydrophobic sol-gel finishing for textiles: Improvement by plasma pre-treatment. *Text. Res. J.* **2013**, *83*, 1190–1200. [CrossRef]
30. Lam, Y.L.; Kan, C.W.; Yuen, C.W. Effect of oxygen plasma pretreatment and titanium dioxide overlay coating on flame retardant finished cotton fabrics. *BioResources* **2011**, *6*, 1454–1474.
31. Usman, J.; Othman, M.H.D.; Ismail, A.F.; Rahman, M.A.; Jaafar, J.; Raji, Y.O.; Gbadamosi, A.O.; el Badawy, T.H.; Said, K.A.M. An overview of superhydrophobic ceramic membrane surface modification for oil-water separation. *J. Mater. Res. Technol.* **2021**, *12*, 643–667. [CrossRef]
32. Caschera, D.; Mezzi, A.; Cerri, L.; de Caro, T.; Riccucci, C.; Ingo, G.M.; Padeletti, G.; Biasiucci, M.; Gigli, G.; Cortese, B. Effects of plasma treatments for improving extreme wettability behavior of cotton fabrics. *Cellulose* **2014**, *21*, 741–756. [CrossRef]
33. Barth-Wehrenalp, G. Inorganic Polymers. *J. Am. Chem. Soc.* **1963**, *85*, 2192–2193. [CrossRef]
34. Bevington, P. *Data Reduction and Error Analysis*; McGraw-Hill: New York, NY, USA, 1969; pp. 56–91. Available online: http://www.spy-hill.net/~{}myers/201/notes/textalk.pdf (accessed on 17 October 2022).
35. Yang, J.; Yin, L.; Tang, H.; Song, H.; Gao, X.; Liang, K.; Li, C. Polyelectrolyte-fluorosurfactant complex-based meshes with superhydrophilicity and superoleophobicity for oil/water separation. *Chem. Eng. J.* **2015**, *268*, 245–250. [CrossRef]
36. Liu, Y.; Wang, X.; Feng, S. Nonflammable and Magnetic Sponge Decorated with Polydimethylsiloxane Brush for Multitasking and Highly Efficient Oil–Water Separation. *Adv. Funct. Mater.* **2019**, *29*, 1902488. [CrossRef]
37. Hoefnagels, H.F.; Wu, D.; de With, G.; Ming, W. Biomimetic superhydrophobic and highly oleophobic cotton textiles. *Langmuir* **2007**, *23*, 13158–13163. [CrossRef]
38. Wang, M.; Long, X.; du, J.; Sun, C.; fu, S.; Xu, C. X-ray photoelectron spectroscopy analysis of cotton treated with the TBCC/H2O2/NaHCO3 system. *Text. Res. J.* **2014**, *84*, 2149–2156. [CrossRef]
39. Jiang, Z.; Li, H.; He, Y.; Liu, Y.; Dong, C.; Zhu, P. Flame retardancy and thermal behavior of cotton fabrics based on a novel phosphorus-containing siloxane. *Appl. Surf. Sci.* **2019**, *479*, 765–775. [CrossRef]
40. Kim, M.G.; Jo, K.I.; Kim, E.; Park, J.H.; Ko, J.W.; Lee, J.H. Preparation of polydimethylsiloxane-modified waterborne polyurethane coatings for marine applications. *Polymers* **2021**, *13*, 4283. [CrossRef]
41. Dakroub, G.; Duguet, T.; Esvan, J.; Lacaze-Dufaure, C.; Roualdes, S.; Rouessac, V. Comparative study of bulk and surface compositions of plasma polymerized organosilicon thin films. *Surf. Interfaces* **2021**, *25*, 101256. [CrossRef]
42. Wakida, T.; Takeda, K.; Tanaka, I.; Takagishi, T. Free Radicals in Cellulose Fibers Treated with Low Temperature Plasma. *Text. Res. J.* **2014**, *59*, 1. [CrossRef]
43. Vaideki, K.; Jayakumar, S.; Thilagavathi, G.; Rajendran, R. A study on the antimicrobial efficacy of RF oxygen plasma and neem extract treated cotton fabrics. *Appl. Surf. Sci.* **2007**, *253*, 7323–7329. [CrossRef]
44. Zhang, H.; Zhang, Z.; Guo, F.; Liu, W. The Influence of Plasma Treatment on the Tribological Properties of Hybrid PTFE/Cotton Fabric/Phenolic Composites. *Polym. Compos.* **2009**, *30*, 1523–1531. [CrossRef]
45. Pappas, D.; Bujanda, A.; Demaree, J.D.; Hirvonen, J.K.; Kosik, W.; Jensen, R.; Mcknight, S. Surface modification of polyamide fibers and films using atmospheric plasmas. *Surf. Coat. Technol.* **2006**, *201*, 4384–4388. [CrossRef]
46. Zhang, Q.; Zhang, C.; Luo, W.; Cui, L.; Wang, Y.J.; Jian, T.; Li, X.; Yan, Q.; Liu, H.; Ouyang, C.; et al. Sequence-Defined Peptoids with –OH and –COOH Groups As Binders to Reduce Cracks of Si Nanoparticles of Lithium-Ion Batteries. *Adv. Sci.* **2020**, *7*, 2000749. [CrossRef]
47. Caschera, D.; Cortese, B.; Mezzi, A.; Brucale, M.; Ingo, G.M.; Gigli, G.; Padeletti, G. Ultra hydrophobic/superhydrophilic modified cotton textiles through functionalized diamond-like carbon coatings for self-cleaning applications. *Langmuir* **2013**, *29*, 2775–2783. [CrossRef]
48. Cho, S.C.; Hong, Y.C.; Cho, S.G.; Ji, Y.Y.; Han, C.S.; Uhm, H.S. Surface modification of polyimide films, filter papers, and cotton clothes by HMDSO/toluene plasma at low pressure and its wettability. *Curr. Appl. Phys.* **2009**, *9*, 1223–1226. [CrossRef]

49. Woźnicka, M. *Synthesis and Properties of Telechelic Polysiloxane and Polyether Homo- and Co-Polymers*; Universität Rostock: Rostock, Germany, 2017. [CrossRef]
50. Groza, A.; Surmeian, A. Characterization of the Oxides Present in a Polydimethylsiloxane Layer Obtained by Polymerisation of Its Liquid Precursor in Corona Discharge. *J. Nanomater.* **2015**, *2015*, 204296. [CrossRef]
51. Vasiljević, J.; Gorjanc, M.; Jerman, I.; Tomšič, B.; Modic, M.; Mozetič, M.; Orel, B.; Simončič, B. Influence of oxygen plasma pre-treatment on the water repellency of cotton fibers coated with perfluoroalkyl-functionalized polysilsesquioxane. *Fibers Polym.* **2016**, *17*, 695–704. [CrossRef]
52. Palamà, I.E.; D'Amone, S.; Arcadio, V.; Caschera, D.; Toro, R.G.; Gigli, G.; Cortese, B. Underwater Wenzel and Cassie oleophobic behaviour. *J. Mater. Chem. A* **2015**, *3*, 3854–3861. [CrossRef]
53. Caschera, D.; Toro, R.G.; Federici, F.; Riccucci, C.; Ingo, G.M.; Gigli, G.; Cortese, B. Flame retardant properties of plasma pre-treated/diamond-like carbon (DLC) coated cotton fabrics. *Cellulose* **2015**, *22*, 2797–2809. [CrossRef]
54. Stanton, M.M.; Ducker, R.E.; MacDonald, J.C.; Lambert, C.R.; McGimpsey, W.G. Super-hydrophobic, highly adhesive, polydimethylsiloxane (PDMS) surfaces. *J. Colloid Interface Sci.* **2012**, *367*, 502–508. [CrossRef] [PubMed]
55. Wu, Z.; Xanthopoulos, N.; Reymond, F.; Rossier, J.S.; Girault, H.H. Polymer microchips bonded by O2-plasma activation. *Electrophoresis* **2002**, *23*, 782–790. [CrossRef]
56. Li, Y.; Zou, C.; Shao, J.; Li, Y. Fabrication of superhydrophobic cotton fabrics through wrapping silica with plasma-induced grafting polymerization. *Text. Res. J.* **2017**, *89*, 401–410. [CrossRef]
57. Gashti, M.P.; Alimohammadi, F.; Shamei, A. Preparation of water-repellent cellulose fibers using a polycarboxylic acid/hydrophobic silica nanocomposite coating. *Surf. Coat. Technol.* **2012**, *206*, 3208–3215. [CrossRef]
58. Cortese, B.; Caschera, D.; Federici, F.; Ingo, G.M.; Gigli, G. Superhydrophobic fabrics for oil-water separation through a diamond like carbon (DLC) coating. *J. Mater. Chem. A.* **2014**, *2*, 6781–6789. [CrossRef]
59. Li, S.; Huang, J.; Ge, M.; Cao, C.; Deng, S.; Zhang, S.; Chen, G.; Zhang, K.; Al-Deyab, S.S.; Lai, Y. Robust Flower-Like TiO2@Cotton Fabrics with Special Wettability for Effective Self-Cleaning and Versatile Oil/Water Separation. *Adv. Mater. Interfaces* **2015**, *2*, 14. [CrossRef]
60. Liu, X.; Ge, L.; Li, W.; Wang, X.; Li, F. Layered double hydroxide functionalized textile for effective oil/water separation and selective oil adsorption. *ACS Appl. Mater. Interfaces* **2015**, *7*, 791–800. [CrossRef]
61. Hosseinzadeh, H.; Mohammadi, S. Synthesis of a novel hydrogel nanocomposite coated on cotton fabric for water-oil separation. *Water Air. Soil Pollut.* **2014**, *225*, 2115. [CrossRef]
62. Vaidulych, M.; Shelemin, A.; Hanuš, J.; Khalakhan, I.; Krakovsky, I.; Kočová, P.; Mašková, H.; Kratochvíl, J.; Pleskunov, P.; Štěrba, J.; et al. Superwettable antibacterial textiles for versatile oil/water separation. *Plasma Process. Polym.* **2019**, *16*, 1900003. [CrossRef]
63. Singh, A.K.; Singh, J.K. Fabrication of zirconia based durable superhydrophobic-superoleophilic fabrics using non fluorinated materials for oil-water separation and water purification. *RSC Adv.* **2016**, *6*, 103632–103640. [CrossRef]
64. Cheng, Q.Y.; Guan, C.S.; Wang, M.; Li, Y.D.; Zeng, J.B. Cellulose nanocrystal coated cotton fabric with superhydrophobicity for efficient oil/water separation. *Carbohydr. Polym.* **2018**, *199*, 390–396. [CrossRef] [PubMed]
65. Wang, J.; Liu, S.; Guo, S. Calcium ions enhanced mussel-inspired underwater superoleophobic coating with superior mechanical stability and hot water repellence for efficient oil/water separation. *Appl. Surf. Sci.* **2020**, *503*, 144180. [CrossRef]
66. Ao, C.; Hu, R.; Zhao, J.; Zhang, X.; Li, Q.; Xia, T.; Zhang, W.; Lu, C. Reusable, salt-tolerant and superhydrophilic cellulose hydrogel-coated mesh for efficient gravity-driven oil/water separation. *Chem. Eng. J.* **2018**, *338*, 271–277. [CrossRef]
67. Teng, C.; Lu, X.; Ren, G.; Zhu, Y.; Wan, M.; Jiang, L. Underwater Self-Cleaning PEDOT-PSS Hydrogel Mesh for Effective Separation of Corrosive and Hot Oil/Water Mixtures. *Adv. Mater. Interfaces* **2014**, *1*, 1400099. [CrossRef]
68. Zeng, Z.; Wu, X.; Liu, Y.; Chen, C.; Tian, D.; He, Y.; Wang, L.; Yang, G.; Zhang, X.; Zhang, Y. Fabrication of a durable coral-like superhydrophilic MgO coating on stainless steel mesh for efficient oil/water separation. *Chem. Eng. Sci.* **2022**, *248*, 117144. [CrossRef]
69. Phanthong, P.; Reubroycharoen, P.; Kongparakul, S.; Samart, C.; Wang, Z.; Hao, X.; Abudula, A.; Guan, G. Fabrication and evaluation of nanocellulose sponge for oil/water separation. *Carbohydr. Polym.* **2018**, *190*, 184–189. [CrossRef]

materials

MDPI

Article

Lithium Accumulation in *Salvinia natans* Free-Floating Aquatic Plant

Anamaria Iulia Török [1], Ana Moldovan [1], Eniko Kovacs [1,2], Oana Cadar [1], Anca Becze [1], Erika Andrea Levei [1,*] and Emilia Neag [1,*]

[1] Research Institute for Analytical Instrumentation Subsidiary, National Institute for Research and Development for Optoelectronics INOE 2000, 67 Donath Street, 400293 Cluj-Napoca, Romania
[2] Faculty of Horticulture, University of Agricultural Sciences and Veterinary Medicine, 3–5 Manastur Street, 400372 Cluj-Napoca, Romania
* Correspondence: erika.levei@icia.ro (E.A.L.); emilia.neag@icia.ro (E.N.)

Abstract: The new context of the intensive use of lithium-based batteries led to increased production of Li and Li-containing wastes. All these activities are potential sources of environmental pollution with Li. However, the negative impact of Li on ecosystems, its specific role in the plants' development, uptake mechanism, and response to the induced stress are not fully understood. In this sense, the Li uptake and changes induced by Li exposure in the major and trace element contents, photosynthetic pigments, antioxidant activity, and elemental composition of *Salvinia natans* were also investigated. The results showed that *Salvinia natans* grown in Li-enriched nutrient solutions accumulated much higher Li contents than those grown in spring waters with a low Li content. However, the Li bioaccumulation factor in *Salvinia natans* grown in Li-enriched nutrient solutions was lower (13.3–29.5) than in spring waters (13.0–42.2). The plants exposed to high Li contents showed a decrease in their K and photosynthetic pigments content, while their total antioxidant activity did not change substantially.

Keywords: *Salvinia natans*; lithium; bioconcentration factor; photosynthetic pigments

Citation: Török, A.I.; Moldovan, A.; Kovacs, E.; Cadar, O.; Becze, A.; Levei, E.A.; Neag, E. Lithium Accumulation in *Salvinia natans* Free-Floating Aquatic Plant. *Materials* **2022**, *15*, 7243. https://doi.org/10.3390/ma15207243

Academic Editors: Roberta G. Toro, Daniela Caschera and Abeer M. Adel

Received: 16 September 2022
Accepted: 16 October 2022
Published: 17 October 2022

1. Introduction

There is a vast number of both terrestrial and aquatic plant species that can uptake pollutants from water, soil, and even from the air [1–4]. The accumulation of metals in aquatic plants depends on the concentration of the metals in water and plant species and the exposure time. Several processes, such as element mobilization, root uptake, xylem loading, root-to-shoot transport, cellular compartmentation, and sequestration, are involved in the accumulation of metals in plants [5]. However, the specific mechanism of metal accumulation in plants is not entirely understood. Metals in ionic form can attach to the surface of plants very quickly, being integrated into the plant's membrane surface or fixed on the sites used for nutrient uptake [6]. The bioavailability of metals to plants and the changes induced in the plants by different metals as well as their defense mechanisms were previously studied [5]. Chlorophyll is responsible for collecting solar energy during photosynthesis, while carotenoids eliminate the reactive oxygen species in the damaged tissues [7]. Changes in the photosynthetic pigments indicate the presence of stress factors, such as toxic elements. Exposure to toxic elements also stimulates the formation of reactive oxygen species and other highly reactive compounds, damaging the plant's lipids, proteins, and nucleic acids [8]. By simultaneous exposure to various elements, the plants face multiple metal stresses, as well as competition between nutrients (K, Ca, Na, Mg), essential (Cu, Zn) and toxic (Cd) elements. High concentrations of nutrients and essential elements may also activate the protective mechanism of plants [9,10].

Aquatic macrophyte species, grouped in emergent, submerged, and free-floating plants, can accumulate a wide range of pollutants, such as toxic metals, pesticides, and radionuclides from waters and wetlands [11,12]. The pollutant's type and location determine the processes of its removal and breakdown by macrophytes [13]. Free-floating aquatic plants are ideal for phytoextraction and rhizofiltration, as they are easy to grow and harvest [9]. The Salviniaceae aquatic species have a remarkable ability to uptake, tolerate and bioaccumulate toxic inorganic pollutants [14–17], phosphates [18], nitrogen [19], arsenic [20], and amino groups from aqueous solutions [21]. *Salvinia natans* (L.) All. (*S. natans*) is a free-floating fern from the Salviniaceae family and has a high tolerance to a wide range of pollutants and a rapid growth rate [22]. It is native to Europe and widely spread within temperate, tropical, and subtropical regions [16,23,24]. *S. natans* has two floating leaves and a submerged leaf, which functions as a root, assimilating nutrients and pollutants [25]. The leaves are predominantly composed and covered by eggbeater-shaped hairs, making them hydrophilic. In order to determine the accumulation potential of *S. natans*, its tolerance to various heavy metals and trace elements was studied [16,17,26–28].

Among the trace elements, lithium (Li) is a critical element used in energy storage, alloys for military and aerospace applications, lubricating greases, glass and ceramics, antidepressant drugs, and hydrogen storage [29]. The use of Li to produce Li-ion batteries steeply increased in the last decade, triggering an increase in Li production and generating Li-containing wastes. As the large-scale recycling of batteries and other Li-containing wastes is still challenging, the contamination of the environment with Li has an increasing trend [30,31]. Because of the small ion radius, Li is a biotic element that is highly soluble in aqueous solutions, frequently forming free cations with a low charge [32,33]. The concentrations of Li vary depending on the nature of the water (i.e., surface water, groundwater, seawater), being a scarce element in natural water [31,33]. The Li concentration varies widely in surface waters. For example, the average Li value was around 40 μg/ L in the USA [34], while it varied from 0.5 to 15.0 μg/L in the Republic of Macedonia [35] and between 20 and 91 μg/L in Ireland [36]. Except for the Austrian, Argentinean, and Chilean regions, where the Li concentrations were reported to exceed 1000 μg/L, the concentration of Li in groundwater ranged from 0.05 to 150 μg/L [37]. In seawater, Li concentrations between 170–190 μg/L were found [38]. In Romania, there is limited information regarding Li's concentration in water. Dippong et al. [39] reported average concentrations of Li between 1.20 and 114 μg/L in Romanian bottled water brands, while Zsigmond et al. [40] found high Li concentrations in most of the studied Transylvanian mineral waters. Torok et al. [41] reported Li concentrations of 1.40–12.20 μg/L in groundwaters from the karst areas in Romania.

Li is not considered an essential element in plant growth and development, yet some studies revealed its beneficial role in plants. Due to the physicochemical similarities between Li, Na, K, and Ca, these elements can be easily uptaken by the plant's roots through the same transport systems [42]. Li bioavailability, phytotoxicity, and its interaction with other elements, as well as the uptake mechanism and the specific role in plant species development, are not fully elucidated [42,43]. Moreover, the understanding of the dose–response relationship and short- and long-term influence on plant physiology is not currently elucidated.

The present study aimed to evaluate the Li uptake by *S. natans* from mono- and multiple-element solutions, as well as from natural spring waters, and to observe the Li-induced changes in the plant's major and trace element contents. At the same time, the plant's vital signs—as a response to the induced stress—were monitored through the photosynthetic pigments content and total antioxidant activity.

2. Materials and Methods

2.1. Experimental Design

S. natans aquatic plants were obtained from a local aquarium store in Cluj-Napoca, Romania. Prior to the experiments, plant species were grown for 30–40 days under laboratory conditions in the Hoagland nutrient solution containing 1.25 mM KNO_3, 1.25 mM $Ca(NO_3)_2$, 0.5 mM $MgSO_4$, 0.25 mM KH_2PO_4, 10 μM FeEDTA, 11.6 μM H_3BO_3, 4.5 μM $MnCl_2 \cdot 4H_2O$, 0.19 μM $ZnSO_4 \cdot 7H_2O$, 0.12 μM $Na_2MoO_4 \cdot 2H_2O$, and 0.08 μM $CuSO_4 \cdot 5H_2O$ [44].

Metal accumulation experiments were carried out in the Hoagland solution enriched with concentrations of 10 mg-Li/L, 20 mg-Li/L, 50 mg-Li/L, and 20 mg-Li/L, 20 mg-Cu/L, 20 mg-Zn/L, and 20 mg-Cd/L by dissolving $LiSO_4 \cdot H_2O$, $CuSO_4 \cdot 5H_2O$, $ZnSO_4 \cdot 7H_2O$, and $Cd(NO_3)_2 \cdot 5H_2O$, as well as in spring waters with different Li concentrations. Spring water samples were collected in November 2021 from the Banpotoc (W1, Hunedoara county, Romania), Borcut (W2, Maramures county, Romania), and Botiza villages (W3, Maramures county, Romania). The major and trace elements in the spring waters are presented in Table 1.

Table 1. Major (Na, Mg, K, and Ca) and trace (Li, Mn, Fe, Ni, Cu, Zn, Sr, and Ba) elements concentrations expressed as average ± standard deviation (n = 3) in Banpotoc (W1), Borcut (W2) and Botiza (W3) spring waters.

		W1	W2	W3
Major elements (mg/L)	Na	20.2 ± 1.6	484 ± 24	599 ± 30
	Mg	44.8 ± 3.6	52.8 ± 4.2	44.5 ± 3.6
	K	2.54 ± 0.20	7.18 ± 0.36	24.5 ±1.2
	Ca	386 ± 19	126 ± 10	154 ± 12
Trace elements (μg/L)	Li	68.6 ± 3.4	233 ± 12	2090 ± 104
	Mn	104 ± 5	351 ± 18	700 ± 35
	Fe	176 ± 9	35.3 ± 1.8	42.3 ± 2.1
	Ni	15.0 ± 0.7	11.0 ± 0.6	14.3 ± 0.7
	Cu	<0.21	3.49 ± 0.28	3.81 ± 0.72
	Zn	5.04 ± 0.46	11.7 ± 1.1	2.56 ± 0.23
	Sr	658 ± 33	1110 ± 56	1350 ± 67
	Ba	1010 ±51	46.6 ± 22.3	301 ± 15

Amounts of 2.5 g (fresh weight, FW) of *S. natans* were placed in 250 mL of enriched Hoagland solution or spring waters. For the control experiments, the Hoagland nutrient solution was used. Plants were left to grow at room temperature (20–23 °C) under natural light conditions for 7 days and then were harvested and analyzed. The *S. natans'* life span is approximately 130–200 days and includes gametophytic and sporophytic phases [23,45]. Generally, sporophyte development requires about 170 days, while gametophyte development requires approximately 35 days [46]. Previous investigations showed that a 6–7 day growth period is enough for the aquatic macrophytes to uptake and accumulate metals, such as Cu, Zn, and Cd, from multi-metal solutions and to develop responses to the induced stress [9,10,47]. All experiments were performed in triplicate (n = 3). The samples of *S. natans* after Li exposure were denoted as follows: (i) control *S. natans*—SC; (ii) *S. natans* exposed to 10 mg-Li/L—S10; (iii) *S. natans* exposed to 20 mg-Li/L—S20; (iv) *S. natans* exposed to 50 mg-Li/L—S50; (v) *S. natans* exposed to 20 mg-Li/L, 20 mg-Cu/L, 20 mg-Zn/L, and 20 mg-Cd/L concentrations—SM20; (vi) *S. natans* exposed to spring water, SW1–SW3.

2.2. Element Analysis

The plant samples were washed with distilled water, oven-dried at 65 °C, ground using an agate mortar, and passed through a 200-μm mesh sieve to obtain a homogenized powder. An amount of 200 mg of plant material was digested with 5 mL of 65% HNO_3 and 2 mL of 30% H_2O_2 (Merck, Darmstadt, Germany). The digested samples were diluted with ultrapure water to a final volume of 25 mL. The measured metal concentrations were expressed as mg/kg dry weight (DW). The Na, Mg, K, Ca, and Fe concentrations were measured using an Optima 5300 DV inductively coupled plasma–atomic emission spectrometer (ICP–OES, Perkin Elmer, Waltham, MA, USA), while the Li, Mn, Fe, Ni, Cu, Zn, Rb, Sr, Cd, and Ba concentrations were determined using an ELAN DRC II inductively coupled plasma–mass spectrometer (ICP–MS, Perkin Elmer, Waltham, MA, USA). ICP multi-element standard solution IV, 1000 mg/L (Merck, Darmstadt, Germany), and multi-element Calibration Standard 3 (Perkin Elmer Pure Plus, Waltham, MA, USA) were used for the calibration, while the measurement accuracy was tested using the standard reference material, 1643f NIST–Trace elements in water (National Institute of Standards and Technology, Gaithersburg, MD, USA) and NIM–GWB 10019 Apple–Trace elements (Institute of Geophysical and Geochemical Exploration, Langfang City, China) for plants. The average recoveries ranged between 93–104%. Ultrapure water (Elga Veolia, High Wycombe, UK) was used for dilutions and the preparation of the standard solutions. The N, C, H and S contents were determined using a Flash 2000 CHNS/O analyzer (Thermo Fisher Scientific, Waltham, MA, USA).

2.3. Photosynthetic Pigments

The photosynthetic pigments (chlorophyll *a*—C_a, chlorophyll *b*—C_b, and carotenoids—C_{x+c}) were measured in an extract of 0.5 g of *S. natans* fresh weight (FW) in 5 mL of methanol. The plant–methanol mixture was vortexed thoroughly for 1 min and incubated at 70 °C for 3 min. The extracts were centrifuged, and the absorbance of the supernatant was measured at 665, 652, and 470 nm using a Nanodrop One Analyzer (Thermo Fisher Scientific, Waltham, MA, USA). The contents of the C_a, C_b, and C_{x+c} pigments were calculated according to Lichtenthaler (1987), and the results were expressed as μg/g (FW) [48].

2.4. Antioxidant Content

An amount of 5 g of fresh *S. natans* was extracted using 20 mL of methanol. The samples were vortexed for 1 min and then centrifuged at 3500 rpm for 10 min. The supernatant was analyzed using Photochem (AnalytikJena, Jena, Germany). Trolox (ACL kit, AnalytikJena, Jena, Germany) was used as a standard for the quantitative assessment of antioxidant activity. The calibration curve was made using a range of pathlengths from 0.5 to 3 nm. The samples were diluted with methanol at 1:20 in order to be in the concentration range of the calibration curve.

2.5. FT–IR Analysis

The FT–IR spectra were recorded using a Spectrum BX II (Perkin Elmer, Waltham, MA, USA) Fourier–transform infrared spectrometer on pellets containing 1% (*w*/*w*) dried and powdered samples in KBr.

2.6. Bioaccumulation Factor

The metal bioaccumulation factor (BCF) was calculated according to Equation (1) [49,50].

$$BCF = CP/CW \tag{1}$$

where, CP is the Li concentration in the plant (mg/kg) and CW is the initial Li concentration in the Hoagland solutions enriched with Li or in the spring waters (mg/L).

2.7. Data Analysis

The OriginLab (version 2020b) software (Northampton, MA, USA) was used for data analysis and the plotting of the charts. The differences between the studied parameters were tested by comparing the averages of the three replicates using Tukey's test for a significance level of p = 0.05 using the OriginPro (version 2020b) software (OriginLab Corporation, Northampton, MA, USA).

3. Results and Discussion

3.1. Li Uptake by S. natans

S. natans showed an excellent ability to accumulate Li from the enriched Hoagland solutions. The Li content in *S. natans*, grown in the control solution, was 0.57 mg/kg DW and increased considerably in the Li-enriched Hoagland solutions. As shown in Figure 1, the Li uptake in the S10, S20, and S50 plants increased 500–1000 fold with the increased Li concentration in the test solution.

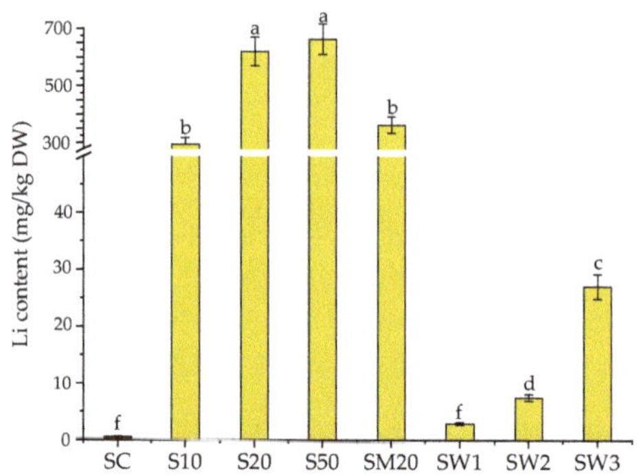

Figure 1. The Li content in *S. natans* grown in control (SC), Hoagland solution enriched with 10 mg-Li/L (S10), 20 mg-Li/L (S20), 50 mg-Li/L (S50), 20 mg-Li/L, 20 mg-Cu/L, 20 mg-Zn/L, and 20 mg-Cd/L (SM20), and Banpotoc (SW1), Borcut (SW2) and Botiza (SW3) spring waters. Li contents marked with different letters are significantly different at the $p < 0.05$ significance level.

A steep increase in the Li content (295 mg/kg DW) was observed in the *S. natans* grown in the solution containing 10 mg-Li/L. This increase was also observed for the plants grown in the enriched Hoagland solution containing 20 mg-Li/L, with Li (621 mg/kg DW) being twice as high as those grown in S10. However, for the *S. natans* grown in 50 mg-Li/L solutions, no further increase in the Li content (664 mg/kg DW) was observed. These results indicated that, up to a concentration of 20 mg-Li/L in the growth medium, the accumulated Li in *S. natans* increased proportionally, but at higher Li concentrations, the accumulated Li remained more or less constant. This could be a consequence of the growth-stimulating effect of low Li concentrations and the toxic effects of high Li contents. Hawrylak-Nowak reported that 50 mg-Li/L (as the mono-element) in the growing medium was toxic for plants and that the toxicity symptoms, as well as the amount of accumulated Li, were species-specific [51,52]. The growth-stimulating effects of low Li concentrations were also reported for maize (5 mg-Li/L) [53] and lettuce (up to 14 mg-Li/L) [42]. Previous studies showed that exposure to the 50 mg-Li/L solutions resulted in the accumulation of 695 mg/kg DW for maize, 3292 mg/kg DW for sunflower, and 1607 and 1544 mg/kg DWs for lettuce [51,52].

The Li content in *S. natans,* grown in SM20, was considerably higher than in the control sample. However, it was almost half lower (364 mg/kg DW) than those grown in S20 (621 mg/kg DW), indicating a possible competition of Li with Cu, Zn and Cd for uptake [47,54]. The Li content in *S. natans* exposed to spring waters was lower (2.90 to 27.1 mg/kg DW) than in those exposed to the Li-enriched growing solution. The low uptake could be explained by the low Li concentration of spring waters. Previous studies have reported that *S. natans* exhibited a high ability to accumulate Cu and Cd from multi-element solutions [46,53].

3.2. Major and Trace Element Contents in S. natans Exposed to Li

Generally, the macro- and micro-element contents in aquatic plants depend on the chemical composition of the aquatic environment, the element of interest, and the plant species [55]. The major and trace element contents in the *S. natans* plants are presented in Figure 2. By simultaneous exposure of *S. natans* to Li, Cu, Zn, and Cd, the accumulation of all major elements decreased (Figure 2a), with the most drastic drop being observed for K (from 23 g/kg DW for the control to 1.64 g/kg DW for the SM20). This fact suggests that K competes with other elements for uptake and that the uptake of Li takes place through K transporters.

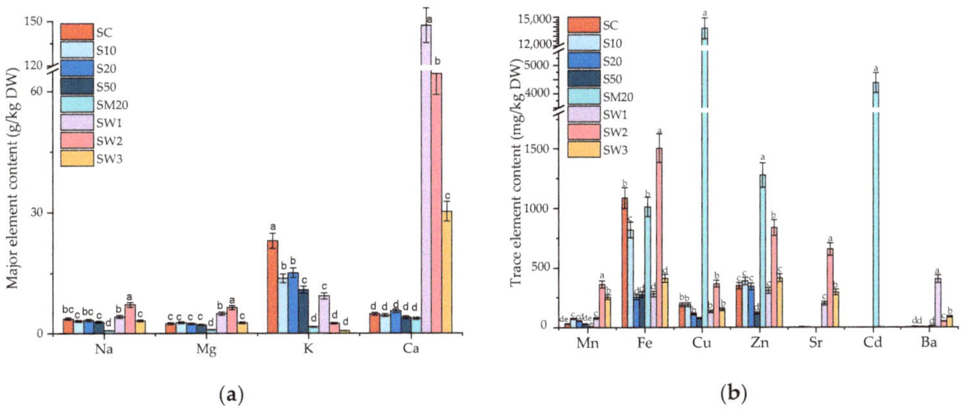

Figure 2. (**a**) Major and (**b**) trace element contents in *S. natans* grown in control (SC), Hoagland solution enriched with 10 mg-Li/L (S10), 20 mg-Li/L (S20), 50 mg-Li/L (S50), 20 mg-Li/L, 20 mg-Cu/L, 20 mg-Zn/L, and 20 mg-Cd/L (SM20), and Banpotoc (SW1), Borcut (SW2) and Botiza (SW3) spring waters. Element contents marked with different letters are significantly different at the $p < 0.05$ significance level.

Compared to the control experiment, *S. natans* exposed to the W1–W3 springs showed several changes in the major and trace element contents. As the spring waters contain 30-, 13-, and 6-fold higher Ca concentrations than in the control sample, Ca was the highest accumulated element. The highest Ca content was measured in SW1 (147 g/kg DW), while the lowest was measured in SW3 (30.2 g/kg DW). The K content in the SW1–SW3 exhibited a decreasing tendency, probably due to the inhibition of the uptake that occurred during exposure to the spring waters. It was previously reported that Ca can inhibit the absorption of Li and normalize the K uptake under Li stress [32]. This fact was also observed in the current study, as the high Ca content in *S. natans* grown in SW1 led to a low Li content compared to those grown in the other two springs, where the Ca concentration was low, and the accumulated Li content was high. Ca and Li uptake is influenced by the initial metal concentrations from the growth medium. As observed in SW2 and SW3, the Li uptake by *S. natans* increased due to the availability of a high Li and low Ca concentration. The K content also decreased after exposure to Li, probably because Li shares the K transport carrier [56]. Hawrylak-Nowak et al. reported an increased K uptake when sunflower plants were exposed to a concentration of 50 mg-Li/L and no differences in K uptake when maize plants were exposed to Li concentrations between 0 to 50 mg-Li/L [51]. In plants, Li interacts with Na, K, and enzymes that require Mg. Therefore, Li can bind into sites that are not occupied by Mg [38].

The metal uptake in *S. natans* occurs in two stages: a fast stage, characterized by a rapid uptake, which can include physical processes (adsorption, ionic exchange, and chelation), and a slow stage, which includes biological processes, such as intracellular uptake, where the metal is transported through plasmalemma into cells [57]. Generally, K has a major role in transporting water and nutrients through the xylem. Thus, when low K contents are found in plants, the translocation of nitrates, phosphates, Ca, Mg, and amino acids decrease [58]. These findings agree with the current results, as the SM20 plants showed lower K and Mg contents, indicating the disturbing effect of Li and other elements on the nutrient transport system. Shahzad et al. reported that Li may replace Mg and Ca in plants during normal metabolic processes [32]. The slight decrease in Ca content indicated the competitive interactions between Ca and Li, as they share binding sites [59].

Exposure to Li led to various changes in the trace element content of *S. natans*. By increasing the Li content in the growth medium, the Fe content decreased from 1088 mg/kg DW in the control plant to 253 mg/kg DW in S20 and 274 mg/kg DW in S50. The decrease in nutrient uptake could be attributed to the changes induced by the high Li concentration in the functioning of essential element transporters [60]. In the case of simultaneous exposure to Li and Cu, Zn, and Cd, *S. natans* accumulated higher contents of Cu and Cd and lower amounts of Zn (Figure 2b). Following the exposure of *S. natans* to the Hoagland solutions enriched with 20 mg-Li/L, 20 mg-Cu/L, 20 mg-Zn/L, and 20 mg-Cd/L initial concentrations (SM20), the Cu, Zn, and Cd contents increased. However, the Li content was lower than in the plants exposed to 20 mg-Li/L (S20). In the Banpotoc, Borcut, and Botiza springs, the initial concentrations of Cu (0.21–3.81 μg-Cu/L) and Zn (2.56–11.7 μg-Zn/L) were much lower than in SM20. The plants exposed to the springs showed lower Cu and Zn contents than the control plant (SC), except for SW2, which showed a 2.8- and 2.4-fold greater Cu and a 2.7- and 2.0-fold greater Zn content. The highest Na, Mn, Fe, Cu, Zn, and Sr trace element contents were obtained in SW2, out of which the Na content (7.11 g/kg DW) was 1.9-fold higher than in the SC. The variation in metal accumulation could be explained based on a differential affinity towards metal and a competition between metal ions during the uptake [17].

Previous studies also reported the potential of *S. natans* to accumulate high levels of metals from water, though the capacity for metal accumulation varies for each treatment. Dhir et al. [61] observed a higher accumulation rate for Cr, Fe, Ni, Cu, Pb, and Cd than for Co, Zn, and Mn. Polechońska et al. reported *S. natans* as Cu, Fe, Ni, and Zn accumulator and Mn hyperaccumulator, while Reeves and Baker as Cd, Ni, Cu, Co, Pb, Cr, Zn, and Mn hyperaccumulator [16,62]. *S. natans* showed a considerable capacity to accumulate metals (Fe, Co, Cr, Pb, Cu, Zn, Cd, and Ni) by increasing the initial concentrations from 30 to 50 mg/L, but the tolerance index decreased, suggesting higher toxicity [17]. The variation in the metal absorption intensity by plants due to the synergistic or antagonistic effects between metals was previously reported in the literature [63]. Cu is an essential element of plants, acting as a cofactor for the enzymes involved in respiration and photosynthesis. However, in high concentrations, Cu is toxic. Therefore, plants regulate their Cu uptake through the metallochaperones (Cu-binding proteins) that deliver it to the plant cells [64]. Zn is a key element in the functioning of the enzymes involved in plant growth and development, such as polyphenol oxidase, ascorbic acid oxidase, and cytochrome oxidase [64,65]. Cu and Zn uptake may be cooperative and could be transported together in the plant cell via similar transporters and channels [64]. The higher uptake of essential elements, such as Zn and Cu, could be explained by their involvement in the enzymatic activity of plants, which may control the absorption intensity of toxic elements. For example, Zn may alleviate the toxic effects of Cd and Pb and, thus, enhance their uptake [66]. The similar properties of Zn and Cd also contribute to an increase in Cd uptake by using the same uptake mechanisms as Zn. The reduction of Mn and Cu uptake in the presence of Zn and Cd was reported for *Melissa officinalis*. This reduction could be attributed to the antagonistic effects between elements that use the same transporters [63,67].

In the present study, the influence of both essential and trace elements on Li uptake was observed. The accumulation of high concentrations of essential (Cu, Zn) and toxic (Cd) elements have antagonistic effects on the Li uptake in *S. natans*. As the Li-specific binding sites seem to lack in *S. natans*, the competition between the biologically active ions and Li uptake is regulated through highly selective channels, as well as active ion transporters. The influence of essential and toxic elements on Li uptake is not fully understood and deserves more research.

3.3. Bioaccumulation Factor

Trace element accumulation (Cr, Co, Zn, Fe, Ni, Cd, Zn, and Mn) by *S. natans* was previously investigated [26]. Comparatively, the absorption of Li gained little attention. The bioaccumulation factor is an important index used to show the ability of a plant to accumulate a pollutant in its tissues [68]. The highest BCF value (Figure 3) was found in SW1 (42.2). Comparable BCF values were also obtained for SW2 (32.1), S20 (31.0), and S10 (29.5), while for S50, SM20, and SW3, the BCF decreased to half (13–18.2). This fact indicates that bioaccumulation is higher in the case of plants exposed to low Li contents than to high Li contents. BCF values, ranging from 1.16 to 1.96, were reported for *Apocynum pictum* grown in soils contaminated with 50–400 mg/kg Li [54]. High Li BCF values were also observed in *Lolium* sp., cultivated in natural conditions, where the Li content in the groundwater and soil samples ranged from 2.2 to 5.6 µg/L and 9.95 to 10 mg/kg, respectively [37].

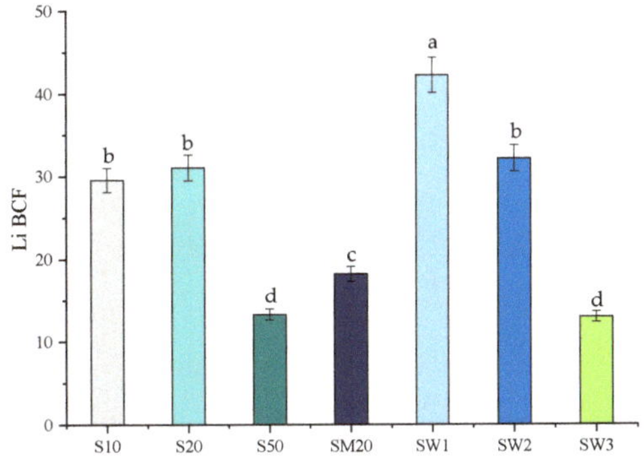

Figure 3. Bioaccumulation factor (BCF) of Li in *S. natans* plants exposed to Hoagland solutions enriched with 10 mg-Li/L (S10), 20 mg-Li/L (S20), 50 mg-Li/L (S50), 20 mg-Li/L, 20 mg-Cu/L, 20 mg-Zn/L, and 20 mg-Cd/L (SM20), and to Banpotoc (SW1), Borcut (SW2) and Botiza (SW3) spring waters. Bioaccumulation factors marked with different letters are significantly different at the $p < 0.05$ significance level.

3.4. Photosynthetic Pigments and Antioxidant Capacity of S. natans after Li Stress Conditions

Exposure to Li led to changes in the photosynthetic pigment content and antioxidant capacity (AC, expressed as µg/mg FW) of *S. natans* (Figure 4). The C_a slowly decreased from 271.8 µg/g FW in the SC to 208.6 and 208.1 µg/g FW in S10 and S20, respectively. The effect of Li on the C_a content can be explained by the interaction between Li and Mg, which have similar physicochemical properties [69]. The C_a concentration in S50 (202.3 µg/g FW) was similar to that of S10 and S20. The lowest value of the C_a concentration was observed in SM20 (79.1 µg/g FW), probably due to the additional stress effects of Cu, Zn, and Cd. Li also had a negative effect on the C_b content in S10, S20, and S50, as well as in SM20. The C_{x+c} concentration slowly decreased with an increase in the Li concentration from S10 to S50. The lowest carotenoid concentration was obtained in the SM20 sample (1.1 µg/g FW).

Noticeably, the concentrations of the photosynthetic pigments in SW1, SW2, and SW3 were lower than the values obtained for the control sample. The C_a concentration decreased from 271.8 µg/g FW in the control sample to 226.1, 211.8, and 95.1 µg/g FW in SW2, SW3, and SW1. The carotenoid concentration was low in SW1 (23.5 µg/g FW), while in SW2 (50.5 µg/g FW) and SW3 (52.2 µg/g FW), it was close to the control sample (65.0 µg/g FW). Generally, in stress conditions, the carotenoid content increases, but under prolonged stress, their content decreases, probably due to damage to the carotenoid biosynthetic pathway [69]. According to the carotenoid content, the W1 spring negatively affected the aquatic plants due to higher Ca, Fe, and Ba concentrations than the W2 and W3. The decrease in the chlorophyll content in the *Apocynum venetum* L. and *Brassica carinata* seedling plants exposed to Li was also reported [70,71]. Aside from Li, other essential and non-essential elements can affect the chlorophyll levels of aquatic plants. A previous study reported decreased chlorophyll and increased carotenoid contents in *Salvinia biloba* after a short time (48 h) exposure to 50 and 100 µM of Cu and Pb. In the same study, exposure to Cd and Zn did not reduce the chlorophyll content [72]. The increase in the carotenoids was attributed to the plant's response to the induced stress through the quench of excess energy protecting the chlorophyll molecules from oxidative damage [72]. In the present study, in SW1, the chlorophyll and carotenoid levels decreased more visibly compared to SW2 and SW3. A possible explanation for this sharp decrease could be that *S. natans* was exposed to a low Li content and high Ca and Ba contents. The reduction in the plant's photosynthetic activity following exposure to high Ba contents was also reported by Dridi et al. [73]. Similarly, a decrease in the photosynthetic pigment level in sunflower (*Helianthus annuus*) and soybean (*Glycine max*) exposed to Ba was reported [74,75]. Soybeans exposed to Ba also showed reduced photosynthetic activity and inhibited K uptake [74]. Therefore, more studies need to be conducted to clarify the competition between Ca, Ba, and Li, and their inhibition effect on the photosynthetic pigment levels.

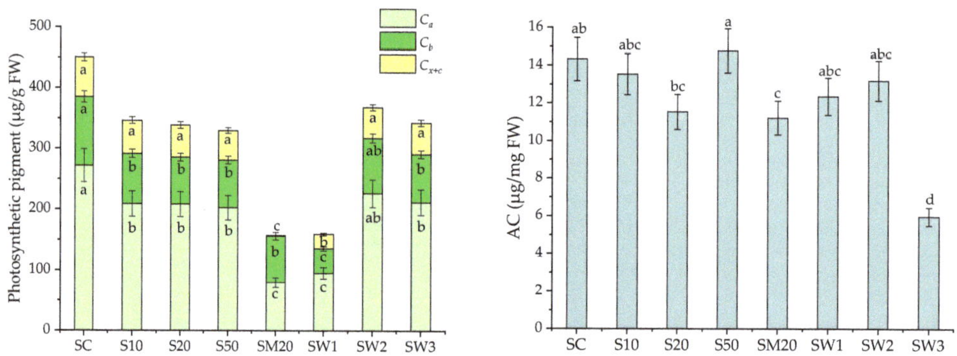

(a) (b)

Figure 4. (**a**) Chlorophyll a (C_a), chlorophyll b (C_b), and carotenoids (C_{x+c}) photosynthetic pigments concentration and (**b**) antioxidant capacity (AC) in *S. natans* grown in control (SC), and Hoagland solution enriched with 10 mg-Li/L (S10), 20 mg-Li/L (S20), 50 mg-Li/L (S50), 20 mg-Li/L, 20 mg-Cu/L, 20 mg-Zn/L, and 20 mg-Cd/L (SM20), and to Banpotoc (SW1), Borcut (SW2) and Botiza (SW3) spring waters. Photosynthetic pigment contents and antioxidant capacity marked with different letters are significantly different at the $p < 0.05$ significance level.

The alterations in the photosynthetic pigment contents could be due to: (i) the low efficiency of the enzymes involved in chlorophyll biosynthesis; (ii) a decrease in the availability of Fe and Mn ions by the inhibition of their uptake in the presence of heavy metals; (iii) the peroxidation of chloroplast membranes during heavy metal stress; (iv) the formation of metal-substituted chlorophylls [61].

No notable differences were observed in the AC between the control (14.3 µg/mg FW) and S10–S50 samples (11.2–14.8 µg/mg FW). In contrast, a notable decrease in the AC was observed in the plants exposed to W3 (2.4-fold smaller than the SC value of 5.98 µg/mg FW). Previously, Upadhyay et al. demonstrated the negative effects of Cu, Zn, and Cd on the AC of *Potamogeton pectinatus* L. and *Potamogeton crispus* L. aquatic plants [76]. According to the obtained results, *S. natans* exposed to the enriched Hoagland solutions and spring waters exhibited small changes in the AC, suggesting the important role of the AC in the abatement of multi-element stress by scavenging free radicals and, hence, protecting the cells from oxidative damage [8]. The negative effect of Li on the AC of *S. natans* was not observed.

3.5. Elemental Composition

The changes in the elemental composition of *S. natans* exposed to different enriched Hoagland solutions and spring waters are shown in Figure 5.

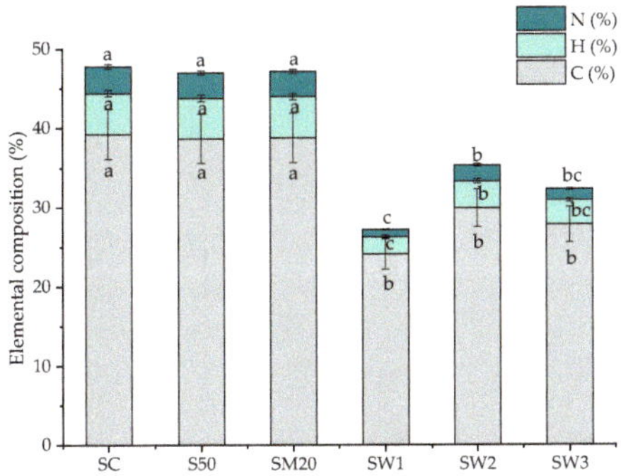

Figure 5. N, C, and H (%) contents in *S. natans* grown in control (SC) and Hoagland solution enriched with 10 mg-Li/L (S10), 20 mg-Li/L (S20), 50 mg-Li/L (S50), 20 mg-Li/L, 20 mg-Cu/L, 20 mg-Zn/L, and 20 mg-Cd/L (SM20), and to Banpotoc (SW1), Borcut (SW2) and Botiza (SW3) spring waters. N, C, and H contents marked with different letters are significantly different at the $p < 0.05$ significance level.

The C, N, and H (%) contents remained unaffected in plants grown on S50 and SM20 compared to the control plants, although a decrease was observed in *S. natans* exposed to spring waters. The S content was below 0.01% in all the studied samples.

Changes in the elemental composition could be attributed to a photosynthetic efficiency reduction that determines an increase in non-photochemical quenching or damage of the reaction center molecules [17] and a decrease in biomass production. The differences observed in the elemental composition of *S. natans* plants exposed to the Hoagland solutions and spring waters could be due to a high amount of non-essential metal ions influencing metal uptake [77].

3.6. FT–IR Spectra

The FT–IR spectra of *S. natans* exposed to Li are presented in Figure 6. The broad band at 3500–3000 cm^{-1} was attributed to the O–H and N–H stretching vibrations [78,79], while the intense peak, around 2920 cm^{-1}, and the shoulder, around 2850 cm^{-1}, are specific to the CH_3 and CH_2 stretching vibrations [78]. The peaks in the 1800–1200 cm^{-1} region are specific to the C=O stretching vibration, indicating the presence of ester-containing compounds commonly found in membrane lipid and cell wall pectin [78].

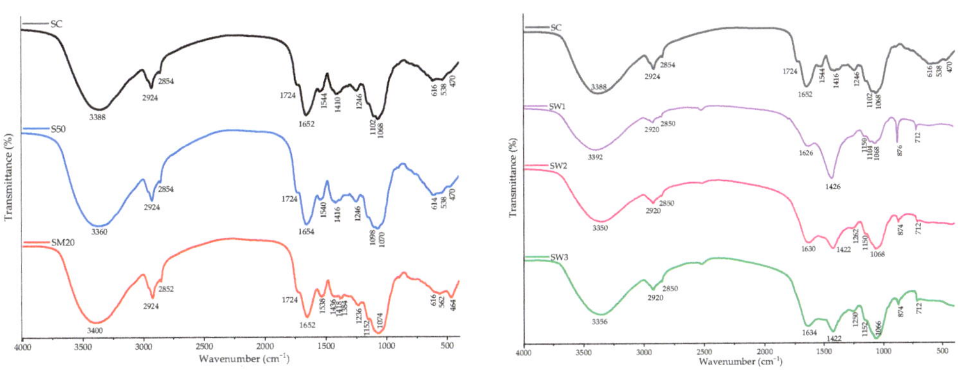

(**a**) (**b**)

Figure 6. FT–IR spectra of *S. natans* grown in control (SC) and Hoagland solution enriched with 50 mg-Li/L (S50), 20 mg-Li/L, 20 mg-Cu/L, 20 mg-Zn/L, and 20 mg-Cd/L (SM20) (**a**) and Banpotoc (SW1), Borcut (SW2) and Botiza (SW3) spring waters (**b**).

The peaks at 1654–1652 cm^{-1}, 1580–1510 cm^{-1}, and 1400–1200 cm^{-1} may also be attributed to amides [78,80,81]. The increase in these peaks could be an indication of the plant's response to stress conditions, as amides are known to have an important role in the plant's defense mechanism [82]. The peak at 1426 cm^{-1} was assigned to the C–H groups [83]. The peaks around 1250 cm^{-1} correspond to the C–O stretch in carboxylic acids, while those at 1130–1000 cm^{-1} correspond to the vibration of C–O–C and O–H in polysaccharides [84]. The peak around 1067 cm^{-1}, observed in the SC, SW1, SW2, and SW3, can be assigned to the C–O stretching of alcohols and carboxylic acids [83]. In the *S. natans* grown in spring waters, a strong peak appeared around 880 cm^{-1}, indicating the linkage of β–glycosides in the molecular structure of polysaccharides [79]. The bands below 800 cm^{-1} are the fingerprint region of the phosphate and sulphur functional groups [83,84].

The FT–IR spectra of S50 were similar to the control, suggesting that the plant that was exposed to Li did not determine the changes in the specific bands. At the simultaneous exposure to Li and Cu, Zn, and Cd, a supplementary band appeared at 1384 cm^{-1}, being assigned to the amides. A peak at 1102 cm^{-1}, attributed to the presence of C–O–C and O–H in the polysaccharides, disappeared, while the intensity of the peak at 464 cm^{-1}, attributed to the phosphate and sulphate functional groups increased, indicating a plant response to the presence of environmental stress and the interactions between Li and the phosphate and sulphate functional groups.

4. Conclusions

S. natans is able to take up and accumulate Li from the nutrient solutions containing low or high Li concentrations and from spring waters. Exposure to a high Li content decreased K and Fe; however, it did not change the Na, Ca, and Mg concentrations. The Li uptake increased with the increase in the Li concentration in the growth medium, up to 20 mg-Li/L, and remained more or less constant at higher concentrations. The accumulation of Li was about two times lower in the case of simultaneous exposure to Li and Cu, Zn, and Cd than in the case of Li exposure. The Li bioaccumulation factor was higher for the plants exposed to a low Li content than for those exposed to high Li contents. Exposure to Li led to a decrease in antioxidant activity and photosynthetic pigments. The photosynthetic pigment content reflected the induced stress effect; therefore, the most affected plants were those simultaneously exposed to Li, Cu, Zn, and Cd.

Author Contributions: Conceptualization, A.I.T., E.A.L. and E.N.; methodology, A.I.T., A.M. and E.N.; analysis, A.I.T., A.M., E.A.L., O.C., A.B. and E.N.; writing—original draft preparation, A.I.T., A.M., E.K., E.A.L. and E.N.; writing—review and editing, A.I.T., E.K., E.N., E.A.L. and O.C. All authors have read and agreed to the published version of the manuscript.

Funding: This work was funded by the Romanian National Core Program, project no. PN 19-18.01.01 (contract no. 18N/8 February 2019).

Institutional Review Board Statement: Not applicable.

Informed Consent Statement: Not applicable.

Data Availability Statement: Not applicable.

Conflicts of Interest: The authors declare no conflict of interest. The funders had no role in the design of the study; in the collection, analyses, or interpretation of data; in the writing of the manuscript; or in the decision to publish the results.

References

1. Rezania, S.; Ponraj, M.; Talaiekhozani, A.; Mohamad, S.E.; Md Din, M.F.; Taib, S.M.; Sabbagh, F.; Sairan, F.M. Perspectives of phytoremediation using water hyacinth for removal of heavy metals, organic and inorganic pollutants in wastewater. *J. Environ. Manag.* **2015**, *163*, 125–133. [CrossRef]
2. Jacob, J.M.; Karthik, C.; Saratale, R.G.; Kumar, S.S.; Prabakar, D.; Kadirvelu, K.; Pugazhendhi, A. Biological approaches to tackle heavy metal pollution: A survey of literature. *J. Environ. Manag.* **2018**, *217*, 56–70. [CrossRef] [PubMed]

3. Han, Y.; Lee, J.; Haiping, G.; Kim, K.-H.; Wanxi, P.; Bhardwaj, N.; Oh, J.-M.; Brown, R.J.C. Plant-based remediation of air pollution: A review. *J. Environ. Manag.* **2022**, *301*, 113860. [CrossRef] [PubMed]
4. Wani, R.A.; Ganai, B.A.; Shah, M.A.; Uqab, B. Heavy metal uptake potential of aquatic plants through phytoremediation technique—A review. *J. Bioremediat. Biodegrad.* **2017**, *8*, 404. [CrossRef]
5. Yan, A.; Wang, Y.; Tan, S.N.; Mohd Yusof, M.L.; Ghosh, S.; Chen, Z. Phytoremediation: A promising approach for revegetation of heavy metal-polluted land. *Front. Plant Sci.* **2020**, *11*, 359. [CrossRef] [PubMed]
6. Khellaf, N.; Djelal, H.; Amrane, A. An Overview of the valorization of aquatic plants in effluent depuration through phytoremediation processes. *Appl. Microbiol.* **2022**, *2*, 309–318. [CrossRef]
7. Emiliani, J.; Oyarce, W.G.L.; Salvatierra, L.; Novo, L.A.B.; Pérez, L.M. Evaluation of Cadmium Evaluation of cadmium bioaccumulation-related physiological effects in Salvinia biloba an insight towards its use as pollutant bioindicator in water reservoirs. *Plants* **2021**, *10*, 2679. [CrossRef]
8. Dhir, B.; Sharmila, P.; Pardha Saradhi, P.; Nasim, S.A. Physiological and antioxidant responses of Salvinia natans exposed to chromium-rich wastewater. *Ecotoxicol. Environ. Saf.* **2009**, *72*, 1790–1797. [CrossRef]
9. Horvat, T.; Vidakovic-Cifrek, Z.; Orescanin, V.; Tkalec, M.; Pevalek-Kozlina, B. Toxicity assessment of heavy metal mixtures by *Lemna minor* L. *Sci. Total Environ.* **2007**, *384*, 229–238. [CrossRef]
10. Torok, A.; Gulyas, Z.; Szalai, G.; Kocsy, G.; Majdik, C. Phytoremediation capacity of aquatic plants is associated with the degree of phytochelatin polymerization. *J. Hazard. Mater.* **2015**, *299*, 371–378. [CrossRef]
11. Dhir, B.; Sharmila, P.; Saradhi, P.P. Potential of Aquatic Macrophytes for Removing Contaminants from the Environment. *Crit. Rev. Environ. Sci. Technol.* **2009**, *39*, 754–781. [CrossRef]
12. Ali, S.; Abbas, Z.; Rizwan, M.; Zaheer, I.E.; Yavaş, İ.; Ünay, A.; Abdel-DAIM, M.M.; Bin-Jumah, M.; Hasanuzzaman, M.; Kalderis, D. Application of floating aquatic plants in phytoremediation of heavy metals polluted water A. Review. *Sustainability* **2020**, *12*, 1927. [CrossRef]
13. Xing, W.; Wu, H.; Hao, B.; Liu, G. Metal accumulation by submerged macrophytes in eutrophic lakes at the watershed scale. *Environ. Sci. Pollut. Res. Int.* **2013**, *20*, 6999–7008. [CrossRef] [PubMed]
14. Espinoza-Quiñones, F.R.; Módenes, A.N.; Thomé, L.P.; Palácio, S.M.; Trigueros, D.E.G.; Oliveira, A.P.; Szymanski, N. Study of the bioaccumulation kinetic of lead by living aquatic macrophyte Salvinia auriculata. *Chem. Eng. J.* **2009**, *150*, 316–322. [CrossRef]
15. Prado, C.; Rosa, M.; Pagano, E.; Prado, F. Metabolic interconnectivity among alternative respiration, residual respiration, carbohydrates and phenolics in leaves of Salvinia minima exposed to Cr(VI). *Environ. Exp. Bot.* **2013**, *87*, 32–38. [CrossRef]
16. Polechońska, L.; Klink, A.; Dambiec, M. Trace element accumulation in Salvinia natans from areas of various land use types. *Environ. Sci. Pollut. Res.* **2019**, *26*, 30242–30251. [CrossRef]
17. Dhir, B.; Srivastava, S. Heavy metal tolerance in metal hyperaccumulator plant, Salvinia natans. *Bull. Environ. Contam. Toxicol.* **2013**, *90*, 720–724. [CrossRef]
18. Rahman, M.A.; Hasegawa, H.; Ueda, K.; Maki, T.; Rahman, M.M. Influence of phosphate and iron ions in selective uptake of arsenic species by water fern (*Salvinia natans* L.). *Chem. Eng. J.* **2008**, *145*, 179–184. [CrossRef]
19. Jampeetong, A.; Brix, H. Nitrogen nutrition of Salvinia natans: Effects of inorganic nitrogen form on growth, morphology, nitrate reductase activity and uptake kinetics of ammonium and nitrate. *Aquat. Bot.* **2009**, *90*, 67–73. [CrossRef]
20. Mukherjee, S.; Kumar, S. Adsorptive uptake of arsenic (V) from water by aquatic fern Salvinia natans. *J. Water Supply Res. Technol.* **2005**, *54*, 47–53. [CrossRef]
21. Kumari, M.; Tripathi, B.D. Effect of aeration and mixed culture of Eichhornia crassipes and Salvinia natans on removal of wastewater pollutants. *Ecol. Eng.* **2014**, *62*, 48–53. [CrossRef]
22. Rápó, E.; Posta, K.; Csavdári, A.; Vincze, B.É.; Mara, G.; Kovács, G.; Haddidi, I.; Tonk, S. Performance comparison of Eichhornia crassipes and Salvinia natans on azo-dye (Eriochrome Black T) phytoremediation. *Crystals* **2020**, *10*, 565. [CrossRef]
23. Szmeja, J.; Gałka, A. Survival and reproduction of the aquatic fern *Salvinia natans* (L.) All. during expansion in the Vistula Delta, south Baltic Sea coast. *J. Freshw. Ecol.* **2013**, *28*, 113–123. [CrossRef]
24. Cui, R.; Nam, S.-H.; An, Y.-J. Salvinia natans: A potential test species for ecotoxicity testing. *Environ. Pollut.* **2020**, *267*, 115650. [CrossRef] [PubMed]
25. Wang, T.; Dou, H.; Liu, C.; Yu, D. Decoupling between plant growth and functional traits of the free-floating fern Salvinia natans under shifted water nutrient stoichiometric regimes. *Flora* **2021**, *281*, 151876. [CrossRef]
26. Dhir, B.; Srivastava, S. Heavy metal removal from a multi-metal solution and wastewater by Salvinia natans. *J. Ecol. Eng.* **2011**, *37*, 893–896. [CrossRef]
27. Mandal, C.; Ghosh, N.; Maiti, S.; Das, K.; Gupta, S.; Dey, N.; Adak, M.K. Antioxidative responses of Salvinia (Salvinia natans Linn.) to aluminium stress and it's modulation by polyamine. *Physiol. Mol. Biol. Plants* **2013**, *19*, 91–103. [CrossRef]
28. Liu, N.; Zhong, G.; Zhou, J.; Liu, Y.; Pang, Y.; Cai, H.; Wu, Z. Separate and combined effects of glyphosate and copper on growth and antioxidative enzymes in *Salvinia natans* (L.) All. *Sci. Total Environ.* **2019**, *655*, 1448–1456. [CrossRef]
29. Parola, V.L.; Liveri, V.T.; Todaro, L.; Lombardo, D.; Bauer, E.M.; Dell'Era, A.; Longo, A.; Caschera, D.; de Caro, T.; Toro, R.G.; et al. Iron and lithium-iron alkyl phosphates as nanostructured material for rechargeable batteries. *Mater. Lett.* **2018**, *220*, 58–61. [CrossRef]

30. Martins, A.; da Silva, D.D.; Silva, R.; Carvalho, F.; Guilhermino, L. Long-term effects of lithium and lithium-microplastic mixtures on the model species Daphnia magna: Toxicological interactions and implications to 'One Health'. *Sci. Total Environ.* **2022**, *838*, 155934. [CrossRef]
31. Bolan, N.; Hoang, S.A.; Tanveer, M.; Wang, L.; Bolan, S.; Sooriyakumar, P.; Robinson, B.; Wijesekara, H.; Wijesooriya, M.; Keerthanan, S.; et al. From mine to mind and mobiles—Lithium contamination and its risk management. *Environ. Pollut.* **2021**, *290*, 118067. [CrossRef] [PubMed]
32. Shahzad, B.; Tanveer, M.; Hassan, W.; Shah, A.N.; Anjum, S.A.; Cheema, S.A.; Ali, I. Lithium toxicity in plants: Reasons, mechanisms and remediation possibilities—A review. *Plant Physiol. Biochem.* **2016**, *107*, 104–115. [CrossRef] [PubMed]
33. Sobolev, O.I.; Gutyj, B.V.; Darmohray, L.M.; Sobolieva, S.V.; Ivanina, V.V.; Kuzmenko, O.A.; Karkach, P.M.; Fesenko, V.F.; Bilkevych, V.V.; Mashkin, Y.O.; et al. Lithium in the natural environment and its migration in the trophic chain. *Ukr. J. Ecol.* **2019**, *9*, 195–203.
34. Kszos, L.A.; Stewart, A.J. Review of lithium in the aquatic environment: Distribution in the United States, toxicity and case example of groundwater contamination. *Ecotoxicology* **2003**, *12*, 439–447. [CrossRef]
35. Bauer, B.; Kavrakovski, Z.; Kostik, V. Lithium content in potable water, surface water, ground water, and mineral water on the territory of Republic of Macedonia. *Int. J. Environ. Res. Public Health* **2014**, *4*, 189–193. [CrossRef]
36. Kavanagh, L.; Keohane, J.; Cleary, J.; Garcia Cabellos, G.; Lloyd, A. Lithium in the natural waters of the south east of Ireland. *Int. J. Environ. Res. Public Health* **2017**, *14*, 561. [CrossRef]
37. Liaugaudaite, V.; Mickuviene, N.; Raskauskiene, N.; Naginiene, R.; Sher, L. Lithium levels in the public drinking water supply and risk of suicide: A pilot study. *J. Trace Elem. Med. Biol.* **2017**, *43*, 197–201. [CrossRef]
38. Aral, H.; Vecchio-Sadus, A. Toxicity of lithium to humans and the environment-a literature review. *Ecotoxicol. Environ. Saf.* **2008**, *70*, 349–356. [CrossRef]
39. Dippong, T.; Hoaghia, M.A.; Mihali, C.; Cical, E.; Calugaru, M. Human health risk assessment of some bottled waters from Romania. *Environ. Pollut.* **2020**, *267*, 115409. [CrossRef]
40. Zsigmond, A.R.; Kékedy-Nagy, L.; Cordoş, E.A.; Măruţoiu, C. Assessment of lithium in transylvanian mineral waters using the platinum-wire loop faes technique. *Environ. Eng. Manag. J.* **2012**, *11*, 1561–1566. [CrossRef]
41. Török, A.I.; Moldovan, A.; Levei, E.A.; Cadar, O.; Tănăselia, C.; Moldovan, O.T. Assessment of lithium, macro- and microelements in water, soil and plant samples from karst areas in Romania. *Materials* **2021**, *14*, 4002. [CrossRef]
42. Naeem, A.; Aslam, M.; Saifullah; Mühling, K.H. Lithium: Perspectives of nutritional beneficence, dietary intake, biogeochemistry, and biofortification of vegetables and mushrooms. *Sci. Total Environ.* **2021**, *798*, 149249. [CrossRef] [PubMed]
43. Shakoor, N.; Adeel, M.; Azeem, I.; Ahmad, M.A.; Zain, M.; Abbas, A.; Zhou, P.; Li, Y.; Ming, X.; Rui, Y. Responses of Agricultural plants to Lithium pollution: Trends, Meta-Analysis, and Perspectives. *bioRxiv* **2022**. [CrossRef]
44. Hoagland, D.R.; Arnon, D.I. The Water-Culture method for growing plants without soil. In *California Agricultural Experiment Station*; University of California, Agricultural Experiment Station: Berkeley, CA, USA, 1938; pp. 29–32.
45. Babenko, L.; Vasheka, O.; Shcherbatiuk, M.; Romanenko, P.; Voytenko, L.; Kosakivska, I. Biometric characteristics and surface microstructure of vegetative and reproductive organs of heterosporous water fern Salvinia natans. *Flora* **2019**, *252*, 44–50. [CrossRef]
46. Gałka, A.; Szmeja, J. Phenology of the aquatic fern *Salvinia natans* (L.) All. in the Vistula Delta in the context of climate warming. *Limnologica* **2013**, *43*, 100–105. [CrossRef]
47. Buta, E.; Török, A.; Zongo, B.; Cantor, M.; Buta, M.; Majdik, C. Comparative studies of the phytoextraction capacity of five aquatic plants in heavy metal contaminated water. *Not. Bot. Horti Agrobot. Cluj-Napoca* **2014**, *42*, 173–179. [CrossRef]
48. Lichtenthaler, H.K. Chlorophylls and carotenoids: Pigments of photosynthetic biomembranes. In *Methods in Enzymology*; Academic Press: Cambridge, MA, USA, 1987; Volume 148, pp. 350–382.
49. Pachura, P.; Ociepa-Kubicka, A.; Skowron-Grabowska, B. Assessment of the availability of heavy metals to plants based on the translocation index and the bioaccumulation factor. *Desalination Water Treat.* **2015**, *57*, 1469–1477. [CrossRef]
50. Zayed, A.; Gowthaman, S.; Terry, N. Phytoaccumulation of Trace Elements by Wetland Plants: I. Duckweed. *J. Environ. Qual.* **1998**, *27*, 715–721. [CrossRef]
51. Hawrylak-Nowak, B.; Kalinowska, M.; Szymańska, M. A Study on selected physiological parameters of plants grown under lithium supplementation. *Biol. Trace Elem. Res.* **2012**, *149*, 425–430. [CrossRef]
52. Kalinowska, M.; Hawrylak-Nowak, B.; Szymańska, M. The influence of two lithium forms on the growth, l-ascorbic acid content and lithium accumulation in lettuce plants. *Biol. Trace Elem. Res.* **2013**, *152*, 251–257. [CrossRef] [PubMed]
53. McStay, N.G.; Rogers, H.H.; Anderson, C.E. Effects of lithium on *Phaseolus vulgaris* L. *Sci. Total Environ.* **1980**, *16*, 185–191. [CrossRef]
54. Das, S.; Mazumdar, K. Phytoremediation potential of a novel fern, Salvinia cucullata, Roxb. Ex Bory, to pulp and paper mill effluent: Physiological and anatomical response. *Chemosphere* **2016**, *163*, 62–72. [CrossRef] [PubMed]
55. Favas, P.J.C.; Pratas, J.; Rodrigues, N.; D'Souza, R.; Varun, M.; Paul, M.S. Metal(loid) accumulation in aquatic plants of a mining area: Potential for water quality biomonitoring and biogeochemical prospecting. *Chemosphere* **2018**, *194*, 158–170. [CrossRef] [PubMed]
56. Kabata-Pendias, A.; Mukherjee, A.B. Trace Elements of Group 1 (Previously Group Ia). In *Trace Elements from Soil to Human*; Kabata-Pendias, A., Mukherjee, A.B., Eds.; Springer Berlin Heidelberg: Berlin/Heidelberg, Germany, 2007; pp. 87–104. [CrossRef]

57. Suñe, N.; Sánchez, G.; Caffaratti, S.; Maine, M.A. Cadmium and chromium removal kinetics from solution by two aquatic macrophytes. *Environ. Pollut.* **2007**, *145*, 467–473. [CrossRef] [PubMed]
58. Ranade-Malvi, U. Interaction of micronutrients with major nutrients with special reference to potassium. *Karnataka J. Agric. Sci.* **2011**, *24*, 106–109.
59. Hemantaranjan, A. *Advances in Plant Physiology (Vol. 18)*; Scientific Publishers: Jodhpur, India, 2019; p. 365.
60. Robinson, B.H.; Yalamanchali, R.; Reiser, R.; Dickinson, N.M. Lithium as an emerging environmental contaminant: Mobility in the soil-plant system. *Chemosphere* **2018**, *197*, 1–6. [CrossRef] [PubMed]
61. Dhir, B.; Sharmila, P.; Pardha Saradhi, P.; Sharma, S.; Kumar, R.; Mehta, D. Heavy metal induced physiological alterations in Salvinia natans. *Ecotoxicol. Environ. Saf.* **2011**, *74*, 1678–1684. [CrossRef] [PubMed]
62. Reeves, R.D.; Baker, A.J.M. Metal-Accumulating Plants. In *Phytoremediation of Toxic Metals-Using Plants to Clean Up the Environment*; Raskin, I., Ensley, B.D., Eds.; John Wiley and Sons: New York, NY, USA, 2000; pp. 193–229.
63. Adamczyk-Szabela, D.; Lisowska, K.; Romanowska-Duda, Z.; Wolf, W.M. Combined cadmium-zinc interactions alter manganese, lead, copper uptake by Melissa officinalis. *Sci. Rep.* **2020**, *10*, 1675. [CrossRef]
64. Li, J.; Yu, H.; Luan, Y. Meta-analysis of the copper, zinc, and cadmium absorption capacities of aquatic plants in heavy metal-polluted water. *Int. J. Environ. Res. Public Health* **2015**, *12*, 14958–14973. [CrossRef] [PubMed]
65. Samecka-Cymerman, A.; Kempers, A.J. Concentrations of heavy metals and plant nutrients in water, sediments and aquatic macrophytes of anthropogenic lakes (former open cut brown coal mines) differing in stage of acidification. *Sci. Total Environ.* **2001**, *281*, 87–98. [CrossRef]
66. Zhang, J.; Yang, N.; Geng, Y.; Zhou, J.; Lei, J. Effects of the combined pollution of cadmium, lead and zinc on the phytoextraction efficiency of ryegrass (*Lolium perenne* L.). *RSC Adv.* **2019**, *9*, 20603–20611. [CrossRef] [PubMed]
67. Dong, Q.; Hu, S.; Fei, L.; Liu, L.; Wang, Z. Interaction between Cd and Zn on Metal Accumulation, Translocation and Mineral Nutrition in Tall Fescue (*Festuca arundinacea*). *Int. J. Mol. Sci.* **2019**, *20*, 3332. [CrossRef] [PubMed]
68. Krayem, M.; Khatib, S.E.; Hassan, Y.; Deluchat, V.; Labrousse, P. In search for potential biomarkers of copper stress in aquatic plants. *Aquat. Toxicol.* **2021**, *239*, 105952. [CrossRef] [PubMed]
69. Liliana, C.; Inga, Z.; Ludmila, R.; Tatiana, C.; Ana, P.; Andrei, A.; Svetlana, D.; Larisa, G.; Decebal, I. Biomass of Arthrospira platensis enriched with lithium by bioaccumulation and biosorption process. *Food Biosci.* **2021**, *41*, 100950. [CrossRef]
70. Jiang, L.; Wang, L.; Mu, S.-Y.; Tian, C.-Y. Apocynum venetum: A newly found lithium accumulator. *Flora* **2014**, *209*, 285–289. [CrossRef]
71. Li, X.; Gao, P.; Gjetvaj, B.; Westcott, N.; Gruber, M.Y. Analysis of the metabolome and transcriptome of Brassica carinata seedlings after lithium chloride exposure. *Plant Sci.* **2009**, *177*, 68–80. [CrossRef]
72. Emiliani, J.; Llatance Oyarce, W.G.; Bergara, C.D.; Salvatierra, L.M.; Novo, L.A.B.; Pérez, L.M. Variations in the phytoremediation efficiency of metal-polluted water with salvinia biloba: Prospects and toxicological impacts. *Water* **2020**, *12*, 1737. [CrossRef]
73. Dridi, N.; Bouslimi, H.; Duarte, B.; Caçador, I.; Sleimi, N. Evaluation of physiological and biochemical parameters and some bioindicators of barium tolerance in *Limbarda crithmoides* and *Helianthus annuus*. *Int. J. Plant Biol.* **2022**, *13*, 115–131. [CrossRef]
74. Suwa, R.; Jayachandran, K.; Nguyen, N.T.; Boulenouar, A.; Fujita, K.; Saneoka, H. Barium toxicity effects in soybean plants. *Arch. Environ. Contam. Toxicol.* **2008**, *55*, 397–403. [CrossRef]
75. Marisamy, K.; Duraipandian, M.; Sevugaperumal, R.; Ramasubramanian, V. Barium toxicity induced changes on growth, photosynthetic, biochemical and antioxidative enzymes level in *helianthus annuus* (L.). *Int. J. Recent Sci. Res.* **2016**, *7*, 9781–9787.
76. Upadhyay, A.K.; Singh, N.K.; Rai, U.N. Comparative metal accumulation potential of *Potamogeton pectinatus* L. and *Potamogeton crispus* L.: Role of enzymatic and non-enzymatic antioxidants in tolerance and detoxification of metals. *Aquat. Bot.* **2014**, *117*, 27–32. [CrossRef]
77. Dhir, B.; Srivastava, S. Disposal of metal treated Salvinia biomass in soil and its effect on growth and photosynthetic Efficiency of wheat. *Int. J. Phytoremediation* **2012**, *14*, 24–34. [CrossRef] [PubMed]
78. Ertani, A.; Francioso, O.; Ferrari, E.; Schiavon, M.; Nardi, S. Spectroscopic-chemical fingerprint and biostimulant activity of a protein-based product in solid form. *Molecules* **2018**, *23*, 1031. [CrossRef]
79. Wu, W.; Zhu, Y.; Zhang, L.; Yang, R.; Zhou, Y. Extraction, preliminary structural characterization, and antioxidant activities of polysaccharides from Salvia miltiorrhiza Bunge. *Carbohydr. Polym.* **2012**, *87*, 1348–1353. [CrossRef]
80. Sadat, A.; Joye, I.J. Peak Fitting Applied to Fourier Transform Infrared and Raman Spectroscopic Analysis of Proteins. *Appl. Sci.* **2020**, *10*, 5918. [CrossRef]
81. Chen, Y.; Chen, J.; Chang, C.; Chen, J.; Cao, F.; Zhao, J.; Zheng, Y.; Zhu, J. Physicochemical and functional properties of proteins extracted from three microalgal species. *Food Hydrocoll.* **2019**, *96*, 510–517. [CrossRef]
82. Kumar, V.; Bhatt, V.; Kumar, N. Chapter 9—Amides From Plants: Structures and Biological Importance. In *Studies in Natural Products Chemistry*; Atta ur, R., Ed.; Elsevier: Amsterdam, Netherlands, 2018; Volume 56, pp. 287–333.
83. Cajamarca, F.A.B.; Corazza, M.Z.; Prete, M.C.; Dragunski, D.C.; Rocker, C.; Caetano, J.; Gonçalves, J.A.C.; Tarley, C.R.T. Investigation on the performance of chemically modified aquatic macrophytes—Salvinia molesta for the micro-solid phase preconcentration of Cd(II) on-line coupled to FAAS. *Bull. Environ. Contam. Toxicol.* **2016**, *97*, 863–869. [CrossRef]
84. Lima, L.K.S.; Silva, M.G.C.; Vieira, M.G. Study of binary and single biosorption by the floating aquatic macrophyte Salvinia natans. *Braz. J. Chem. Eng.* **2016**, *33*, 649–660. [CrossRef]

materials

MDPI

Article

Mechanical Performance of Fly Ash Based Geopolymer (FAG) as Road Base Stabilizer

Liyana Ahmad Sofri [1,2,*], Mohd Mustafa Al Bakri Abdullah [2,3], Andrei Victor Sandu [4,5,6,*], Thanongsak Imjai [7], Petrica Vizureanu [4,8], Mohd Rosli Mohd Hasan [9], Mohammad Almadani [10], Ikmal Hakem Ab Aziz [2,3] and Farahiyah Abdul Rahman [1]

1 Faculty of Civil Engineering & Technology, Universiti Malaysia Perlis (UniMAP), Arau 02600, Malaysia
2 Centre of Excellence Geopolymer and Green Technology, (CEGeoGTech), Universiti Malaysia Perlis (UniMAP), Arau 01000, Malaysia
3 Faculty of Chemical Engineering & Technology, Universiti Malaysia Perlis (UniMAP), Arau 01000, Malaysia
4 Faculty of Material Science and Engineering, Gheorghe Asachi Technical University of Iasi, 41 D. Mangeron St., 700050 Iasi, Romania
5 Romanian Inventors Forum, Str. Sf. P. Movila 3, 700089 Iasi, Romania
6 National Institute for Research and Development for Environmental Protection INCDPM, 294 Splaiul Independentei, 060031 Bucharest, Romania
7 Center of Excellence in Sustainable Disaster Management, School of Engineering and Technology, Walailak University, Nakhonsithammarat 80161, Thailand
8 Technical Sciences Academy of Romania, Dacia Blvd. 26, 030167 Bucharest, Romania
9 School of Civil Engineering, Engineering Campus, Universiti Sains Malaysia, Penang 14300, Malaysia
10 Department of Civil Engineering, Faculty of Engineering—Rabigh Branch, King Abdulaziz University, Jeddah 21589, Saudi Arabia
* Correspondence: liyanasofri@unimap.edu.my (L.A.S.); sav@tuiasi.ro (A.V.S.)

Citation: Sofri, L.A.; Abdullah, M.M.A.B.; Sandu, A.V.; Imjai, T.; Vizureanu, P.; Hasan, M.R.M.; Almadani, M.; Aziz, I.H.A.; Rahman, F.A. Mechanical Performance of Fly Ash Based Geopolymer (FAG) as Road Base Stabilizer. *Materials* **2022**, *15*, 7242. https://doi.org/10.3390/ma15207242

Academic Editors: Roberta G. Toro, Daniela Caschera and Abeer M. Adel

Received: 25 August 2022
Accepted: 13 October 2022
Published: 17 October 2022

Abstract: This study examines the strength development of fly ash-based geopolymer (FAG) as a stabilizer for road base material for pavement construction. In the last decade, there has been a rapid development of conventionally treated bases, such as cement-treated bases. However, a major problem with this kind of application is the shrinkage cracking in cement-treated bases that may result in the reflection cracks on the asphalt pavement surface. This study explores the effects of FAG on base layer properties using mechanistic laboratory evaluation and its practicability in pavement base layers. The investigated properties are flexural strength (FS), unconfined compressive strength (UCS), shrinkage, and resilient modulus (RM), as well as indirect tensile strength (ITS). The findings showed that the mechanical properties of the mixture enhanced when FAG was added to 80–85% of crushed aggregate, with the UCS being shown to be a crucial quality parameter. The effectiveness of FAG base material can have an impact on the flexible pavements' overall performance since the base course stiffness directly depends on the base material properties. As a stabilizing agent for flexible pavement applications, the FAG-stabilized base appeared promising, predicated on test outcomes.

Keywords: geopolymer; fly ash; base course material; road base stabilization; unconfined compressive strength

1. Introduction

The world has seen the tremendous expansion of road networks in emerging countries during the last several years. However, this rapid expansion has a serious implication since constructing roads consumes many crushed or virgin aggregates, especially for the development of base layers for roads [1]. A cement-treated base is an example of a stabilized base layer [2]. It consists of compacted mixtures of granular materials, Portland cement, and water. In current years, applying a cement-stabilized base has led to pavement damage, induced by the cement hydration process, as well as moisture loss [3]. Every year, cement output increases globally, and there are no signs that it will ever decline [4]. Since 1950, global cement production has multiplied 30 times, and nearly four-fold since 1990, making it the third largest source of anthropogenic carbon dioxide (CO_2) emissions, after fossil fuels and land-use changes [5]. The production of cement requires enormous amounts of natural resources, for example, limestone, natural gas, fossil fuels, and electricity. High temperatures are necessary to manufacture cement in power plants, which results in 8 to 10 percent of carbon dioxide (CO_2) being emitted into the environment [6]. The usage of cement may be reduced by utilizing several pozzolanic raw materials, including palm oil, rice husk ash, fly ash (FA) [7], fuel ash, bagasse ash [8], and phosphogypsum [9].

Geopolymer is a well-known environmentally friendly substance. It has been shown to have good characteristics. A geopolymer precursor material is composed of silicon (Si) and aluminum (Al) [10]. After a few steps, it would react with extremely alkaline activator solutions to produce a geopolymer binder [11]. With an alkaline activator solution, materials including blast furnace slag, metakaolin, and FA may be activated, leading to the development of geopolymerization [12]. Raw pozzolanic materials, for instance, FA, are widely used in geopolymer cement and concrete as aluminosilicate raw materials [13–18]. It can be applied as a replacement for or in addition to regular Portland cement in a variety of ways. FA application in cement provides benefits from an environmental and economic point of view since it possesses a zero-cost industrial waste that must be disposed of [19].

The most reliable primary raw material for developing alkali-activated materials is FA, whose activation has been the subject of extensive study over the past few decades [20]. It would positively impact the reduction of cement production and the development of sustainable building materials, as FA is a hazardous waste of electrical steam generated by burning coal [21]. Other widely used raw materials are kaolin, metakaolin, and bottom ash [22]. Previous studies have indicated that FA can also be utilized as a soil stabilizer for the sub-base layer in constructing roads [23]. FA is also excellent material for producing geopolymer concrete as a replacement for cement, and is suitable to be used as a road base stabilizer.

Studies have been carried out continuously to develop FA-based geopolymer (FAG) as an alternative binder. The strength, chemistry, morphological, and adhesion properties of FAG used as an asphalt binder modifier were investigated by Atmaja et al. [24]. Class F of FA was mixed with alkaline solutions (i.e., sodium hydroxide (NaOH) and sodium silicate (Na_2SiO_3)) to prepare the geopolymer. The modified bitumen's impact values were higher than the base bitumen after adding the geopolymer. Overall, it produced the best results, and the ideal concentration for bitumen modification was 5% geopolymer. Dayal & Soundarapandian [25] investigated the properties of geopolymer-coated aggregates made of FA and their impact on the properties of bituminous mixes. The addition of a geopolymer paste based FA to a natural aggregate improved the physical and mechanical qualities of marginal aggregates, allowing them to fulfill the pavement construction requirements. According to the findings, geopolymers made with high calcium (Ca) content FA resulted in significantly enhanced aggregate strength properties in bituminous concrete mixtures produced with geopolymer-coated aggregates formed with FA.

In a separate study by Hu et al. [26], class F fly ash and red mud were combined with a sodium hydroxide and sodium silicate solution to generate a geopolymer stabilized base. According to the findings of the tests, both geopolymer stabilized bases demonstrated potential for use in flexible pavement applications at room temperature. The unconfined compressive strength of the fly ash geopolymer base material improved by 23 percent, from 13.1 MPa to 17.1 MPa. Sermsak et al. [27] reported that the 7 day unconfined compressive strength of recycled concrete aggregate, fly ash geopolymer improved as the fly ash (FA) content increased. This is because FA silica and alumina can react with calcium, resulting in pozzolan and geopolymerization products, which met the requirements for a low-traffic road. The SEM images also revealed that spherical FA and recycled concrete aggregate particles are covalently bonded, resulting in a dense matrix.

Based on the information presented above, an investigation was performed to evaluate the effects of FAG on mechanical properties as a road base stabilizer. Although several studies have been undertaken to evaluate the mechanical properties of this stabilizer, there is a lack of investigation comprising flexural strength (FS), shrinkage, unconfined compressive strength (UCS), resilient modulus (RM), and indirect tensile strength (ITS). The purpose of this study was to determine whether it would be possible and practical to make use of this material, as well as to examine the probable existence of this material in the FAG.

2. Materials and Methods

2.1. Fly Ash (FA)

Fly ash (FA) employed in the research was acquired from Manjung Coal-Fired Power Station located in Lumut, Perak, Malaysia. The FA's color was brown, and the fundamental chemical composition of FA was SiO_2 (30.8%), Al_2O_3 (13.1%), Fe_2O_3 (22.99%), and CaO (22.3%), which were identified via X-ray fluorescence (XRF). The FA microstructure diagram and chemical composition have been demonstrated in Figure 1 and Table 1 accordingly.

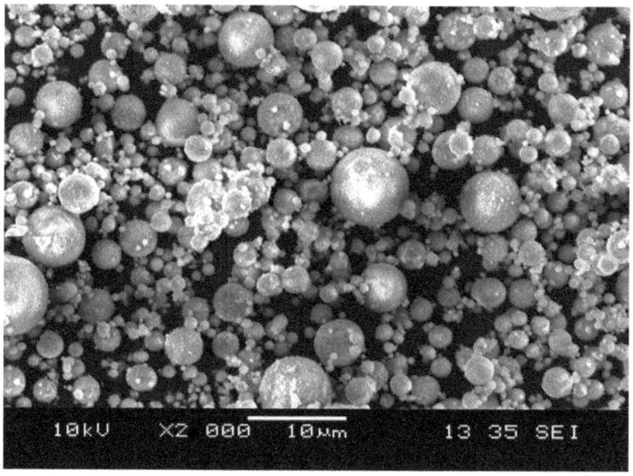

Figure 1. Morphology of the FA particles.

Table 1. The FA chemical composition (%).

Composition	Weight (%)
SiO_2	30.80
CaO	22.30
Fe_2O_3	22.99
Al_2O_3	13.10
MgO	4.00
K2O	1.60
TiO_2	0.89
SO_3	2.67
MnO	0.21
LOI	1.44

2.2. Crushed Aggregate

Crushed aggregate was obtained from Pens Industries Perlis, Malaysia. Figure 2 shows the crushed aggregate's gradation within the upper and lower limit and meets the requirement for pavement base material by the Public Work Department (JKR), Malaysia. Note that crushed aggregate typically had nominal aggregate particles ranging roughly from quarry dust to 38 mm in size. Crushed aggregate typically had nominal aggregate particles ranging roughly from quarry dust to 38 mm in size. Los Angeles (LA) abrasion test was also conducted on a crushed aggregate sample following ASTM C131 [28]. The LA abrasion value was 25.8%. Meanwhile, the aggregate crushing value and aggregate crushing value test are 10.3% and 18.5%, respectively. The water absorption of crushed aggregate is 1.65%. These values indicated that the crushed aggregate meets the requirement for stabilized base materials of the JKR Malaysia.

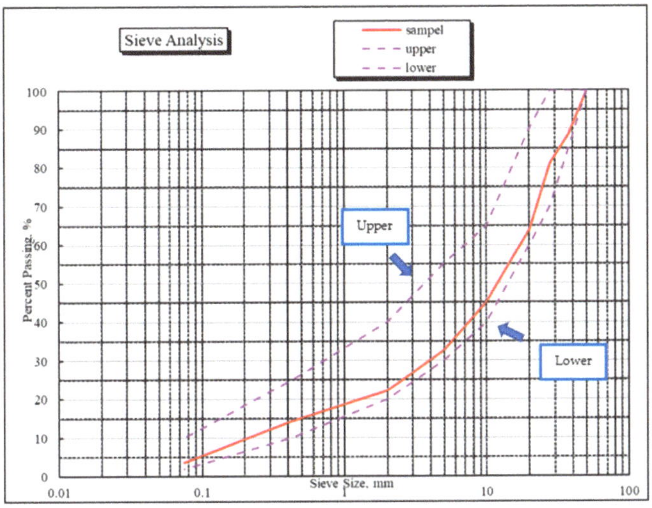

Figure 2. Upper and lower limit for crushed aggregate.

2.3. Specimen Preparation

Prior to combining with FA, an alkaline activator, including sodium hydroxide (NaOH) and sodium silicate (Na_2SiO_3), was produced. In an automated mixer, the alkaline activator and FA were combined to form a homogenous paste. The technique began with a 5 min mixture of Na_2SiO_3 and NaOH solution. Next, FA was added to the basin and stirred for another 5 min. The Na_2SiO_3 to NaOH ratio was 2.5 and was fixed from previous research by Mustafa et al. [29] while the solid to liquid (S/L) ratio was 2.0. NaOH molarities of 10 M. For geopolymer sample preparation, alkali solutions with a concentration of 10 M were utilized. These concentrations were chosen based on past work and other works of literature [30].

2.4. Mixing Process of Fly Ash Geopolymer Concrete

The fly ash, alkali activator, and crushed aggregates were mixed until homogeneous based on the selected ratio of geopolymer paste. The FA-based geopolymer concrete was molded and compacted in a mold for several tests. The coarse and fine aggregates were taken as 80% and 85% of the whole mixture by mass. The coarse and fine aggregates utilized in this research are 45% and 55%, correspondingly, of the total aggregate. Subsequently, the remainder is the combined volume of geopolymer binder, FA, and alkaline solution. The mixed design details with different aggregate percentages of 80% (A80/G20) and 85% (A85/G15) are presented in Table 2. According to the Portland Cement Association, the amount of total aggregate used in the mixture is typically greater than 80% by mass. To obtain the maximum dry density for 95% compaction (i.e., a commonly used compaction requirement for road pavement construction), 0.5%, and 3.5% water were added to the mixtures of A80/G20 and A85/G15, respectively, for preparing the samples of unconfined compressive strength (UCS), flexural strength (FS), shrinkage, indirect tensile strength (ITS), and resilient modulus (RM).

Table 2. Mix proportion of FAG concrete having different aggregate percentages.

Sample Name	NaOH Concentration (M)	S/L Ratio	Na_2SiO_3/NaOH Ratio	Percentage of Aggregate (A)	Percentage of FAG (G)
A80/G20	10 M	2.0	2.5	80	20
A85/G15			2.5	85	15

2.5. Unconfined Compressive Strength (UCS) Test

The compressive strength of a mixture is assessed using the UCS test. The mixture was made using a cylindrical plastic mold that was 38 mm (diameter) × 76 mm (height) for the paste, and 101.60 mm (diameter) × 116.4 mm (height) for the FAG concrete, as per ASTM D 1633 [31]. Utilizing a hydraulic compressive strength machine, a load was continuously applied within the range of 140 to 70 kPa/s as required by ASTM D 1633 to measure the average UCS of the specimens post 7, 28, and 90 days of curing time. In addition, to simulate the exact ambient temperature in a pavement application, samples curing at room temperature were assessed in this study. To obtain the Maximum Dry Density for 95% compaction (a commonly used compaction requirement for road construction), 0.5% and 3.5% of water were added to the mixtures of A80/G20 and A85/G15, respectively, for preparing the samples of UCS and other testing in this research.

2.6. Flexural Strength (FS) Test

The standard test procedure ASTM C78 was used to determine the FS, which is represented as the modulus of rupture [32]. FS is measured using the usual dimensions of specimens (400 × 100 × 100) mm^3. At a one-third distance from both beam supports, equal weights are applied. Note that FS is the maximum tensile stress attained when loading increases if a fracture occurs within the center third of the beam.

2.7. Shrinkage Test

The shrinkage experiment is performed at room temperature following ASTM C490 [33]. The beam specimens were chosen for the experiment, about (300 × 75 × 75) mm^3. Correspondingly, a strain gauge is pasted on both sides of the specimen to test the straining change. Dial gauges with 0.01 mm revolutions were used to collect measurements for dimensional changes, which were then divided by the height of the specimens to obtain shrinkage or swelling values. The initial reading, known as the zero reading, was used as a standard for subsequent readings, and dial gauges were adjusted to reflect this reading. Anticlockwise readings indicate shrinking. Measurements of shrinkage values were obtained at various intervals, with more readings taken early in the curing period.

2.8. Indirect Tensile Strength (ITS) Test

The ITS test was carried out following ASTM D6931 [34]. A standard cylinder specimen with 101.6 mm (diameter) × 65 mm (height) is placed horizontally between the loading surfaces of the compression testing machine. Stabilized specimens underwent indirect tensile testing with an enforced strain rate of 1 mm/min. In this test, a compressive loading condition is applied along two axial lines to the cylindrical specimen. A constant rate of the load is used to cause failure in tension. Figure 3 shows that the load is spread out by two bearing strips to stop multiple cracks and crush at the point of load.

Figure 3. Indirect Tensile Strength Test.

2.9. Resilient Modulus (RM) Test

The RM test, as portrayed in Figure 4, has the potential to examine the materials relative quality to get a result for pavement analysis and design. Temperature, loading rate, rest intervals, and frequencies all have an impact on this test procedure. A cylindrical metal specimen that has 63.5 ± 2.5 mm height $\times$ 101.60 mm internal diameter was utilized for the mixture's preparation. The RM of the mixtures was determined using the ASTM D 4123 [35] repeated-load indirect tension test by applying a loading frequency of 1.0 Hz and 2000 N compressive loads having a waveform at 25 °C (the appropriate load range might be 10–50% of the ITS).

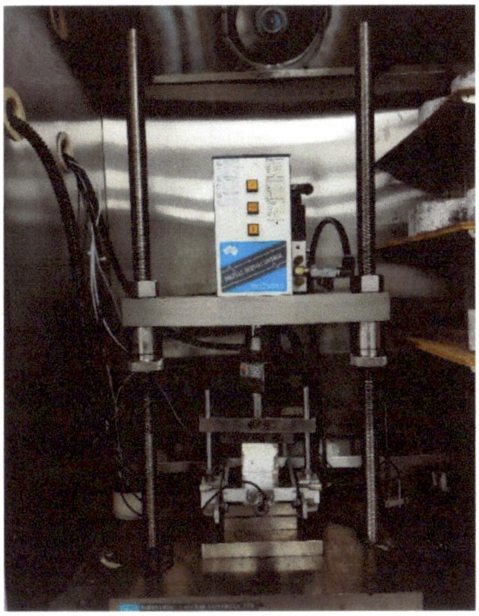

Figure 4. Resilient Modulus.

3. Results and Discussions

3.1. Unconfined Compressive Strength (UCS)

The traffic loads transferred from the asphalt surface course are sustained by the base course layer, which serves as the primary load-bearing course in pavement constructions. Therefore, strength is the primary characteristic of the base layer. It is commonly acknowledged that the unconfined compressive strength (UCS) is a crucial signal for determining the tensile and shear characteristics of the base layer. The UCS of FAG concrete is measured at aging days of 7, 28, and 90, as presented in Figure 5. The graph presents FAG concrete with different total aggregate content, named A80/G20 and A85/G15. UCS values for 7, 28, and 90 days cured specimens were determined to be within 10.3 to 34.9 MPa for both mixtures.

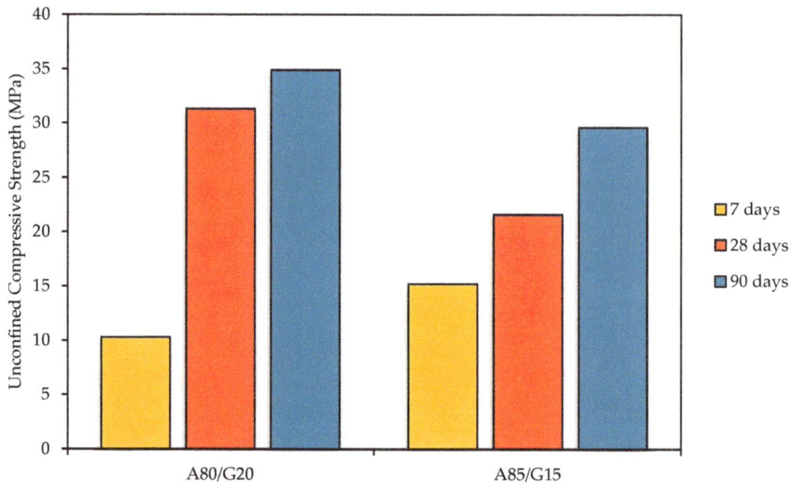

Figure 5. Unconfined Compressive Strength of FAG A80/G20 and A85/G15 mixtures at 7, 28, and 90 days.

The lowest UCS (10.3 MPa) was for A80/G20 at 7 days, which has 80% crushed aggregate at the geopolymer content of 20%. In contrast, the mixture with the same content has the highest value of UCS (34.9 MPa) at 90 days. A mixture of A85/G15 has a range of strength values of 15.2 MPa at 7 days to 29.6 MPa at 90 days. The minimum requirement to be fulfilled is 5 MPa for 7 days and 10 MPa for 28 days, according to Public Work Department (JKR), Malaysia for a stabilized base [36]. The results indicated that the strength value at A85/G15 was lower than A80/G20 at 28 days and 90 days.

It is believed the result had related to the presence of additional water of 3.5% from the compaction test to the mixtures of A85/G15 and had caused a reduction in the availability of geopolymer matrix. From this figure, it is evident that adding water content significantly reduces the strength. This result was supported by Aliabdo et al. [37]. The mixtures containing 15% (A85/G15) alkaline activator with the addition of water exhibited comparatively less strength than those that contained 20% (A80/G20) of the alkaline activator with the addition of water. When more water was added to the A85/G15 mixes, it raised the water-to-solid ratio (w/s) and reduced the alkaline activator solution's concentration, resulting in a drop in strength.

The UCS increased at both mixtures due to increased chemical reactions during the geopolymerization process. FA contains high silica (Si) and alumina (Al) content, as discussed in the chemical composition analysis. In geopolymerization, FA serves as a source of reactive Si and Al to form silicate and aluminate hydrates, which are important for developing strength. The high presence of Si and Al in FA reacted with calcium

(Ca) and generated calcium aluminate silicate hydrate (C-A-S-H) eventually improve its mechanical properties.

The improvement in UCS with crushed aggregate in the mixture is due to improved load transmission in the road base between the mixture's particles. More precisely, greater aggregate percentages generate higher degrees of friction and interlocking processes between aggregate particles and the geopolymer binder. As a result, the pavement's capacity to withstand strong traffic loads has been much improved, both in the magnitude and frequency of wheel loads.

The UCS of FAG concrete is around 50% greater than that reported by Dong et al. [38] who indicated that the UCS of cement-stabilized base materials varied from 5 to 7 MPa after 7 days.

3.2. Flexural Strength (FS)

The capacity of a material to withstand deformation under applied stress is measured by its flexural strength (FS) (or rupture modulus). The test findings of FS are presented in Figure 6. Other than that, the FS of the mixtures changed substantially under three different curing times of 7, 28, and 90 days, as indicated by the data displayed. It has been discovered that the pattern of variation in UCS and the pattern of variation in FS are largely consistent. The FS of FAG concrete base materials constantly rises as curing times increase, although at varying rates.

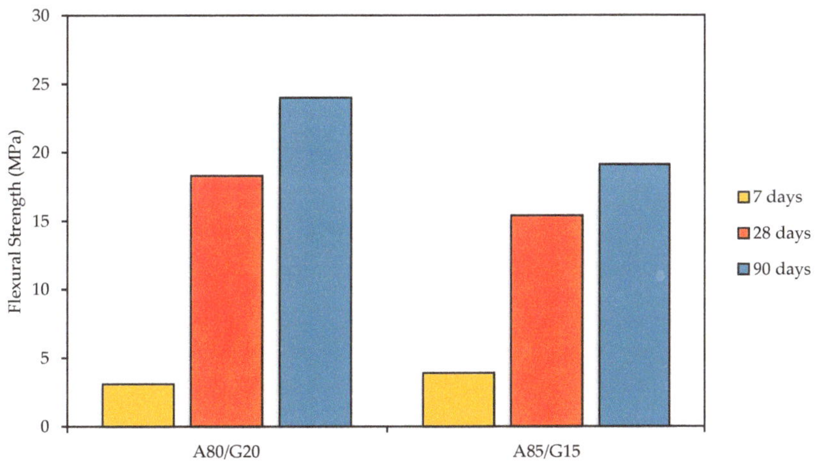

Figure 6. Flexural Strength for FAG A80/G20 and A85/G15 mixtures at 7, 28, and 90 days.

For different curing days, the strength of A80/G20 mixes ranges from 3.1 MPa to 24.0 MPa, whereas the strength of A85/G15 mixtures ranges from 3.9 MPa to 19.1 MPa. FS boosted promptly, specifically in the 7 to 28 day curing periods, while it most likely increased slowly throughout the curing periods of 28 to 90 days. In the standard procedure of stabilized base from the JKR Malaysia, the FS requirement is 2.0 MPa at 28 days [36].

The FS value of A85/G15 mixtures was lower than A80/G20 mixtures due to an increase in aggregate content. The possible reason is the weak interface bond strength between geopolymer binders and crushed aggregate for A85/G15 compared to A80/G20 mixtures, as seen in Figure 6. From the presented graph, the age effect of FAG concrete on the FS is due reactivity of the aluminosilicate precursors of the geopolymer gel, which is resulted from the substantially fast geopolymerization process.

A mixture of A85/G15 at 28 and 90 days is related to the presence of 3.5% additional water and has caused the reduction of the availability of the geopolymer matrix. The strength gradually diminished as a result of its increased water-to-solids ratio (w/s) and decreased alkaline activator solution concentration. Even with just 0.5% of water and an alkaline activator, the strength increased significantly. A comparatively less strength was displayed by the mixtures comprising 15% alkaline activator and 3.5% water compared to those with 20% alkaline activator and 0.5% water.

Furthermore, the rapid increase in flexural strength of FAG was approximately 83% after curing between 7 and 14 days. Additionally, it was demonstrated that aging time has a significant impact on how well a geopolymer develops its strength. This is caused by the FA's high iron (Fe_2O_3) concentration, which is about 22.99%. In the geopolymer system, the reaction between Fe_2O_3 FA and the alkali solution takes some time before the iron silicate binder gel is produced. Compared to Si and Al, iron has a large atomic diameter and a high atomic mass. Therefore, the geopolymer's resilience is aided by ferrosialate in addition to the sodium aluminosilicate gel. As a result of the prolonged curing time, the strength of concrete supplied by iron oxide has increased [39].

The treated layers in the pavement system are subject to flexural stress when a load is applied. Hence, the FS of FAG concrete mixtures is a crucial quality. The treated layers must be considered for fatigue cracking, much like the repetitive load application in the pavement system. The assessment of fatigue life for such mixtures is heavily influenced by flexural stiffness. Note that FS is an essential performance parameter for assessing the crack resistance of a cementitious-stabilized layer subjected to loading.

3.3. Shrinkage Analysis

Shrinkage increases rapidly during the first 7 days, and the increase rate is slower in the subsequent days. From Figure 7, it is reasonable to conclude that the drying shrinkage shows an increment in the drying period for the type of blend mixtures that were tested, ranging from 0 to 90 days.

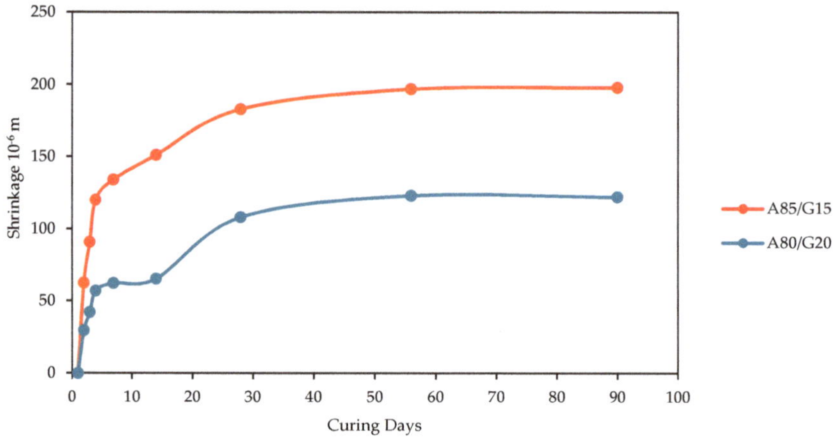

Figure 7. Shrinkage value for A80/G20 and A85/G15 mixtures.

The evaluation reveals that the presence of crushed aggregate material, and consequently, geopolymer binder, has a substantial impact on the potential for shrinkage. For example, the mixture blend that was prepared with 80% crushed aggregate material and treated with 20% geopolymer content (A80/G20) showed a lower drying shrinkage value, which is up to 122 $\times$ 10^{-6} m at 90 days of curing compared to the mixture A85/G15 value of 198 $\times$ 10^{-6} m at the same number of curing days. This is because the A80/G20 mixture was treated with 20% geopolymer content. However, beginning on day 56, the shrinkage value remained constant for both mixes.

The available water consumption in the hydration reaction of the mixture is the primary cause of the increment in drying shrinkage that occurs in conjunction with the presence of geopolymer content. Therefore, an increase in the aggregate content of the materials for stability may result in a high moisture consumption and, as a direct consequence of this, a significant degree of shrinkage, followed by cracking of the material. For example, the mixture of A85/G15 had a high-water absorption percentage compared to A80/G20 because the crushed aggregate content of A85/G15 was greater than A80/G20; thus, there is an excess of moisture during evaporation of A85/G15 that leads to high shrinkage compared to A80/G20 mixtures. The other reason of the lowest shrinkage is the water added to the mixture of A80/G20 was lower than mixture A85/G15 which is 0.5% and 3.5% respectively.

The moisture content is an important component in reducing the degree of shrinkage of stabilized materials since it is mostly induced by moisture loss due to hydration or evaporation. Increasing the moisture content causes more moisture to evaporate, hence a greater magnitude of shrinkage stresses. This was related to an excess of moisture that was not required for hydration when the higher moisture content was used. Therefore, it is required to control the moisture content and preferably to use the Optimum Moisture Content in the compaction test. When compared to the traditional base materials, the FAG base material's dry shrinkage value was found to be slightly lesser, suggesting that the usage of geopolymer as a stabilizing agent potentially helps alleviate reflective cracks.

3.4. Indirect Tensile Strength (ITS)

Under traffic loads, the base course of pavement constructions is subjected to tensile stresses and strains. Figure 8 illustrates the indirect tensile strength (ITS) test findings for the specimens prepared in the range of A80/G20 and A85/G15 (crushed aggregate and geopolymer content) and cured for the period of 7 and 28 days before testing. For mixture A80/G20, the increasing percentage is about 49.7% for curing 7 to 28 days. Meanwhile, about a 14% increase percentage was observed for mixture A85/G15 in 7 to 28 days of curing.

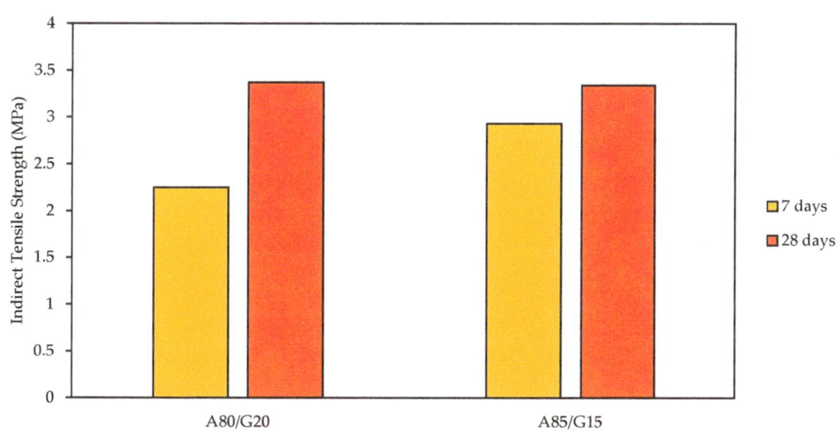

Figure 8. Indirect Tensile Strength for A80/G20 and A85/G15.

The development rate of indirect tensile strength, in general, is proportional to the amount of geopolymer used. This is because high Ca in FA content in the mixture can contribute to the strength and binding effect. With increasing volumes of crushed aggregate and curing time, the value of ITS increased consistently. Like the UCS, the reduction in ITS is more significant at greater geopolymer levels than at low ones. In general, the ITS values for the different examined specimens remained within a range of 12 to 16% of the UCS values. The addition of FA enhanced the indirect tensile strength, and so did the decrease of porosity. Reducing the porosity increases the contact between the particles, boosting their bonding strength.

3.5. Resilient Modulus (RM)

The important mechanical parameter utilized to examine the pavement materials' performance is its resilient modulus (RM), which is essential for calculating stresses, strains, and displacements within pavement structure layers susceptible to traffic-induced loads. The resilient modulus test outcomes are shown in Figure 9 for mixtures cured for 7 and 28 days. The figure shows that the A80/G20 combination had the greatest RM (32,881 MPa) and the lowest RM (18,819.3 MPa) at 28 days and 7 days, respectively, with A85/G15 mixtures in between. This strong connection between these two mixtures is consistent with the results of flexural strength.

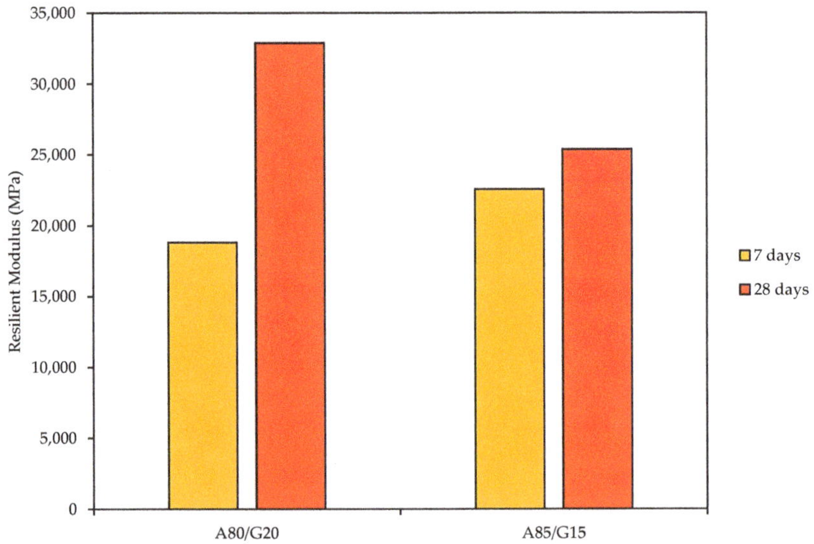

Figure 9. Two Resilient Modulus for A80/G20 and A85/G15 mixtures.

Strong support for the upper asphalt courses and even distribution of vehicle wheel loads in the lower subgrade are provided by the high stiffness stabilized base layer. While materials having high stiffness have a tendency to become brittle, which raises the tensile condition of the layers and leads to early failure, excessive base stiffness may have a detrimental influence on the service life of the pavement. Therefore, the effect of the base layer modulus on the performance of the pavement has been studied by numerous academics [40]. The findings showed that boosting the base layer's modulus could significantly reduce tensile stress at the base of asphalt courses and pavement deflection, resulting in a more significant economic benefit by lowering the overall thickness of pavement structures.

4. Conclusions

In this research, mechanical parameters were used to evaluate how well FAG performed as a road base, for instance, flexural strength (FS), indirect tensile strength (ITS), resilient modulus (RM), shrinkage, and unconfined compressive strength (UCS) with different percentages of crushed aggregates. The mixture of A80/G20 and A85/G15 was used to determine the performance evaluation of the pavement base. The following conclusion can be summarized.

- The major finding is that only 80% and 85% of crushed aggregates are appropriate for usage as a road base material as per the compaction test result. The increase in strength value is due to the friction and interlocking actions between the aggregate particles and the geopolymer binder, which can enhance the load transfer in the road base layer. From the results obtained, the UCS values for 7, 28, and 90-days cured specimens were in the range of 10.3–34.9 MPa for all mixtures.

- The flexural strength of FAG concrete base grew continuously as curing times increased, and the strength findings followed the same pattern.
- FAG base material was discovered to have a slightly less dry shrinkage value than conventional base materials, suggesting that using geopolymer as a stabilizing agent may assist in improving reflective cracks. Additionally, it is one of the elements influencing how well the pavement performs.
- For the results of the indirect tensile strength of the A80/G20 mixture, the percentage increase was about 49.7% for the 7 to 28 day curing period. The growth percentage of the A85/G15 mixture increased by 14% during the 7 to 28-day curing period.
- The A80/G20 mixture obtained the highest resilient modulus of 32,881 MPa, and the same mixture also recorded the lowest is 18,819.3 MPa. These two mixtures magnitudinal relationship reflects the findings of the UCS test.

The above conclusions have verified that incorporating FAG as a road base stabilizer is conducive to promoting environmental sustainability. However, further research is essential to examine the properties and performance of geopolymer road bases using the geopolymerization method using different raw materials.

Author Contributions: Conceptualization, writing—original draft preparation and supervision, L.A.S. and M.M.A.B.A.; methodology and investigation, T.I., M.R.M.H., M.A. and I.H.A.A.; validation, data curation and writing—review and editing, A.V.S. and P.V.; formal analysis and visualusation F.A.R. All authors have read and agreed to the published version of the manuscript.

Funding: This study was supported by the Center of Excellence Geopolymer and Green Technology (CEGeoGTECH), Faculty of Civil Engineering & Technology, Universiti Malaysia Perlis and School of Civil Engineering, Universiti Sains Malaysia. This research was also supported by TUIASI from the University Scientific Research Fund (FCSU).

Institutional Review Board Statement: Not applicable.

Informed Consent Statement: Not applicable.

Data Availability Statement: Not applicable.

Acknowledgments: Special thanks to those who have indirectly or directly contributed to this paper project.

Conflicts of Interest: The authors declare no conflict of interest.

References

1. Hoy, M.; Horpibulsuk, S.; Arulrajah, A. Strength Development of Recycled Asphalt Pavement-Fly Ash Geopolymer as A Road Construction Material. *Constr. Build. Mater.* **2016**, *117*, 209–219. [CrossRef]
2. Li, H.; Cui, C.; Cai, J.; Zhang, M.; Sheng, Y. Utilization of Steel Slag in Road Semi-Rigid Base: A Review. *Coatings* **2022**, *12*, 994. [CrossRef]
3. Li, Z. Study on Performance of Retarded Composite Semi-Rigid Base Mixed with Rubber Powder. *Materials* **2022**, *15*, 4683. [CrossRef]
4. Fazli, H.; Yan, D.; Zhang, Y.; Zeng, Q. Effect of Size of Coarse Aggregate on Mechanical Properties of Metakaolin-Based Geopolymer Concrete and Ordinary Concrete. *Materials* **2021**, *14*, 3316. [CrossRef]
5. De Souza, F.B.; Yao, X.; Gao, W.; Duan, W. Graphene Opens Pathways to a Carbon-Neutral Cement Industry. *Sci. Bull.* **2022**, *67*, 5–8. [CrossRef]
6. Abbas, S.N.; Qureshi, M.I.; Abid, M.M.; Zia, A.; Ur, M.A.; Tariq, R. An Investigation of Mechanical Properties of Fly Ash Based Geopolymer and Glass Fibers Concrete. *Sustainability* **2022**, *14*, 10489. [CrossRef]
7. Russo, F.; Veropalumbo, R.; Biancardo, S.A.; Oreto, C.; Scherillo, F.; Viscione, N. Reusing Jet Grouting Waste as Filler for Road Asphalt Mixtures of Base Layers. *Materials* **2021**, *14*, 3200. [CrossRef]
8. Topark-Ngarm, P.; Chindaprasirt, P.; Sata, V. Setting Time, Strength, and Bond of High-Calcium Fly Ash Geopolymer Concrete. *J. Mater. Civ. Eng.* **2015**, *27*, 04014198. [CrossRef]
9. Haneklaus, N.; Barbossa, S.; Basallote, M.D.; Bertau, M.; Bilal, E.; Chajduk, E.; Chernysh, Y.; Chubur, V.; Cruz, J.; Dziarczykowski, K.; et al. Closing the Upcoming EU Gypsum Gap with Phosphogypsum. *Resour. Conserv. Recycl.* **2022**, *182*, 106328. [CrossRef]

10. Tahir, M.F.M. Mechanical and Durability Analysis of Fly Ash Based Geopolymer with Various Compositions for Rigid Pavement Applications. *Materials* **2022**, *15*, 3458. [CrossRef]
11. Aziz, I.H.; Abdullah, M.M.A.B.; Heah, C.Y.; Liew, Y.M. Behaviour Changes of Ground Granulated Blast Furnace Slag Geopolymers at High Temperature. *Adv. Cem. Res.* **2019**, *32*, 465–475. [CrossRef]
12. Abdullah., A. Article The Effects Of Various Concentrations of Naoh on The Inter-Particle Gelation of a Fly Ash Geopolymer Aggregate. *Materials* **2021**, *14*, 1111. [CrossRef]
13. Barbhuiya, S.; Pang, E. Strength and Microstructure of Geopolymer Based on Fly Ash and Metakaolin. *Materials* **2022**, *15*, 3732. [CrossRef]
14. Burduhos Nergis, D.D.; Vizureanu, P.; Corbu, O. Synthesis and Characteristics of Local Fly Ash Based Geopolymers Mixed with Natural Aggregates. *Rev. Chim.* **2019**, *70*, 1262–1267. [CrossRef]
15. Shahedan, N.F.; Abdullah, M.M.A.; Mahmed, M.; Kusbiantoro, A.; Tammas-Williams, S.; Li, L.Y.; Azia, I.H.; Vizureanu, P.; Wyslocki, J.J.; Bloch, K.; et al. Properties of a New Insulation Material Glass Bubble in Geopolymer Concrete. *Materials* **2021**, *14*, 809. [CrossRef]
16. Nergis, D.D.B.; Vizureanu, P.; Ardelean, I.; Sandu, A.V.; Corbu, O.C.; Matei, E. Revealing the Influence of Microparticles on Geopolymers' Synthesis and Porosity. *Materials* **2020**, *13*, 3211. [CrossRef]
17. Azimi, E.A.; Abdullah, M.M.A.; Vizureanu, P.; Salleh, M.A.A.M.; Sandu, A.V.; Chaiprapa, J.; Yoriya, S.; Hussin, K.; Aziz, I.H. Strength Development and Elemental Distribution of Dolomite/Fly Ash Geopolymer Composite under Elevated Temperature. *Materials* **2020**, *13*, 1015. [CrossRef]
18. Faris, M.A.; Abdullah, M.M.A.; Sandu, A.V.; Ismail, K.N.; Moga, L.M.; Neculai, O.; Muniandy, R. Assessment of Alkali Activated Geopolymer Binders as an Alternative of Portlant Cement. *Mater. Plast.* **2017**, *54*, 145–154. [CrossRef]
19. Jo, M.; Soto, L.; Arocho, M.; St John, J.; Hwang, S. Optimum Mix Design of Fly Ash Geopolymer Paste and Its Use in Pervious Concrete for Removal of Fecal Coliforms and Phosphorus in Water. *Constr. Build. Mater.* **2015**, *93*, 1097–1104. [CrossRef]
20. Kalombe, R.M.; Ojumu, V.T.; Eze, C.P.; Nyale, S.M.; Kevern, J.; Petrik, L.F. Fly Ash-Based Geopolymer Building Materials for Green and Sustainable Development. *Materials* **2020**, *13*, 5699. [CrossRef]
21. Ziegler, D.; Formia, A.; Tulliani, J.M.; Palmero, P. Environmentally-Friendly Dense And Porous Geopolymers Using Fly Ash And Rice Husk Ash As Raw Materials. *Materials* **2016**, *9*, 466. [CrossRef]
22. Ismail, A. Laboratory Investigation on The Strength Characteristics of Cement- Treated Base. *Appl. Mech. Mater.* **2014**, *507*, 353–360. [CrossRef]
23. Nordin, N.; Abdullah, M.M.A.B.; Tahir, M.F.M.; Sandu, A.V.; Hussin, K. Utilization of Fly Ash Waste as Construction Material. *Int. J. Conserv. Sci.* **2016**, *7*, 161–166.
24. Atmaja., S. Investigation of The Chemical, Strength, Adhesion and Morphological Properties of Fly Ash Based Geopolymer-Modified Bitumen. *Constr. Build. Mater.* **2020**, *255*, 119364.
25. Dayal, S.; Soundarapandian, N. Effect of Fly-Ash Based Geopolymer Coated Aggregate on Bituminous Mixtures. *Gra Devinar* **2018**, *70*, 187–199.
26. Hu, W.; Nie, Q.; Huang, B.; Su, A.; Du, Y.; Shu, X.; He, Q. Mechanical Property and Microstructure Characteristics Of Geopolymer Stabilized Aggregate Base. *Constr. Build. Mater.* **2020**, *191*, 1120–1127. [CrossRef]
27. Tiyasangthong, S.; Yoosuk, P.; Krosoongnern, K.; Sakdinakorn, R.; Tabyang, W.; Phojan, W.; Suksiripattanapong, C. Stabilization of Recycled Concrete Aggregate Using High Calcium Fly Ash Geopolymer as Pavement Base Material. *Infrastructures* **2022**, *7*, 117. [CrossRef]
28. *C131/C131M-14*; Standard Test Method for Resistance to Degradation of Small-Size Coarse Aggregate by Abrasion and Impact in the Los Angeles Machine. ASTM: West Conshohocken, PA, USA, 2014.
29. Al Bakri, A.M.; Kamarudin, H.; Bnhussain, M.; Rafiza, A.R.; Zarina, Y. Effect of Na_2SiO_3/NaOH Ratios and NaOH Molarities on Compressive Strength of Fly-Ash-Based Geopolymer. *ACI Mater. J.* **2012**, *109*, 503.
30. Sofri, L.A.; Abdullah, M.M.A.B.; Hasan, M.R.M.; Huang, Y. Effect of Naoh Concentration and Fly Ash/Alkaline Activator Ratio on The Compressive Strength Of Road Base Material. *AIP Conf. Proc.* **2018**, *2030*, 020293.
31. *ASTM D1633-00(2007)*; Standard Test Methods for Compressive Strength of Molded Soil-Cement Cylinders. ASTM: West Conshohocken, PA, USA, 2007; pp. 1–15.
32. *ASTM C78/C78M-18*; Standard Test Method for Flexural Strength of Concrete (Using Simple Beam with Third-Point Loading. ASTM: West Conshohocken, PA, USA, 2002; pp. 1–3.
33. *ASTM C490/C490M-08*; Standard Practice for Use of Apparatus for the Determination of Length Change of Hardened Cement Paste, Mortar, and Concrete. ASTM: West Conshohocken, PA, USA, 2008; pp. 1–5.
34. *ASTM D6931-12*; Standard Test Method for Indirect Tensile (IDT) Strength of Bituminous Mixtures. ASTM: West Conshohocken, PA, USA, 2007; pp. 1–5.
35. *ASTM D4123*; Standard Test Method for Indirect Tension Test for Resilient Modulus of Bituminous. ASTM: West Conshohocken, PA, USA, 1995; Volume 82, pp. 2–5.
36. *JKR/SPJ/2008-S4*; Standard Specification for Road Works Part 4 Flexible Pavement. JKR: Kuala Lumpur, Malaysia, 2008; pp. 1–187.
37. Aliabdo, A.A.; Abd Elmoaty, A.E.M.; Salem, H.A. Effect of Water Addition, Plasticizer and Alkaline Solution Constitution on Fly Ash Based Geopolymer Concrete Performance. *Constr. Build. Mater.* **2016**, *121*, 694–703. [CrossRef]

38. Xuan, D.; Molenaar, A.A.A.; Houben, L.J.M. Compressive and Indirect Tensile Strengths of Cement-Treated Mix Granulates with Recycled Masonry and Concrete Aggregates. *J. Mater. Civ. Eng.* **2012**, *24*, 577–585. [CrossRef]
39. Kaze, R.C. Microstructure and Engineering Properties of $Fe_2O_3(FeO)$-Al_2O_3-SiO_2 Based Geopolymer Composite. *J. Clean Prod.* **2018**, *199*, 849–859. [CrossRef]
40. Grilli, A.; Bocci, M.; Tarantino, A.M. Experimental Investigation on Fibre-Reinforced Cement-Treated Materials Using Reclaimed Asphalt. *Constr. Build. Mater.* **2013**, *38*, 491–496. [CrossRef]

materials

MDPI

Article

Equilibrium and Kinetic Modeling of Crystal Violet Dye Adsorption by a Marine Diatom, *Skeletonema costatum*

Mohamed Ashour [1,*], Ahmed E. Alprol [1,*], Mohamed Khedawy [1], Khamael M. Abualnaja [2] and Abdallah Tageldein Mansour [3,4,*]

1 National Institute of Oceanography and Fisheries (NIOF), Cairo 11516, Egypt
2 Department of Chemistry, College of Science, Taif University, P.O. Box 11099, Taif 21944, Makkah, Saudi Arabia
3 Animal and Fish Production Department, College of Agricultural and Food Sciences, King Faisal University, P.O. Box 420, Hofuf 31982, Al-Ahsa, Saudi Arabia
4 Fish and Animal Production Department, Faculty of Agriculture (Saba Basha), Alexandria University, Alexandria 21531, Egypt
* Correspondence: microalgae_egypt@yahoo.com (M.A.); ah831992@gmail.com (A.E.A.); amansour@kfu.edu.sa (A.T.M.)

Abstract: Significant efforts have been made to improve adsorbents capable of eliminating pollutants from aqueous solutions, making it simple and quick to separate from the treated solution. In the current study, the removal of Crystal Violet Dye (CVD) from an aqueous synthetic solution onto a marine diatom alga, *Skeletonema costatum,* was investigated. Different experiments were conducted as a function of different pH, contact time, adsorbent dosage, temperature, and initial CVD concentration. The highest adsorption efficiency (98%) was obtained at 0.4 g of *S. costatum*, pH 3, and a contact time of 120 min, at 25 °C. Furthermore, Fourier-transform infrared spectroscopy (FTIR) results display that binding of CVD on *S. costatum* may occur by electrostatic and complexation reactions. Moreover, the Brunauer–Emmett–Teller surface area analysis (BET) obtained was 87.17 $m^2\ g^{-1}$, which, in addition to a scanning electron microscope (SEM), reveals large pores that could enhance the uptake of large molecules. However, the equilibrium adsorption models were conducted by Halsey, Langmuir, Freundlich, Henderson, and Tempkin isotherm. In addition, multilayer adsorption isotherm best described the uptake of CVD onto *S. costatum*. The maximum monolayer adsorption capacity (q_{max}) was 6.410 mg g^{-1}. Moreover, thermodynamic parameters of the adsorption studies suggested that the uptake of CVD onto *S. costatum* was endothermic and spontaneous. The pseudo-first-order, pseudo-second-order, and intra-particle diffusion kinetic equations were applied to model the adsorption kinetic data. It was seen that the kinetics of the adsorption may be described using pseudo-second-order kinetic equations. Finally, the present work concluded that the marine diatom alga *S. costatum* is suitable as a natural material for the adsorption of CVD.

Keywords: algae; bioremediation; crystal violet dye; modeling; adsorption kinetics

Citation: Ashour, M.; Alprol, A.E.; Khedawy, M.; Abualnaja, K.M.; Mansour, A.T. Equilibrium and Kinetic Modeling of Crystal Violet Dye Adsorption by a Marine Diatom, *Skeletonema costatum*. *Materials* **2022**, *15*, 6375. https://doi.org/10.3390/ma15186375

Academic Editors: Gerald Anthony Murphy, Roberta G. Toro, Daniela Caschera and Abeer M. Adel

Received: 19 August 2022
Accepted: 7 September 2022
Published: 14 September 2022

1. Introduction

Water is a necessary component of life and is utilized by almost all living things on the world's surface [1]. It is also a valuable resource in world economies because agriculture and other commercial uses consume around 70% of freshwater [2,3]. The untreated discharge of synthetic dyes from the industrial sector (textiles, printing, dyeing cosmetics, paints, plastic, rubber, leather, food, etc.) is a major concern for researchers working on wastewater pollution protection [4,5]. Moreover, industrial wastewater mainly contains several organic complicated molecules and dyestuff which are discharged into highly dyed wastewater, which has caused critical environmental concerns globally [6–12]. Intensive technological and industrial development, along with the increase in other human activities, has led to the overuse of water resources and increased water pollution [7,13]. The textile industry is the main industry that utilizes dyes causing the main source of water pollution, which poses a threat to aquatic life and human life [14,15]. Therefore, dye removal from wastewater poses a challenge to prevent dyes from causing harm to the environment [8], especially to aquatic biological systems [16]. To manage the problems of water pollution, numerous physical, chemical, and biological approaches to decomposing many textile dyes were used over the latest years [8]. For example, advanced oxidation processes, nanofiltration, photocatalytic degradation, adsorption, and biological treatment are among the technologies used to remove organic pollutants [2,17–19]. The primary objective of these strategies was to reduce damaging dyes into non-toxic end products, but most of them failed, since the end compounds were more toxic than the primary dyes [20,21]. Among these technologies, adsorption has grown as an effective biological and economical technique for decontamination of organic pollutants from wastewater [22–26]. Sharma et al. [26] reported that adsorption is considered a potential technique because of its ease of operation and relatively low cost of application required for wastewater pollution removal.

There has been a lot of attention recently to converting marine biomass applications into value-added products [27–29], especially sorbents, to reduce the impacts of environmental pollutants [23]. Recently, algal cells (microalgae and/or seaweeds) are one of the most important low-cost, eco-friendly, sustainable, and biologically promising technologies for dye and pollutant removal from aqueous solutions [8,30,31]. Algal cells have a long history of bioremediations of toxic dyes and/or water pollutants [32–34]. Generally, due to their biomolecules, bioactive compounds, and their functional groups, algae are the richest organisms in our world that can serve and be utilized in several bio-industries, such as human food supplements, animal feeds, bio-pharmaceuticals, bio-cosmetics, bio-fertilizers, fine-biochemicals, bio-energy, and bio-plastic, etc. [35–46]. Microalgae have a great prospect of interacting with water pollutants as a major producer in the food chain and a major player in the balance of the marine ecosystem [47–49]. Additionally, the use of dead organisms in adsorption is more beneficial for the treatment of water, since dead microorganisms are not affected by harmful wastes, do not necessitate a constant supply of nutrients, and can regenerate and be reused numerous times [50]. Dead cells can be kept or used for a long time at room temperature without putrefaction taking place. Additionally, it has been demonstrated that dead cells acquire contaminants to an equal or greater degree than developing or resting cells [51]. The use of dead biomass in powdered form in the column does, however, present several challenges, including the difficulty of easily separating the biomass after adsorption, mass loss after regeneration, low strength and density, and small particle size, which make it challenging to utilize them in column applications. Dead biomass can be immobilized in supportive material to address these issues [52].

The marine diatom *Skeletonema costatum* is a chain of microalgal cells extensively used in marine hatcheries as live feed for marine larvae [53–55]. In addition, *S. costatum* species have many bioremediation advantages in adsorbing several pollutants from the aquatic environment, in a process that is simple, easy to observe, and sensitive to toxins [56,57]. The bioremediation process of *S. costatum* is carried out via passive transport, through the cell wall, and an active transport mechanism [3,58]. The *S. costatum* cell wall contains numerous functional groups, such as amine, hydroxyl, carboxyl, phosphate, and thiol groups. These functional groups, mainly negatively charged, can react with the positively charged ions of many pollutants [56,59]. The dye removal, especially by using algae, may be attributed to the accumulation of dye ions on the surface of algal biopolymers and further to the diffusion of the dye molecules from the aqueous phase to the solid phase of the biopolymer [60]. Extracellular polymers consist of surface functional groups, which enhance the sorption of the dye molecules onto the surface of the polymer (floc) during the dye removal process [61].

Crystal violet dye (CVD) is mainly used during the manufacturing process to dye cotton, silk, inks, and paints [62]. The protein CVD is used to detect bloody fingerprints. CVD is also applied throughout many kinds of adhesive taps for latent printing. Gram's stain is frequently employed in the demonstration and basic classification of dyes [22].

CVD is also used to create vibrant dyes [18,63]. It is carcinogenic and has been classified as a recalcitrant molecule, since it is poorly metabolized by microbes, it is non-biodegradable, and it can persist in a variety of environments [64]. Furthermore, biological waste materials, eco-friendly behavior, and the possibility of reusability or regenerating to find low-cost adsorbent materials and reactivity, thus acting as suitable adsorbents, have been considered [65,66].

The objective of the present research was to explore the possibility of using the marine diatom alga *Skeletonema costatum* as an adsorbent for dye elimination, for example, CVD, a widely used dye in the textile-processing industry. The impact of experimental parameters such as pH, contact times, adsorbent dosages, temperatures, and initial dye concentrations on the adsorptive capacity of the adsorbent was examined. On the other hand, thermodynamic properties, equilibrium adsorption isotherms, and adsorption kinetics were examined by using conventional theoretical methods to investigate the adsorption mechanism. Moreover, the industrial wastewater and seawater applications, regeneration, and reusability have been studied. Therefore, in this study, this material was selected to investigate its ability to remove toxic dyes. However, even though there is literature on the marine diatom alga *Skeletonema costatum* as an adsorbent for the removal of heavy metal ions, where the previous studies showed the insufficiency of the use of this material in removing toxic dyes from wastewater, this work shows an experimental study on the removal of dyes, especially CVD, by using this novel material. Since it considerably improved the adsorption effectiveness of the specified dyes, it was chosen for the following investigation to cover the gap in the literature. Therefore, the current work aims to create a high-performance, environmentally friendly powder adsorbent from *Skeletonema costatum* culture for the removal of CVD from an aqueous solution. Furthermore, *Skeletonema costatum* was considered due to its novelty, which may easily be regenerated and recycled, which makes it applicable for treatment processes. Moreover, the highest suitable model was achieved by using error functions.

2. Materials and Methods

2.1. Materials

2.1.1. *Skeletonema costatum*

The marine diatom alga *Skeletonema costatum* was identified and isolated previously from the Eastern Harbor of Alexandria Coast (31°13′48″ N; 29°53′12″ E), close to the National Institute of Oceanography and Fisheries (NIOF), Egypt. An isolated diatom strain, *S. costatum* (cell size from 2 to 5 μm), was cultured at an indoor scale under standard conditions of temperature (22 ± 2 °C), salinity (34 ± 1 ppt), continuous illumination (3500 ± 500 Lux), and continuous aeration. After a batch culture in conical flasks (1 L) filled with F/2 standard Guillard medium [67], the culture cells were harvested by centrifuge (3000× *g*. 10 min^{-1}) at the late exponential phase on day 5. The harvested biomasses were dried (at 55 °C for 48 h) and preserved at −20 °C for further applications.

2.1.2. Crystal Violet Dye (CVD)

The adsorbate, crystal violet (CV) dye, was a basic, acquired from Daystar, Mumbai, India (basic dye, C.I. 42555, max. abs. = 588 nm, molecular formula $C_{25}H_{30}N_3Cl$ and standard purity (97.5% specified)) (Figure 1). Distilled water was used to obtain the desired concentrations. The original pH of the dye solution was 2.0. The maximum absorbance of the dye solutions was monitored using a spectrophotometer to determine their concentrations (Milton Roy, spectronic 2ID).

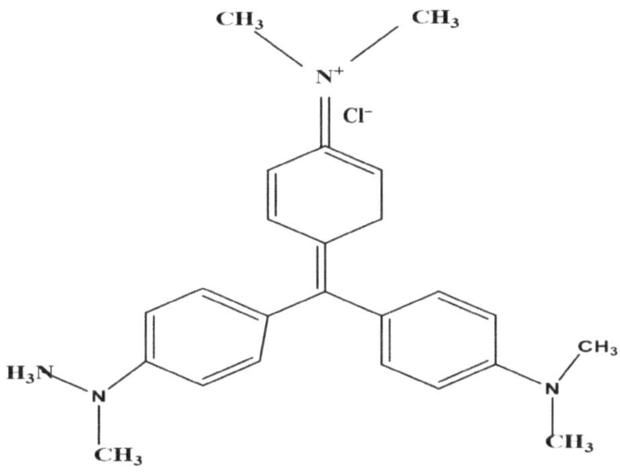

Figure 1. The chemical structure of the crystal violet dye (CVD).

2.1.3. Chemical and Reagents

The chemicals used in the adsorption study, including sodium hydroxide (NaOH, 98.6%) and sulfuric acid (H_2SO_4, 98%), were obtained from Fisher Scientific (Fisher Scientific, Montreal, QC, Canada). All the applied chemicals throughout this study were analytical-grade reagents.

2.2. Methods

Batch Adsorption Process

Batch adsorption trials were done in a shaker at various ranges of parametric conditions. Optimization reactions were achieved for pH factors in the range 3, 5, 7, 9, and 11, initial CVD concentration influence in the range 5, 10, 15, 20, and 40 mg L^{-1}, the contact time parameter (15, 30, 60, 120, and 180 min), biomass amount (0.05, 0.1, 0.2, 0.3, and 0.4 g), and temperature influence at 25, 35, 45, and 55 °C. For these aims, the used bottles were shaken at a controlled temperature (25 °C), using an orbital shaker (Orbital incubator model SI50, Stuart Scientific, Redhill, UK) for the minimal contact time necessary to achieve equilibrium, as indicated by the kinetic data, with an agitation rate of 110 rpm.

The efficiency of the treatment was assessed to determine the concentration using the UV–visible spectroscopic method at 588 nm (Milton Roy, Spectronic 21D, Houston, TX, USA). The impact of pH was evaluated by modifying the reaction mixture to various initial values of pH and evaluating the remaining color at the equilibrium contact time. Liquid solutions of NaOH and H_2SO_4 were utilized to regulate the needed values of pH. The percentages of removal and q_e (mg g^{-1}) of CVD onto *S. costatum* were calculated, as described by Ghoneim, et al. [68], as the following:

$$q_e = \frac{(C_i - C_f) \times V}{W} \quad (1)$$

$$\text{Percentage removal } (\%) = \frac{(C_i - C_f)}{C_i} \times 100 \quad (2)$$

where C_i and C_f are the initial and the final CVD (mg L^{-1}), respectively; W is the quantities of *S. costatum* adsorbent (g); and V is the solution volume (L).

For kinetic observations, 10 mL flasks containing CVD (5 mg L^{-1}) and the adsorbent (0.05 g) were exposed to agitation for numerous contact time intervals. At regular intervals, the flasks were removed from the shakers, and the remaining concentration of CVD solution was calculated.

2.3. *Characterizations*

Before and after the adsorption process, the surface morphology of the adsorbents was characterized via a Scanning Electron Microscope, SEM (JSM-IT200 with an accelerating voltage of 20.0 kV at a magnification ×2000). The samples were coated with platinum to prevent charging. On the other hand, the functional groups of the adsorbents, as well as a binding mechanism (s), were conducted using a Fourier-transform infrared spectrometer (FTIR) (Thermo Nicolet 6700). Furthermore, in this study, quantachrome instruments (cyl. pore) (NLDFT Ads. model) were used to examine the surface area of the raw adsorbent at 77.35 K.

2.4. *Data Analysis*

The normality and homogeneity assumptions were performed before the statistical analysis of the data. The experiments were applied in triplicates and the obtained data were presented in the form of means ± standard deviation (SD, $n = 3$). Before analysis, all results in percentages were arcsine transformed [69].

3. Results and Discussion

3.1. *S. costatum (Adsorbent) Characterization*

3.1.1. SEM

The scanning electron microscope investigations of *S. costatum*, before and after the adsorption process (Figure 2), concluded that the adsorbent (Figure 2A) has an irregular and heterogeneous surface morphology with a well-developed porous structure. Moreover, pores of several sizes and shapes were determined. The presence of these fine particles leads to an increase in the surface area and the porosity of *S. costatum* [70]. After adsorption of CVD (Figure 2B), the remarkable changes were investigated in the morphological structure of CVD-loaded *S. costatum*, revealing an irregular surface with a molecular cloud of the CVD, where the CVD may be adsorbed and trapped. The higher the porosity, the higher the adsorption capacity [17].

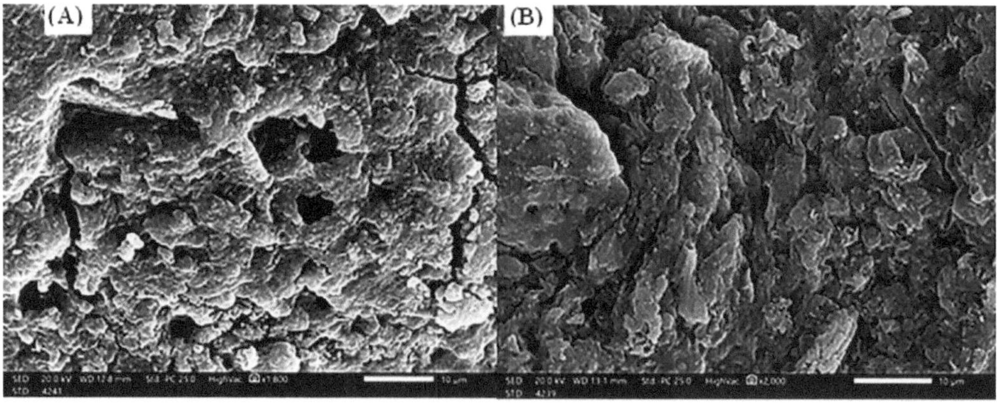

Figure 2. Scanning electron microscope (20.0 kv, ×2000) of *S. costatum* adsorbent before (**A**) and after (**B**) adsorption of CVD.

3.1.2. FTIR

FTIR spectra of the *S. costatum* before and after the adsorption process confirmed that there are several functional groups, as shown in Figure 3.

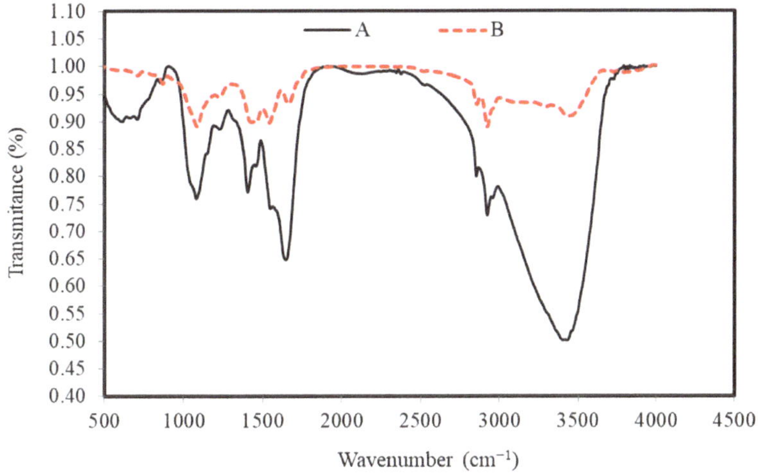

Figure 3. FTIR peaks of the *S. costatum* adsorbent before (A) and after (B) adsorption of CVD.

The wide peaks, which are located at around 3403, 3295, and 3495 cm^{-1}, are typically due to adsorbed water and/or the groups of hydroxyl [64]. The bands that existed around 2922 and 2856 cm^{-1} correspond to C–H vibrations of methyl and methylene groups. However, the band around 1645 and 1662 cm^{-1} generally happened by the stretching vibration of C=O carboxyl groups, while the bands around 1548 and 1542 cm^{-1} are ascribed to the aromatic ring or C=C stretching vibration [71]. In addition, the bands at around 1406 and 1458 cm^{-1} correspond to the C–H in-plane bending vibrations in methylene and methyl groups. The band at 1227–1230 cm^{-1} corresponds to C–C stretching in alkanes and a comparatively intense band at about 1081 and 1081 cm^{-1} can be assigned to C–O stretching vibrations in alcohols. The C–H out-of-plane bending vibrations cause the bands at 853 and 871 cm^{-1}. Disappearance and band shifting from 1406 to 1457 cm^{-1} are determined in the CVD-adsorbed *S. costatum* attributed to the stretching of C=C modes of the benzene ring of lignin. Transmittance at wave number 3429 cm^{-1} is found to be shifted to 3403 cm^{-1} on adsorption and this perhaps assumes the responsibility for the chemical reaction of the CVD with O–H groups on the adsorbent. Several bands were found in the region of 1000–500 cm^{-1}, which is perhaps attributed to the C–H stretching. From FTIR observations, CV dye ions may interact with the active surface of *S. costatum* adsorbents, resulting in the creation of new absorption bands, changes in absorption strength, and a movement in the wavenumber of functional groups. CVD ions bound to the active sites of the *S. costatum* adsorbents through complexation mechanism and electrostatic attraction. In addition, the electrostatic attraction was between carbonate groups and CVD ions. Furthermore, the complexation mechanism engaged in sharing the electron pair between electron donor atoms (N and O). The current work determined that carbonate, carbonyl, hydroxyl, and amine are the primary adsorption sites in *S. costatum* adsorbents. Another study found that Cu adsorption by *S. costatum* has a high percentage of removal due to the same functional groups being responsible for the adsorption process, which were mainly carboxyl, hydroxyl, and amine, as well as, thiol and phosphate groups, as reported by Pratama et al. [56,57,59].

3.1.3. BET

Table 1 shows the surface characteristics of the dry adsorbent employed in this investigation based on adsorbent-specific surface area characterization using the BET technique, according to nitrogen adsorption and desorption isotherms. Table 1 and Figure 4 show the adsorbent's surface area was 87.172 m^2 g^{-1}, as determined by the results. The total pore volume and average pore size of *S. costatum* were also evaluated and found to be 0.103 CC g^{-1} and 3.131 nm, respectively.

Table 1. Physical properties of *S. costatum*.

Properties	Data
BET surface area	87.17 m^2 g^{-1}
Single-point BET	74.10 m^2 g^{-1}
DH desorption	62.89 m^2 g^{-1}
Skeletal density	2.2 g C^{-1}
Average pore sizes	3.131 nm
Average pore volume	0.103 CC g^{-1}

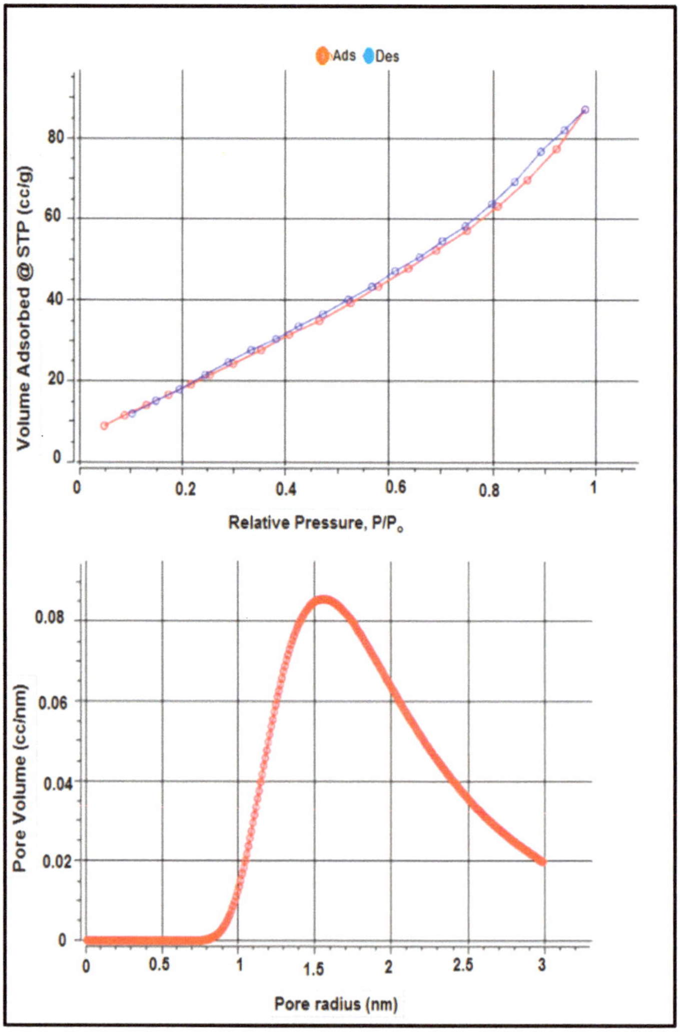

Figure 4. Isotherm–Isotherm isotherms and the pore size distribution.

3.2. Adsorption Studies

3.2.1. pH

The pH value is an important parameter affecting the removal uptake of toxic pollutants from wastewater due to the variation of H^+ and OH^- in the treatment medium [72]. The effect of solution pH on adsorption of CVD on *S. costatum* was studied by varying the pH of the dye solution for an initial concentration of 5 mg L^{-1}, while the adsorbent dosage, contact time, agitation speed, and temperature were fixed at 0.4 g/50 mL, 180 min, 110 rpm, and 25 °C, respectively. The obtained results indicate that the adsorption process of CVD was favorable in an acidic medium, as shown in Figure 5.

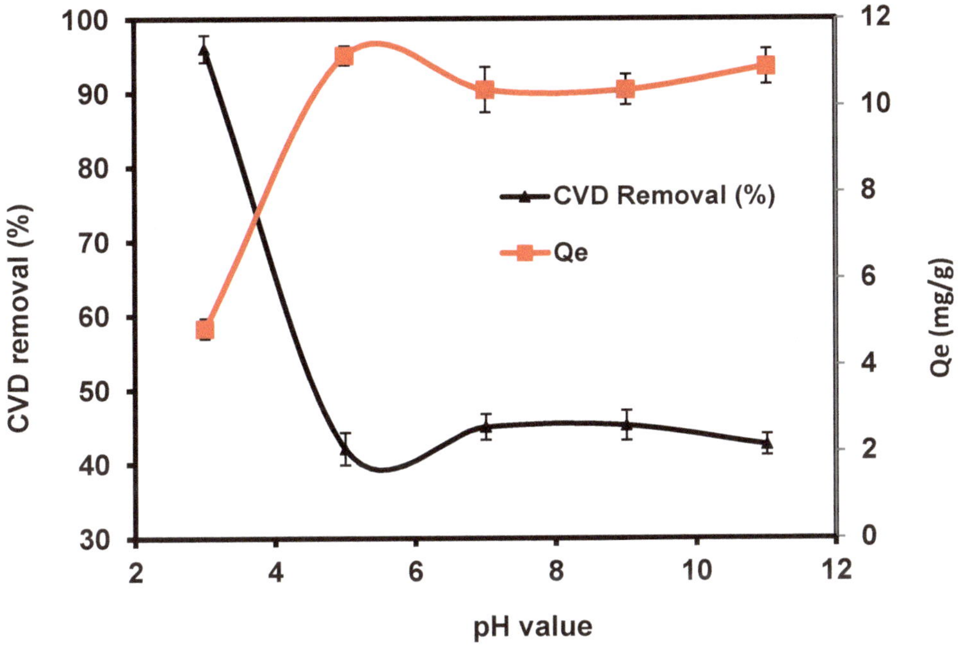

Figure 5. Impact of pH values on the adsorption of CVD onto *S. costatum*.

Dye removal efficiency decreased when the pH increased from 5 to 11, indicating that pH did not significantly affect the percentage of CVD removal, particularly under alkaline conditions. The highest percentage of CVD uptake was obtained at pH 3, with a percentage removal of 96%. CVD is based on chromophores combined with several kinds of reactive groups that react with the active functional groups of *S. costatum*, such as chitin, carboxyl, hydroxyl, acidic polysaccharides, amine, and the other functional groups [57,59]. Higher removal observed at low pH values is perhaps attributed to the electrostatic appeal between negatively charged CVD anions and positively charged *S. costatum* cell surface [73]. Furthermore, as the pH value increases, the adsorption of CVD decreases because of competition between the anionic dye and excess OH^- ions in the solution [74]. According to Wu et al. [75], the amount of adsorption decreased because the negatively charged adsorbent gave fewer effective sites for adsorption as repulsive forces increased. Following that, adsorption experiments were carried out at the optimal pH.

3.2.2. Adsorbent Dose of *S. costatum*

The effect of different *S. costatum* levels (0.05, 0.1, 0.2, 0.3, and 0.4 g/50 mL) were studied at 25 °C, for 180 min. With the increasing doses of *S. costatum*, the quantities of CVD adsorption and the percentage of removal increased from 90.90% to 95.20% and from 0.05 to

0.4g (Figure 6). This phenomenon can be attributed to an increase in the adsorbent-specific surface area and the availability of more adsorption sites [76]. Furthermore, this pattern might be described as a result of partial biomass agglomeration at increasing biomass concentrations, resulting in a reduction in the specific surface area of adsorption [77]. Only about a 3% increase was observed as the adsorbent dose was increased from 0.05 to 0.03 g; overlap of adsorption sites may describe the small increase in percentage adsorption. The decrease in the amount of CVD adsorbed q_e (mg g^{-1}) with increased adsorbent mass is attributed to a division of the flow or a concentration gradient between the solution and the absorbing surface. When *S. costatum* sorbent mass increases, the level of CVD that is adsorbed onto the unit weight of *S. costatum* decreases, resulting in a drop in the q_e value as the adsorbent mass concentration rises [78]. In the present study, the adsorbent dose of 0.4 g/50 mL is the optimal value for the removal of CVD.

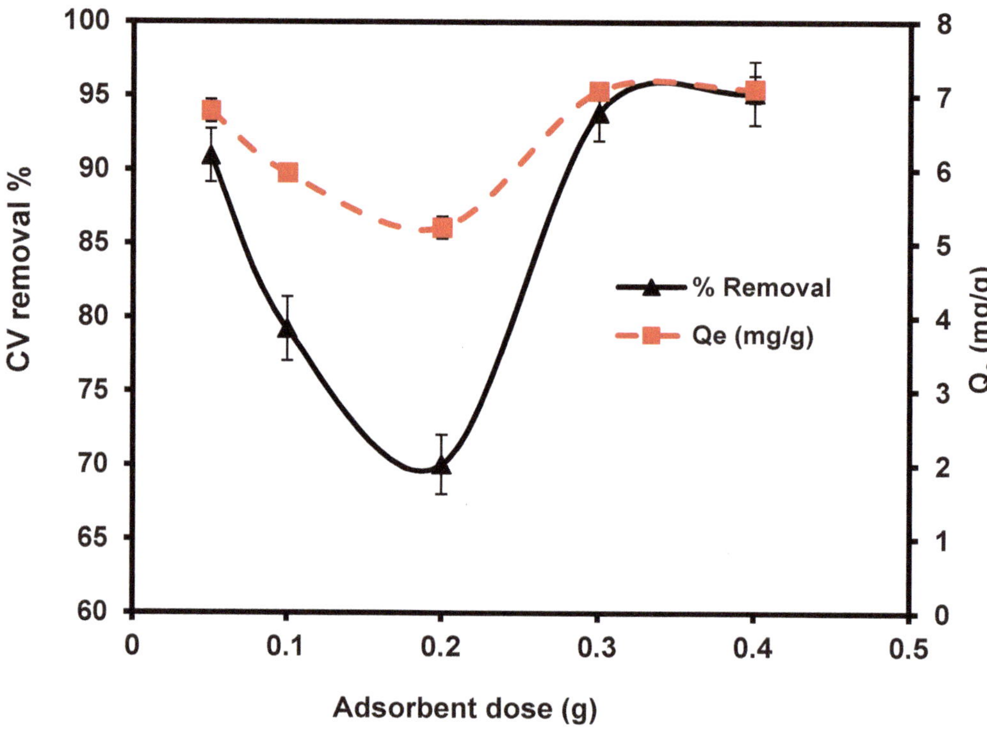

Figure 6. Impact of *S. costatum* concentrations on the adsorption of CVD.

3.2.3. CVD Concentrations

The effect of initial CVD levels (5, 10, 15, 20, and 40 mg L^{-1}) on adsorption onto 0.1 g/ 50 mL of the dried *S. costatum* for a contact time of 3 h at 25 °C was investigated. Figure 7 shows that the uptake capacity and adsorption capacity of CVD by *S. costatum* increased with the increasing initial level of CVD. This increase in the amount of adsorption with increased initial CVD concentration may be due to the increased driving force that dominates the resistance of mass transfer of the CVD between the solid phase and bulk liquid phase [79].

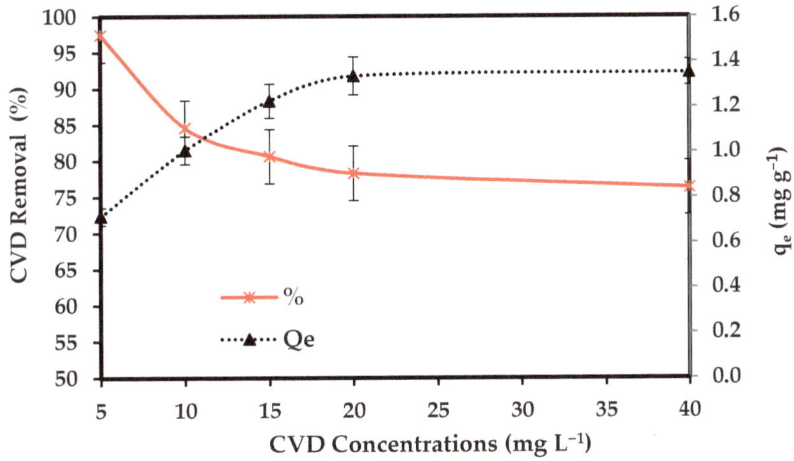

Figure 7. Impact of CVD concentrations on the adsorption process.

Furthermore, the surface of the sorbent is bare in the initial stage; sorption happens quickly and is usually regulated by the diffusion process from the bulk to the surface. Since there are fewer sorption sites accessible at this stage, the sorption is most likely an attachment-controlled process. Increasing the initial concentration results in increased diffusion of more CVD molecules in the bulk solution towards the external surface of the *S. costatum* and enhanced interaction between crystal violet anions and the *S. costatum* surface leading to increased adsorption uptake of CVD. The adsorption typically starts rapidly and then gradually slows because when a significant number of vacant surface sites seem to have been able to adsorb at first, then it is always difficult to occupy the residual vacant surface sites due to repulsive forces between the CVD molecules on the *S. costatum* and the bulk phase [74]. The obtained data indicate that the efficiency of *S. costatum* in the removal of CVD was highly related to the initial concentration of CVD. Similar suggestions were reported by Alprol et al. [80], who indicated that the efficiency of *A. platensis* in the removal of IV2R was highly related to the initial concentration of IV2R. Inventor et al. [81] confirmed that initial quick adsorption is due to dye molecules contacting available surface adsorption sites, while the following gradual adsorption is due to dye molecules' absorption into the pores of the adsorbents.

3.2.4. Contact Times

The practical efficiency of the adsorption technique in solution is largely based on the immersion time in the medium (i.e., contact time). Figure 8 illustrates the amount of CVD adsorbed onto *S. costatum* as a time function at 0.1 g/50 mL. CVD adsorptions were fast initially and gradually reduced over time until equilibrium was attained. Fast adsorption was observed within the first 15 min (87.88%) and then the adsorption rate of dye is high, which may be attributed to abundant adsorption sites, which attracted the dye particles from the bulk solution at the initial stages. The adsorption process becomes less efficient due to the gradual occupancy of these sites. In the figure, it is shown that the adsorption reaction nearly attained equilibrium within 120 min, after which no substantial amount of crystal violet was adsorbed with increasing contact time after equilibrium was achieved, indicating that all available sorption sites had been occupied [68]. Due to increased repulsive forces between dye molecules and the bulk solution, the occupation of the remaining unoccupied spots becomes increasingly difficult as time passes.

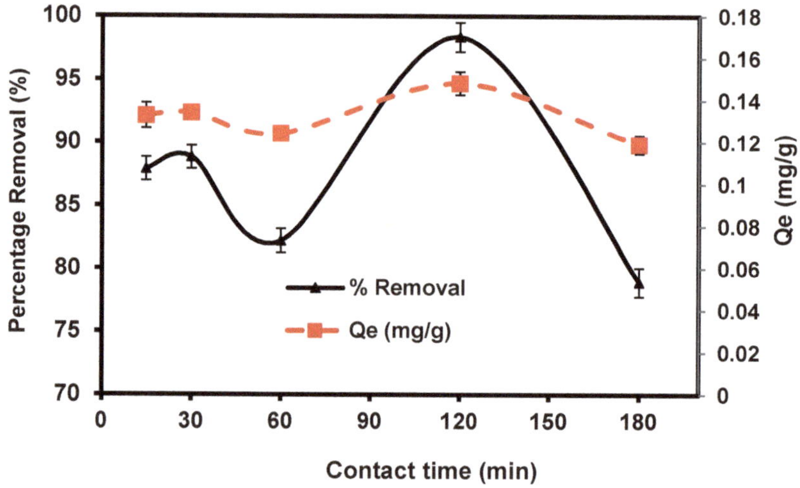

Figure 8. Impact of agitating time in the percentage removal of crystal violet dye.

3.2.5. Temperature

The temperatures of 25, 35, 45, and 55 °C were studied with 0.1/ 50 mL (Figure 9). The maximum percentage removal (79.30%) of CVD was observed at 25 °C. Based on the relation between removal efficiency and temperature, the results reported that the CVD sorption takes place through the aggregation of the dimers and/or monomers of CVD. The dimeric or monomeric dispersed from the cellular membrane are perhaps trapped as an outcome of accumulation, even at a low equilibrium CVD level. The relative increase in dye level in the cell probably leads to aggregation. Aggregation is most likely caused by a relative rise in CVD concentration in the cell. Hence, the reduction of sorbed quantities at higher temperatures, perhaps due to deaggregation, is based on the relative increase in thermal motions of CVD aggregates [82].

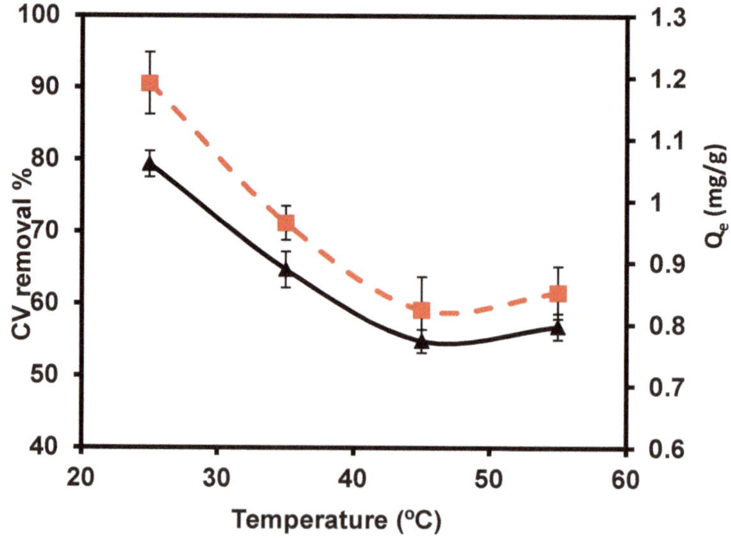

Figure 9. Impact of several temperatures on the removal percentage of CVD.

3.3. Thermodynamics Properties

To investigate the thermodynamic properties and thermal impact on the adsorption process, several temperatures (25, 35, 45, and 55 °C) were conducted with 0.05 g/50 mL of *S. costatum*. Different parameters of thermodynamics such as entropy change (ΔS), enthalpy change (ΔH), and Gibbs free energy (ΔG) were calculated to determine the natural process of the adsorption, as shown in Table 2.

Table 2. The calculated values of thermodynamic parameters of the adsorption of CVD onto *S. costatum*.

T (°C)	Q_e (mg g^{-1})	ΔG (kJ mol^{-1})	ΔH (kJ mol^{-1})	ΔS (kJ mol^{-1} k)
25	1.190	−5.050	8.636	−0.036
35	0.964	−3.279		
45	0.824	−2.355		
55	0.852	−2.636		

The magnitude of ΔG (kJ mol^{-1}) was conducted using the following equation:

$$\Delta G = -\ RTLnKa \tag{3}$$

where R is the universal gas constant (0.008314 kJ mol^{-1} K); T is the absolute temperature (°K); and Ka is the adsorption equilibrium constant. ΔH (kJ mol^{-1}) was calculated by the following equation:

$$\Delta H = \ \Delta G\ + T\Delta S \tag{4}$$

A plot of ΔG against T was found to be linear and the value of ΔH and ΔS was calculated from the slope and intercept of the plots. The negative values of ΔG for *S. costatum* have confirmed the feasibility of the process and the spontaneous nature of adsorption. However, the values of ΔG were increased from −5.050 to −2.636 kJ mol^{-1} as the temperature increased from 25 to 55 °C. The values of ΔH and ΔS were calculated and are given as 8.636 kJ mol^{-1} and −0.036 kJ mol^{-1} K, respectively. Hossain et al. [83] and Subbareddy et al. [84] reported that the positive value of ΔH indicates that the adsorption reaction was endothermic, and the strong affinity of the *S. costatum* diatom for CVD ions suggests some structural changes in CVD ions and the *S. costatum* [85]. In addition, the negative value of ΔS also suggests that the adsorption was enthalpy-driven and spontaneous [86]. Furthermore, the negative ΔS reflected the affinity of the *S. costatum* for the CVD and the decrease of randomness at the solid/liquid interface occurring in the internal structure during the adsorption process [75,86]. The negative value of ΔS shows a decline in the degree of freedom of the adsorbed CVD ions [87].

3.4. Equilibrium Adsorption Isotherm

Adsorption isotherms are essential in calculating the maximum capacity and the kind of adsorption front that is produced. To characterize the equilibrium nature of *S. costatum*, various adsorption isotherms have been examined [82]. The data were fitted to the Freundlich and Halsey isotherm equations and the fixed parameters of the isotherm equations were calculated.

3.4.1. Langmuir Isotherm

The Langmuir model [88] assumes monolayer adsorption (the adsorbed layer is one molecule in thickness), with adsorption only occurring at a finite (fixed) number of definite localized sites that are identical and equivalent, with no lateral interaction and steric hindrance between the adsorbed molecules, even on adjacent sites. In its derivation, the Langmuir isotherm refers to homogeneous adsorption, in which each molecule possesses constant enthalpies and sorption activation energy (all sites possess an equal affinity for the adsorbate) [89]. The nonlinear form of the Langmuir equation was calculated by the following non-linear form (Lineweaver–Burk linearization of the Langmuir model—Type 2) [90]:

$$q_e = q_{max}\ K_L\ C_e/(1 + K_L C_e) \quad (5)$$

where q_{max} (mg g^{-1}) is the maximum sorption capacity corresponding to the saturation capacity (representing the total binding sites of an adsorbent) and K_L (L mg^{-1}) is a coefficient related to the affinity among the adsorbent and dye ions.

The relationship was calculated from plotting curve $(1/q_e)$ vs. $(1/C_e)$:

$$1/q_e = 1/(K_L\ q_{max}\ C_e) + 1/q_{max} \quad (6)$$

where K_L and q_{max} are determined from the slope and intercept, respectively.

The CVD dye showed a maximal uptake capacity (q_{max}) of 17.76 mg g^{-1} by the *S. costatum*. The observed linear correlation coefficient (R^2 = 0.876) indicates this model does not give a good fit to the adsorption process (Figure 10). The Langmuir constant (K_L), which is related to the heat of adsorption, was found to be 5.06. The Langmuir constant K_L can serve as an indicator of the strength or affinity of the sorbent for the solute [87].

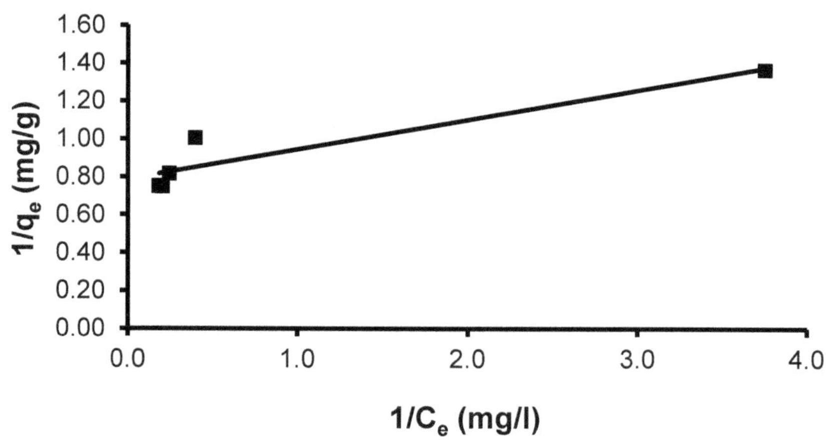

Figure 10. Plot of the Langmuir isotherm model for CVD adsorbed onto *S. costatum*.

The essential characteristics of the Langmuir isotherms can be expressed in terms of a dimensionless constant separation factor or equilibrium parameter, R_L, which is defined by Hall et al. [91]:

$$R_L = 1/(1+ K_L\ C_i) \quad (7)$$

where Ci is the initial adsorbate concentration (mg L^{-1}). The R_L value indicates the shape of the isotherm, as shown in Table 3. R_L values lie between 0 and 1, which indicates a favorable sorption isotherm for CVD. The RL was found to be 0.011.

Table 3. Relationship between RL and the type of isotherm [92].

R_L	Type of Isotherm
$R_L > 1$	Unfavorable
$R_L = 1$	Linear
$0 < R_L < 1$	Favorable
$R_L = 0$	Irreversible

3.4.2. Freundlich Isotherm

The Freundlich model can be applied to multilayer adsorption, with non-uniform distribution of adsorption heat and affinities over the heterogeneous surface [93]. Using the observed data of the *S. costatum* experiment, the feasibility of the empirical Freundlich isotherm was determined depending on sorption on heterogeneous surfaces, by plotting log (q_e) versus log (C_e). The Freundlich adsorption isotherm was expressed in the equation as the following [94]:

$$q_e = K_f \, C_e^{1/n} \tag{8}$$

$$\text{Log } q_e = \text{Log } K_F + 1/n \text{ Log} C_e \tag{9}$$

Freundlich constants K_F and 1/n were determined from the isotherm equation according to Equation (9). Freundlich adsorption isotherm constants and correlation coefficients are shown in Table 4. As its value is closer to zero, the constant 1/n, which represents the intensity of dye adsorption onto the sorbent or surface heterogeneity, becomes more heterogeneous. The n_F value reflects the degree of nonlinearity between solution concentration and adsorption in the following ways: if nF = 1, adsorption is linear; if nF < 1, the adsorption process is chemical; if nF > 1, the adsorption is a beneficial physical process [95]. The value of n was calculated to be 5.102, showing that the adsorption process is chemical. In addition, the value of 1/n (0.196) was found to lie between 0 and 1, indicating that CVD is confidently adsorbed by the *S. costatum* diatom adsorbent. Correspondingly, the obtained results fit with the data of the experiment of the Freundlich isotherm model, which was confirmed by a high correlation coefficient of R^2 = 0.936 for the adsorbent (Figure 11). Therefore, the Freundlich isotherm can be employed to describe heterogeneous systems.

Table 4. Factors of the isotherm models from linear solvation.

Isotherm Model	Isotherm Parameters	Values
Langmuir	Qm (mg g^{-1})	17.76
	R_L	0.011
	K_L	5.06
	R^2	0.876
	X^2	1.013
Freundlich	1/n	0.196
	nF	5.102
	K_F ($mg^{1-1/n}$ $L^{1/n}$ g^{-1})	12.94
	R^2	0.936
	X^2	0.001
Tempkin	A_T	136.38
	B_T	13.38
	b_T	185.16
	R^2	0.895
	X^2	1205.66
Halsey	n	5.1
	K	475,898
	R^2	0.936
	X^2	0.002
Henderson	n_h	1
	K_h	0.006
	R^2	0.78
	X^2	62.093

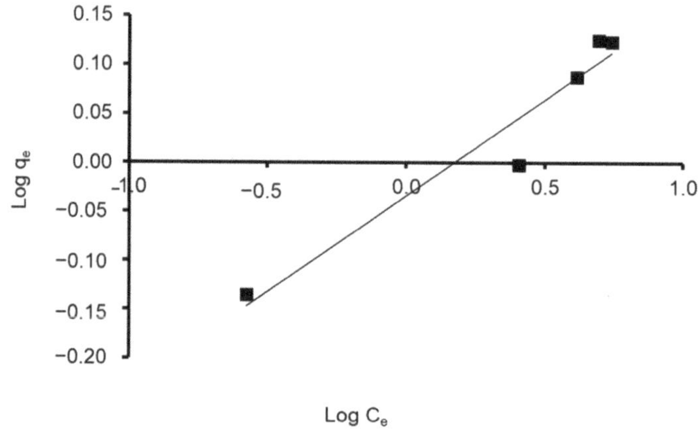

Figure 11. Plot of the Freundlich isotherm model for CVD adsorbed onto *S. costatum*.

3.4.3. Tempkin Isotherm

According to the Tempkin isotherm model, the heat of adsorption for every molecule in the layer falls linearly with coverage as a result of adsorbent–adsorbate interactions. Additionally, the model assumes that adsorption is characterized by a uniform distribution of binding energies up to maximum binding energy [96].

The Tempkin isotherm was calculated from the following form [97]:

$$q_e = B \text{ Ln } A + B \text{ Ln } C_e \tag{10}$$

$$q_e = B \text{ Ln } (AT\ Ce) \tag{11}$$

where A is the constant of equilibrium binding (L g^{-1}) associated with higher binding energy, B = (RT/b) (J mol^{-1}) is the Tempkin constant and is related to the heat of adsorption and the gas constant (R = 8.314 J mol^{-1} K^{-1}), and b (J mol^{-1}) is a constant corresponding to the heat of sorption.

A plot of q_e vs. Ln C_e enables the determination of isotherm constants A and b_T. Table 4 shows the values of the parameters. The data revealed that the Tempkin isotherm model applies to the CVD adsorption onto *S. costatum* adsorbent, as shown by the low regression correlation coefficient (R^2 = 0.895). This can be obvious when the estimated q_e values differ from those derived from the experimental study, as in Figure 12.

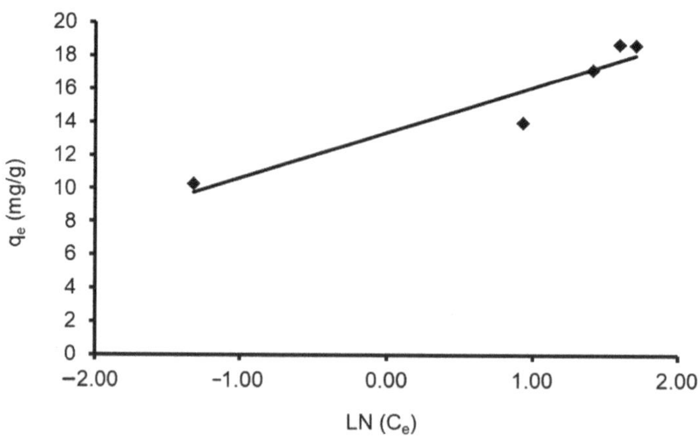

Figure 12. Plot of the Tempkin isotherm models for CVD adsorbed onto *S. costatum*.

3.4.4. Halsey and Henderson Isotherm

These models are suitable for heterosporous solids and the multilayer sorption technique [98]. The Halsey model was calculated using the following equation [99]:

$$\text{Ln } q_e = \frac{1}{n}\text{Ln K} + \frac{1}{n}\text{ Ln } C_e \tag{12}$$

Meanwhile, the Henderson model was obtained from the following equation [97]:

$$\text{Ln }[-\text{Ln}(1 - C_e)] = \text{Ln K} + \frac{1}{n}\text{ Ln } q_e \tag{13}$$

where n and K are Halsey constants for both Halsey and Henderson isotherms. A plot of ln q_e versus ln C_e for Halsey and ln $[-\ln(1-C_e)]$ versus ln q_e for Henderson. The isotherm constants and correlation coefficients are summarized in Table 4. Halsey showed a correlation coefficient of $R^2 = 0.936$ (Figure 13), while Henderson showed an R^2 of 0.780 (Figure 14). The results obtained display that the Halsey model is applicable for the adsorption of CVD onto the adsorbent.

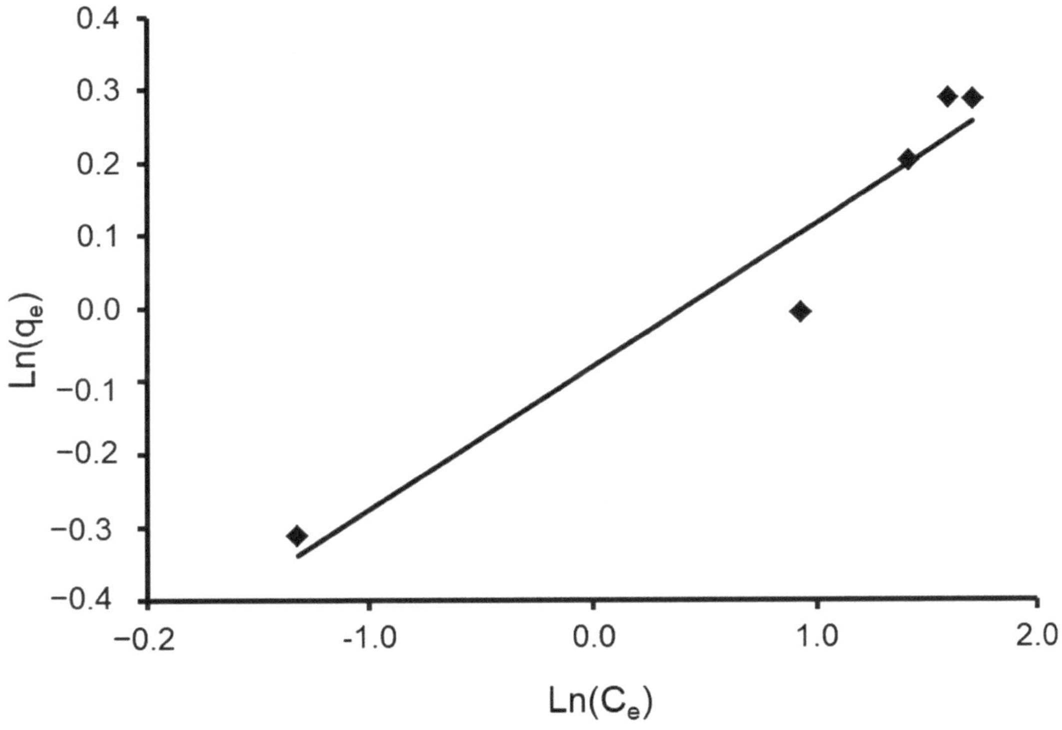

Figure 13. Plot of the Halsey isotherm model for CVD adsorbed onto *S. costatum*.

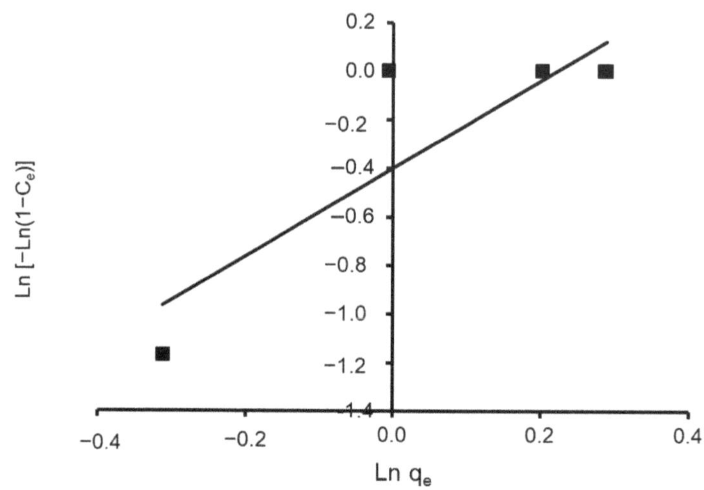

Figure 14. Plot of the Henderson isotherm model for CVD adsorbed onto *S. costatum*.

3.4.5. Error Analyses

This study used isotherm models to evaluate the materials' removal effectiveness, and the linear correlation coefficient was used to determine the relationship between the experimental data and the model under study (R^2). The Chi-square (X^2) test was additionally used as a standard to ensure the fitting quality was the non-linear regression. This statistical analysis is based on the sum of the squares of the differences between experimentally determined data and model-calculated data, where each squared difference was divided by the corresponding data produced by modeling, as calculated by the following Equation (14).

$$X^2 = \frac{\left(q_{e,isotherm} - q_{e,calc}\right)^2}{q_{e,isotherm}} \quad (14)$$

Based on the coefficient of determination (R^2) value and the smallest Chi-square test (X^2) result, the best fit model for the adsorption isotherm was selected. Halsey and Freundlich isotherm equations were found to best match the equilibrium data acquired for CVD onto *S. costatum* after analysis of R^2 values and the data were fitted. When compared to the other isotherm forms, the Langmuir, Henderson, and Tempkin forms provided the worst fit to the experimental data. Additionally, the Chi-square test of both the Halsey and Freundlich isotherm yielded the lowest value, indicating that the $q_{e,\ cal}$ of the Halsey and Freundlich isotherms and the $q_{e,}$ exp, as summarized in Table 4, are the isotherms that are most similar to one another. Therefore, with consideration of the R^2 and the error functions, it is concluded that the Halsey and Freundlich isotherm best fitted the experimental data. The R_L and n_F values from the experiment indicate that the adsorption process is favorable and chemical.

3.5. Adsorption Kinetic

To examine the regulatory mechanism of the adsorption technique, for example, mass transfer and chemical reaction, Lagergren pseudo-first-order and pseudo-second-order kineticss in addition to the intraparticle diffusions were applied to the investigation of the experimental data of CVD adsorption by *S. costatum* in the batch examinations (Figures 11–13).

3.5.1. Pseudo-First–Order Kinetic

The pseudo-first-order rate expression of Lagergren depends on the sorption capacity of the adsorbent and is generally expressed as follows:

$$Dq/d_t = K_1 (q_e - q_t) \quad (15)$$

where q_e is the quantity of CVD adsorbed at equilibrium (mg g^{-1}), q_t is the quantity of CVD adsorbed at time t (mg g^{-1}), and K_1 is expressed as a pseudo-first-order rate constant (min^{-1}). The integrating equation was assessed [100] as follows:

$$Log (q_e/q_e - q_t) = K_1t/2.303 \quad (16)$$

The pseudo-first-order equation is given by the following formula in an equation:

$$Log (q_e - q_t) = Log\ q_e - K_1t/2.303 \quad (17)$$

The plot of log ($q_e - q_t$) agents t provides a straight line for the first-order adsorption kinetics. The value of the first-order rate constant k_1 was given from the slope of the straight line. The K_1 correlation coefficients, R^2 values, and predicted and experimental q_e values for different *S. costatum*/CVD combinations are presented in Table 5. Figure 15 shows a plot of the form of the pseudo-first-order model at 25 °C. However, the correlation coefficients for the first-order kinetic model were low. In addition, the theoretical q_e values found from the first-order kinetic model did not give reasonable values (R^2 = 0.752). This suggests that this adsorption process does not follow a first-order reaction.

Table 5. The calculated values of kinetic parameters of the adsorption of CVD onto *S. costatum*.

Kinetic Models	Parameters	Values
The pseudo-first-order	q_e (calc.) (mg g^{-1})	63.09
	k_1 (min^{-1})	0.00414
	R^2	0.752
	q_e (exp.) (mg g^{-1})	0.1316
The pseudo-second-order	q_e (calc.) (mg g^{-1})	0.12
	k_2 (mg g^{-1} min^{-1})	2.16
	h_o	0.03
	R^2	0.978
	q_e (exp.) (mg g^{-1})	0.1316
The intraparticle diffusion	K_{dif} (mg g^{-1} $min^{-0.5}$)	0.0583
	C (calc.) (mg g^{-1})	0.01
	R^2	0.438

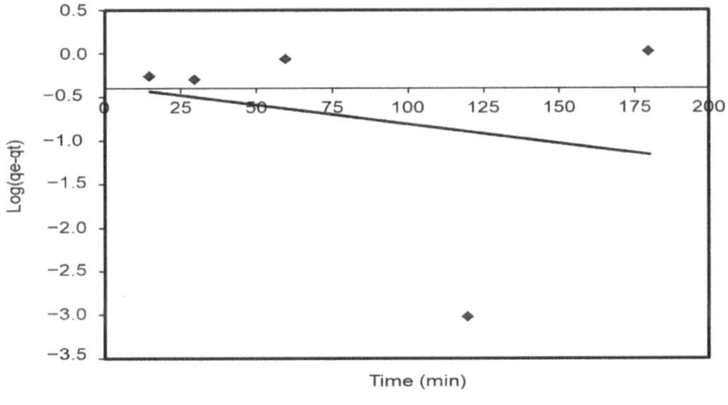

Figure 15. Pseudo-first-order plots for the adsorption of CVD onto *S. costatum*.

3.5.2. Pseudo-Second–Order Kinetic

The pseudo-second-order equations depend on the sorption efficiency of the solid phase, which, contrary to the other model, predicts the behavior over the whole range of concentrations. The pseudo-second-order kinetic rate equation is expressed as:

$$dq_t / d_t = K_2 (q_e - q_t)^2 \quad (18)$$

where K_2 is the constant of the second-order rate (g m^{-1} min^{-1}). The integrated equation is presented as follows:

$$1/(q_e - q_t) = 1/q_e + K_2 \quad (19)$$

A form of the pseudo-second-order equation was obtained as described by Yang, et al. [101] as follows:

$$t/q_t = 1/K_2\, q_e^2 + t/q_e \quad (20)$$

Plots of (t/q_t) against (t) would provide a relationship (Figure 10), and the values of the q_e and K_2 parameters can be calculated from the slope and intercept, respectively.

Meanwhile, the initial adsorption rate h_o (mg g^{-1} min^{-1}) at the time was calculated by the following [102]:

$$h_o = K_2 q_e^2 \quad (21)$$

The results indicate that the values of qe for second-order kinetics are closer to the estimated values (Table 5). The correlation coefficient (R^2 = 0.978) value for the pseudo-second order is also closer to unity when compared to the first order (Figure 16 and Table 5). The near values between the predicted values of q_e from the pseudo-second-order model and the experimental q_e, in contrast to the pseudo-first-order, predicted q_e values, which significantly differed from experimental q_e, serve as more evidence for this. This shows that the adsorption process follows a pseudo-second-order model and that it appears to be governed by chemical adsorption, including valency forces, by sharing or exchange of electrons among dye anions and the adsorbent, which provides the best correlation of the data for the dye [73].

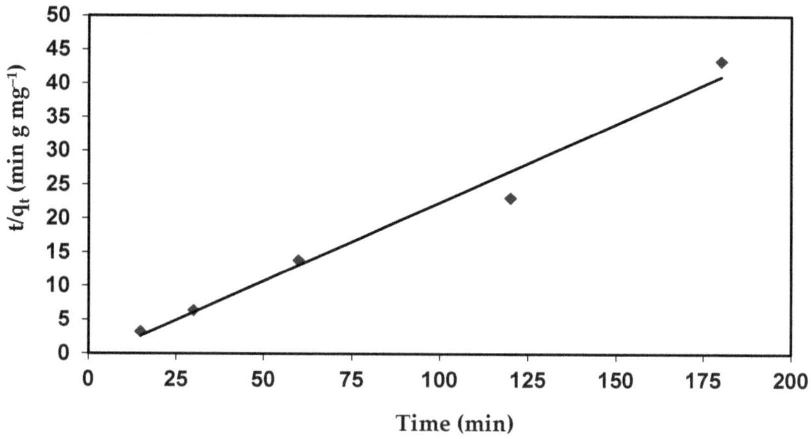

Figure 16. Pseudo-second-order plots for the adsorption of CVD onto *S. costatum*.

3.5.3. Intra-Particle Diffusion Equation

Intra-particle diffusion is a key rate-controlling phase in the first phase of adsorption, and outside mass transfer or boundary layer diffusion could be defined by the initial rate of solute sorption. The acquired data were examined using mass transfer and kinetic models [103]. Adsorption kinetics are obtained by the following stages. The intra-particle diffusion model was calculated as follows [73,104]:

$$q_t = K_{dif}\, t^{1/2} + C \quad (22)$$

where C is the intercept and K_{dif} (mg g^{-1} $min^{-0.5}$) reflects the intra-particle diffusion rate constant, which is estimated through the slope of the regression line. The intra-particle diffusion model showed no fit with the experimental data (R^2 = 0.438). Moreover, the data assigned that, in the intra-particle diffusion plots for adsorption of CVD, it appeared that none of the regions has C values equal to zero, which did not pass throughout the origin (Figure 17), suggesting that the pore diffusion is not the rate-controlling step of the investigated adsorption process (Table 5).

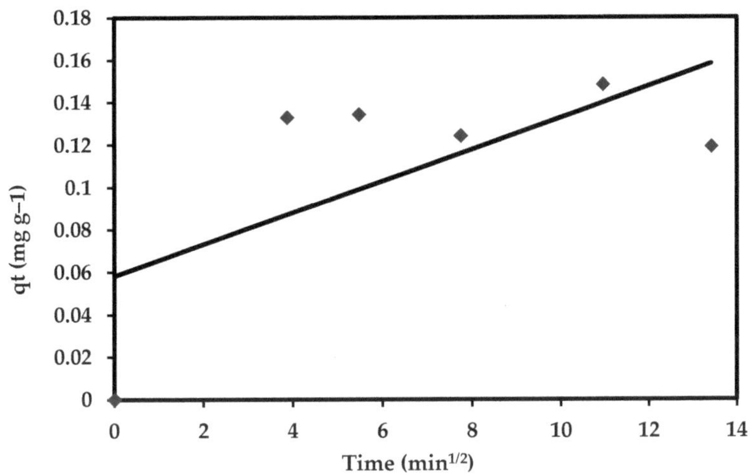

Figure 17. The intra-particle diffusion plots for the adsorption of CVD onto *S. costatum*.

3.6. Industrial Wastewater and Seawater Applications

A wastewater sample was collected from the El-Naboria drain, which belongs to Alexandria, Egypt, and comprises numerous industrial sewage and agricultural wastes. To assess the adsorbent validity of *S. costatum*, real wastewater mixed with simulated dye samples was collected to examine the removal of CVD by the adsorbent under optimal conditions (Figure 18). Owing to the low CVD concentrations, an amount of CVD was added to obtain 5 mg L^{-1} of this dye. The characteristics of the wastewater used are as follows: values of salinity (PPT) are 186.5 and pH is 8.44; the dissolved oxygen (mg L^{-1}) value was 11.01, nitrite (135.45 μg L^{-1}), nitrate (622.7 μg L^{-1}), ammonia (468.5 μg L^{-1}), and phosphate (52.37 μg L^{-1}). The CVD solution was filtered to remove precipitates and suspended the matter. The presence of salt did not affect the removal of CVD by *S. costatum*, which leads us to deduce that there was no interaction between the salt and the *S. costatum*, nor between the salt and the CVD. Moreover, the high concentration of CVD ions may make them more preferably adsorbed by *S. costatum*, since the effect of salt cannot influence the active groups of the adsorbent. The results show that adding saltwater to the dye solution increases the adsorption capacity of the CVD, and the reason for this is that the interactions between solvent molecules and NaCl ions are fewer than that of dyes, which leads to an increase in the solubility of dyes in the aqueous solution, and thus leads to a desire in the ability to adsorb it on the surface. Furthermore, it is possible to explain this flow based on the process of imbibition. This is because this process leads to swelling of the adsorbent material and the removal of the osmotic pressure of the saline solution, which leads to a change in the geometry of the pores of the adsorbent material, and thus leads to a decrease or increase in adsorption according to changes of the shape of the surface [105]. The application of *S. costatum* to remove the CVD from wastewater was carried out under optimized conditions (pH 3 for 180 min at 25 °C and 0.05 g of adsorbents in the solution). Deionized water comprising a similar concentration of CVD was applied as a control to estimate the effect of *S. costatum* on CVD. The obtained results reveal that changing the type of wastewater had a substantial impact on the greatest removal, with deionized water having the least impact on the adsorption method, with a color removal of 90.9% at pH 3 after 180 min.

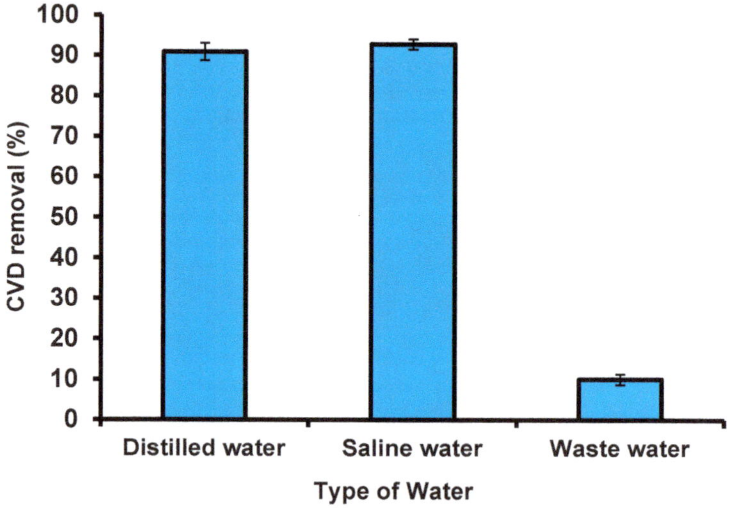

Figure 18. Removal of CVD from different solutions using *S. costatum*.

By contrast, real wastewater comprises very high concentrations of interfering ions from numerous pollutants, which had a significant influence on the elimination efficiency of CVD, with a color removal of 10.08 % after 180 min at pH 3. Regardless, the results reveal that *S. costatum* adsorbents can be successfully used as a low-cost adsorbent to remove CVD from aqueous mixtures and wastewater. Furthermore, changing the same kind of CVD solution did not affect the maximal adsorption capabilities.

3.7. Regeneration and Reusability Study

The regeneration and reuse of the sorbent material are essential for its economic viability [8]. To evaluate the reusability of *S. costatum*, adsorption–desorption experiments were conducted using 5 mg L^{-1} of CVD solution, shaken at 110 rpm for 3 h, as presented in Figure 19. The current results reveal that the removal efficiency of the *S. costatum* for CVD uptake remained nearly unchanged for the three consecutive series. The initial removal efficiency was 90.10% in 180 min, which was reduced to 69.30% after the first round. Hence, it could be concluded that the sorption capacity of the adsorbent remains unaffected with extended regeneration rounds. This decrease shows that the adsorbent represents a good material that can be used in industrial operations. It can be easily known that desorption efficiency decreases with increasing cycle numbers due to the decrease of adsorption capacity. This indicates that, while biomass-adsorbing sites were still available, they became more difficult to locate. Changes in the chemistry and structure of the adsorbent, as well as changes in the mass transport conditions, may have an influence on adsorption performance after the third cycle of use [106].

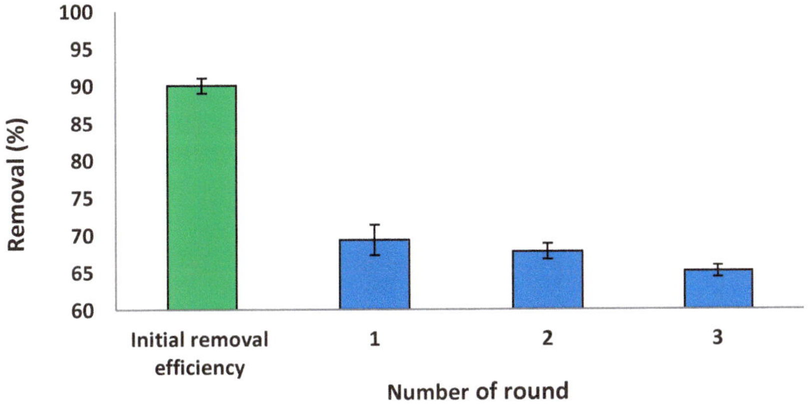

Figure 19. Reusability study of the *S. costatum* after adsorption of CVD.

3.8. Comparison with Other Studies

Table 6 presents the comparison amongst the maximum values of CVD adsorption capacities of the prepared adsorbent, and some adsorbents, comprising some activated carbons, modified biomasses, nanoparticles, and zeolite. In the present work, the amount of maximum monolayer uptake capacity of CVD by *S. costatum* was obtained at 6.41 mg g^{-1}; this amount has been compared with q_m achieved from other studies on the removal of CVD. Table 6 shows the maximum uptake capacity from the Langmuir model by various sorbents.

Table 6. The comparisons of the maximum uptake capacity of CVD by different adsorbents.

Material	q_e (mg g^{-1})	Refs.
Grapefruit peel	254.16	[22]
Lignin-rich Isolate from Elephant Grass	24.99	[102]
Pod-inspired MXene/porous carbon microspheres	750.00	[107]
Fe_3O_4 nanoparticles	269.70 to 282.50	[107]
Graphene oxide cross-linked hydrogel nanocomposites of xanthan gum	371.29	[108]
Soil-Ag nanoparticles	1.92	[109]
Cellulose-based biosorbent	153.85	[110]
Lemongrass leaf fibers incorporated with cellulose acetate	36.10	[111]
Unused solid waste of Rosewater extraction	168.8	[112]
Chitin	420.10	[113]
Zeolite from bottom ash	17.60	[114]
Natural zeolite	106.67	[115]
Magnetic Zeolite	0.97	[116]
Leaf biomass of *Calotropis procera*	4.14	[117]
Blue-green alga, *Spirulina* sp.	101.87	[118]
Marine diatom, *Skeletonema costatum*	6.41	Present study

3.9. Suggested Mechanism

The adsorption mechanism could be affected by a wide variety of factors. Adsorption is the outcome of the interaction between ions, molecules, and the surface of the adsorbent. The surface of the adsorbent carries different kinds of functional groups, such as amino, carbonyl, phosphate, carboxylate, carboxyl, and hydroxyl, as reported by FTIR analysis, which are responsible for the confiscation of harmful materials from industrial wastewater [119]. However, the dye remediation process is mainly comprised of two mechanisms: active uptake and passive uptake [65]. The active uptake is the exchange of monovalent and divalent ions such as Na, Mg, and Ca in the cell walls, which are replaced by ions of dyes. Passive uptake is the formation of a complex between the ions of dyes with functional groups located on the cell wall; this process is alternating and fast, while active uptake is a process that occurs when dye ions bind to the cell [65]. This mechanism occurs in combination with the usage of dye ions by microorganisms and intracellular accumulation of dye ions; dyes can also be deposited in the metabolism and excretion process on a second level. pH and the presence of other ions are factors that influence the uptake passive process. pH and the presence of other ions influence the passive uptake process, whereas pH and temperature influence the active uptake process [120].

On the other hand, Soedarti et al. [66] found that *Skeletonema* sp., including microalgae, are aquatic organisms containing molecular mechanisms that may discriminate non-essential toxic metals from metal ions that are required for growth. *Skeletonema costatum* can absorb CVD ions in two ways: absorption and adsorption, according to this study. *S. costatum* has cell walls, which allow for adsorption. *S. costatum* has cellulose-based cell walls. The functional groups in the cellulose inside cell walls, such as hydroxyl, can bond with ions in dye [66]. *S. costatum* also absorbs nutrients because it produces phytochelatins [59]. *S. costatum is* like other microalgae that also produces phytochelatins, namely, peptides metallothionein class III (Mt-III), to detoxify heavy metals, which can be induced by the existence of heavy metal ions, as reported by Perales-Vela et al. [121] and Shaw [122].

4. Conclusions

The diatom *Skeletonema costatum* was found to be effective in the uptake of Crystal Violet Dye (CVD). Maximum dye adsorption was achieved by a batch technique at pH 3, adsorbent dose 0.4 g, contact time 120 min, and 5 mg L^{-1} initial dye concentration at 25 °C. The percentage removal of the *S. costatum* diatom rose to 95.2% for an adsorbent dosage of 0.4 g. Freundlich and Halsey's isotherm models describe the adsorption of CVD onto *S. costatum*, suggesting that adsorption was onto a non-uniform site and that the CVD adsorption data followed the pseudo-second-order kinetic model. Thermodynamic data indicate the endothermic nature of adsorption systems. The calculated R_L value was less than 1, while the value of n was greater than 1, hence the adsorption process was favorable. Values of q_e agreed with the values of the current experiment. However, initial rapid adsorption was onto a uniform site, thus maximum monolayer adsorption capacity was obtained to be 6.410 mg g^{-1}. The *S. costatum* displayed significant regeneration possibility and application on real wastewater. Finally, the current work concluded that the marine diatom *S. costatum* is significantly used to eliminate the textile toxic dye from wastewater, since it is an effective, eco-friendly, and locally available adsorbent.

Author Contributions: Conceptualization, A.E.A. and M.A.; methodology, A.E.A., M.A. and M.K.; software, A.E.A., A.T.M. and M.A.; validation, A.E.A., M.A., M.K. and A.T.M.; formal analysis, A.E.A. and M.A.; investigation, A.E.A., M.A. and M.K.; resources, A.E.A., M.A., K.M.A. and A.T.M.; data curation, A.E.A., M.A., M.K. and A.T.M.; writing—original draft preparation, A.E.A., M.A. and A.T.M.; writing—review and editing, A.E.A., M.A. and A.T.M.; visualization, A.E.A., M.A. and A.T.M.; supervision, A.E.A., M.A. and A.T.M.; project administration, A.E.A., M.A. and A.T.M.; funding acquisition, A.E.A., M.A., K.M.A. and A.T.M. All authors have read and agreed to the published version of the manuscript.

Funding: This research was funded by Taif University Researchers Supporting Project number TURSP-2020/267, Taif University, Taif, Saudi Arabia.

Institutional Review Board Statement: Not applicable.

Informed Consent Statement: Not applicable.

Data Availability Statement: Not applicable.

Acknowledgments: The authors gratefully appreciate Taif University Researchers Supporting Project number TURSP-2020/267, Taif University, Taif, Saudi Arabia for support publication.

Conflicts of Interest: The authors declare no conflict of interest.

References

1. Sharma, A.; Choudhry, A.; Rathi, G.; Tara, N.; Abdulla, N.K.; Sajid, M.; Chaudhry, S.A. Ferrite based magnetic nanocomposites for wastewater treatment through adsorption. In *Contamination of Water*; Elsevier: Amsterdam, The Netherlands, 2021; pp. 449–460.
2. Mangla, D.; Sharma, A.; Ikram, S. Critical review on adsorptive removal of antibiotics: Present situation, challenges and future perspective. *J. Hazard. Mater.* **2022**, *425*, 127946. [CrossRef] [PubMed]
3. Choudhry, A.; Sharma, A.; Khan, T.A.; Chaudhry, S.A. Flax seeds based magnetic hybrid nanocomposite: An advance and sustainable material for water cleansing. *J. Water Process Eng.* **2021**, *42*, 102150. [CrossRef]
4. Jabeen, A.; Bhatti, H.N. Adsorptive removal of reactive green 5 (RG-5) and direct yellow 50 (DY-50) from simulated wastewater by Mangifera indica seed shell and its magnetic composite: Batch and Column study. *Environ. Technol. Innov.* **2021**, *23*, 101685. [CrossRef]
5. Karaman, C.; Karaman, O.; Show, P.-L.; Orooji, Y.; Karimi-Maleh, H. Utilization of a double-cross-linked amino-functionalized three-dimensional graphene networks as a monolithic adsorbent for methyl orange removal: Equilibrium, kinetics, thermodynamics and artificial neural network modeling. *Environ. Res.* **2022**, *207*, 112156. [CrossRef] [PubMed]
6. El-Bindary, A.A.; Hussien, M.A.; Diab, M.A.; Eessa, A.M. Adsorption of Acid Yellow 99 by polyacrylonitrile/activated carbon composite: Kinetics, thermodynamics and isotherm studies. *J. Mol. Liq.* **2014**, *197*, 236–242. [CrossRef]
7. Abualnaja, K.M.; Alprol, A.E.; Abu-Saied, M.A.; Ashour, M.; Mansour, A.T. Removing of Anionic Dye from Aqueous Solutions by Adsorption Using of Multiwalled Carbon Nanotubes and Poly (Acrylonitrile-styrene) Impregnated with Activated Carbon. *Sustainability* **2021**, *13*, 7077. [CrossRef]
8. Abualnaja, K.M.; Alprol, A.E.; Abu-Saied, M.A.; Mansour, A.T.; Ashour, M. Studying the Adsorptive Behavior of Poly(Acrylonitrile-*co*-Styrene) and Carbon Nanotubes (Nanocomposites) Impregnated with Adsorbent Materials towards Methyl Orange Dye. *Nanomaterials* **2021**, *11*, 1144. [CrossRef]
9. Abualnaja, K.M.; Alprol, A.E.; Ashour, M.; Mansour, A.T. Influencing Multi-Walled Carbon Nanotubes for the Removal of Ismate Violet 2R Dye from Wastewater: Isotherm, Kinetics, and Thermodynamic Studies. *Appl. Sci.* **2021**, *11*, 4786. [CrossRef]
10. Hunge, Y.M.; Uchida, A.; Tominaga, Y.; Fujii, Y.; Yadav, A.A.; Kang, S.-W.; Suzuki, N.; Shitanda, I.; Kondo, T.; Itagaki, M. Visible light-assisted photocatalysis using spherical-shaped bivo4 photocatalyst. *Catalysts* **2021**, *11*, 460. [CrossRef]
11. Zhang, M.; Sun, X.; Wang, C.; Wang, Y.; Tan, Z.; Li, J.; Xi, B. Photocatalytic degradation of rhodamine B using $Bi_4O_5Br_2$-doped ZSM-5. *Mater. Chem. Phys.* **2022**, *278*, 125697. [CrossRef]
12. Yadav, A.; Kang, S.-W.; Hunge, Y. Photocatalytic degradation of Rhodamine B using graphitic carbon nitride photocatalyst. *J. Mater. Sci. Mater. Electron.* **2021**, *32*, 15577–15585. [CrossRef]
13. Gupta, V.; Agarwal, A.; Singh, M.; Singh, N. Removal of Red RB dye from aqueous solution by belpatra bark charcoal (BBC) adsorbent. *J. Mater. Environ. Sci.* **2017**, *8*, 3654–3665.
14. Alprol, A.E. Study of environmental concerns of dyes and recent textile effluents treatment technology: A Review. *Asian J. Fish. Aquat. Res.* **2019**, *3*, 1–18. [CrossRef]
15. Karimi-Maleh, H.; Darabi, R.; Shabani-Nooshabadi, M.; Baghayeri, M.; Karimi, F.; Rouhi, J.; Alizadeh, M.; Karaman, O.; Vasseghian, Y.; Karaman, C. Determination of D&C Red 33 and Patent Blue V Azo dyes using an impressive electrochemical sensor based on carbon paste electrode modified with ZIF-8/g-C_3N_4/Co and ionic liquid in mouthwash and toothpaste as real samples. *Food Chem. Toxicol.* **2022**, *162*, 112907. [PubMed]
16. Alprol, A.E.; Heneash, A.M.M.; Soliman, A.M.; Ashour, M.; Alsanie, W.F.; Gaber, A.; Mansour, A.T. Assessment of Water Quality, Eutrophication, and Zooplankton Community in Lake Burullus, Egypt. *Diversity* **2021**, *13*, 268. [CrossRef]
17. Khan, M.M.; Khan, A.; Bhatti, H.N.; Zahid, M.; Alissa, S.; El-Badry, Y.A.; Hussein, E.E.; Iqbal, M. Composite of polypyrrole with sugarcane bagasse cellulosic biomass and adsorption efficiency for 2,4-dicholrophonxy acetic acid in column mode. *J. Mater. Res. Technol.* **2021**, *15*, 2016–2025. [CrossRef]
18. Noreen, S.; Tahira, M.; Ghamkhar, M.; Hafiz, I.; Bhatti, H.N.; Nadeem, R.; Murtaza, M.A.; Yaseen, M.; Sheikh, A.A.; Naseem, Z. Treatment of textile wastewater containing acid dye using novel polymeric graphene oxide nanocomposites (GO/PAN, GO/PPy, GO/PSty). *J. Mater. Res. Technol.* **2021**, *14*, 25–35. [CrossRef]
19. Yadav, A.; Hunge, Y.; Kang, S.-W. Porous nanoplate-like tungsten trioxide/reduced graphene oxide catalyst for sonocatalytic degradation and photocatalytic hydrogen production. *Surf. Interfaces* **2021**, *24*, 101075. [CrossRef]
20. Khan, M.A.N.; Siddique, M.; Wahid, F.; Khan, R. Removal of reactive blue 19 dye by sono, photo and sonophotocatalytic oxidation using visible light. *Ultrason. Sonochem.* **2015**, *26*, 370–377. [CrossRef]
21. Tahir, N.; Bhatti, H.N.; Iqbal, M.; Noreen, S. Biopolymers composites with peanut hull waste biomass and application for Crystal Violet adsorption. *Int. J. Biol. Macromol.* **2017**, *94*, 210–220. [CrossRef]
22. Saeed, A.; Sharif, M.; Iqbal, M. Application potential of grapefruit peel as dye sorbent: Kinetics, equilibrium and mechanism of crystal violet adsorption. *J. Hazard. Mater.* **2010**, *179*, 564–572. [CrossRef] [PubMed]
23. Ferreira, B.C.S.; Teodoro, F.S.; Mageste, A.B.; Gil, L.F.; de Freitas, R.P.; Gurgel, L.V.A. Application of a new carboxylate-functionalized sugarcane bagasse for adsorptive removal of crystal violet from aqueous solution: Kinetic, equilibrium and thermodynamic studies. *Ind. Crops Prod.* **2015**, *65*, 521–534. [CrossRef]
24. Gopi, S.; Pius, A.; Thomas, S. Enhanced adsorption of crystal violet by synthesized and characterized chitin nano whiskers from shrimp shell. *J. Water Process Eng.* **2016**, *14*, 1–8. [CrossRef]

25. Alprol, A.E.; Abd El Azzem, M.; Amer, A.; El-Metwally, M.E.; Abd El-Hamid, H.T.; El-Moselhy, K.M. Adsorption of cadmium (II) ions from aqueous solution onto mango leaves. *Asian J. Phys. Chem. Sci.* **2017**, *2*, 1–11. [CrossRef]
26. Sharma, A.; Chaudhry, S.A. Adsorption of Pharmaceutical Pollutants Using Lignocellulosic Materials. In *Green Materials for Wastewater Treatment*; Springer: Berlin/Heidelberg, Germany, 2020; pp. 277–289.
27. Abbas, E.M.; Ali, F.S.; Desouky, M.G.; Ashour, M.; El-Shafei, A.; Maaty, M.M.; Sharawy, Z.Z. Novel Comprehensive Molecular and Ecological Study Introducing Coastal Mud Shrimp (*Solenocera Crassicornis*) Recorded at the Gulf of Suez, Egypt. *J. Mar. Sci. Eng.* **2020**, *9*, 9. [CrossRef]
28. Metwally, A.S.; El-Naggar, H.A.; El-Damhougy, K.A.; Bashar, M.A.E.; Ashour, M.; Abo-Taleb, H.A.H. GC-MS analysis of bioactive components in six different crude extracts from the Soft Coral (*Sinularia maxim*) collected from Ras Mohamed, Aqaba Gulf, Red Sea, Egypt. *Egypt. J. Aquat. Biol. Fish.* **2020**, *24*, 425–434. [CrossRef]
29. Ashour, M.; Mabrouk, M.M.; Abo-Taleb, H.A.; Sharawy, Z.Z.; Ayoub, H.F.; Van Doan, H.; Davies, S.J.; El-Haroun, E.; Goda, A.M.S.A. A liquid seaweed extract (TAM®) improves aqueous rearing environment, diversity of zooplankton community, whilst enhancing growth and immune response of *Nile tilapia, Oreochromis niloticus,* challenged by *Aeromonas hydrophila. Aquaculture* **2021**, *543*, 736915. [CrossRef]
30. Wang, S.; Hu, S.; Shang, H.; Barati, B.; Gong, X.; Hu, X.; Abomohra, A.E.-F. Study on the co-operative effect of kitchen wastewater for harvest and enhanced pyrolysis of microalgae. *Bioresour. Technol.* **2020**, *317*, 123983. [CrossRef]
31. Tu, R.; Jin, W.; Xi, T.; Yang, Q.; Han, S.-F.; Abomohra, A.E.-F. Effect of static magnetic field on the oxygen production of *Scenedesmus obliquus* cultivated in municipal wastewater. *Water Res.* **2015**, *86*, 132–138. [CrossRef]
32. Han, S.; Jin, W.; Chen, Y.; Tu, R.; Abomohra, A.E.-F. Enhancement of lipid production of *Chlorella pyrenoidosa* cultivated in municipal wastewater by magnetic treatment. *Appl. Biochem. Biotechnol.* **2016**, *180*, 1043–1055. [CrossRef]
33. Tadda, M.A.; Li, C.; Gouda, M.; Abomohra, A.E.-F.; Shitu, A.; Ahsan, A.; Zhu, S.; Liu, D. Enhancement of nitrite/ammonia removal from saline recirculating aquaculture wastewater system using moving bed bioreactor. *J. Environ. Chem. Eng.* **2021**, *9*, 105947. [CrossRef]
34. Essa, D.; Abo-Shady, A.; Khairy, H.; Abomohra, A.E.-F.; Elshobary, M. Potential cultivation of halophilic oleaginous microalgae on industrial wastewater. *Egypt. J. Bot.* **2018**, *58*, 205–216. [CrossRef]
35. Ashour, M.; Hassan, S.M.; Elshobary, M.E.; Ammar, G.A.G.; Gaber, A.; Alsanie, W.F.; Mansour, A.T.; El-Shenody, R. Impact of Commercial Seaweed Liquid Extract (TAM®) Biostimulant and Its Bioactive Molecules on Growth and Antioxidant Activities of Hot Pepper (*Capsicum annuum*). *Plants* **2021**, *10*, 1045. [CrossRef] [PubMed]
36. Battah, M.; El-Ayoty, Y.; Abomohra, A.E.-F.; El-Ghany, S.A.; Esmael, A. Optimization of growth and lipid production of the Chlorophyte Microalga *Chlorella vulgaris* as a feedstock for biodiesel production. *World Appl. Sci. J.* **2013**, *28*, 1536–1543.
37. Cao, B.; Sun, Y.; Guo, J.; Wang, S.; Yuan, J.; Esakkimuthu, S.; Bernard Uzoejinwa, B.; Yuan, C.; Abomohra, A.E.F.; Qian, L.; et al. Synergistic effects of co-pyrolysis of macroalgae and polyvinyl chloride on bio-oil/bio-char properties and transferring regularity of chlorine. *Fuel* **2019**, *246*, 319–329. [CrossRef]
38. Ebaid, R.; Elhussainy, E.; El-Shourbagy, S.; Ali, S.; Abomohra, A.E.-F. Protective effect of Arthrospira platensis against liver injury induced by copper nanoparticles. *Orient. Pharm. Exp. Med.* **2017**, *17*, 203–210. [CrossRef]
39. El-Mesery, H.S.; Mao, H.; Abomohra, A.E.F. Applications of non-destructive technologies for agricultural and food products quality inspection. *Sensors* **2019**, *19*, 846. [CrossRef]
40. El-Shouny, W.; Sharaf, M.; Abomohra, A.; Abo-Eleneen, M. Production enhancement of some valuable compounds of *Arthrospira platensis*. *J. Basic Environ. Sci.* **2015**, *2*, 74–83.
41. Shao, W.; Ebaid, R.; El-Sheekh, M.; Abomohra, A.; Eladel, H. Pharmaceutical applications and consequent environmental impacts of *Spirulina (Arthrospira):* An overview. *Grasas Y Aceites* **2019**, *70*, 292. [CrossRef]
42. Huo, S.; Basheer, S.; Liu, F.; Elshobary, M.; Zhang, C.; Qian, J.; Xu, L.; Arslan, M.; Cui, F.; Zan, X.; et al. Bacterial intervention on the growth, nutrient removal and lipid production of filamentous oleaginous microalgae *Tribonema* sp. *Algal Res.* **2020**, *52*, 102088. [CrossRef]
43. Osman, M.E.; Abo-shady, A.M.; Elshobary, M.E. Production and characterization of antimicrobial active substance from some macroalgae collected from Abu-Qir bay (Alexandria) Egypt. *Afr. J. Biotechnol.* **2013**, *12*, 6847–6858.
44. Osman, M.E.H.; Abo-shady, A.M.; Elshobary, M.E. In vitro screening of antimicrobial activity of extracts of some macroalgae collected from Abu-Qir bay Alexandria, Egypt. *Afr. J. Biotechnol.* **2010**, *9*, 7203–7208.
45. Madkour, F.F.; Kamil, A.E.-W.; Nasr, H.S. Production and nutritive value of *Spirulina platensis* in reduced cost media. *Egypt. J. Aquat. Res.* **2012**, *38*, 51–57. [CrossRef]
46. El-Kassas, H.Y.; Mohamed, L.A. Bioremediation of the textile waste effluent by *Chlorella vulgaris*. *Egypt. J. Aquat. Res.* **2014**, *40*, 301–308. [CrossRef]
47. Yang, J.; Zhu, Q.; Chai, J.; Xu, F.; Ding, Y.; Zhu, Q.; Lu, Z.; Khoo, K.S.; Bian, X.; Wang, S. Development of environmentally friendly biological algicide and biochemical analysis of inhibitory effect of diatom *Skeletonema costatum*. *Chin. Chem. Lett.* **2021**, *33*, 1358–1364. [CrossRef]
48. Li, H.; Xu, T.; Ma, J.; Li, F.; Xu, J. Physiological responses of *Skeletonema costatum* to the interactions of seawater acidification and the combination of photoperiod and temperature. *Biogeosciences* **2021**, *18*, 1439–1449. [CrossRef]
49. Shabaka, S.H. Checklist of seaweeds and seagrasses of Egypt (Mediterranean Sea): A review. *Egypt. J. Aquat. Res.* **2018**, *44*, 203–212. [CrossRef]

50. Fu, Y.; Viraraghavan, T. Removal of Congo Red from an aqueous solution by fungus *Aspergillus niger*. *Adv. Environ. Res.* **2002**, *7*, 239–247. [CrossRef]
51. Sulaymon, A.H.; Ebrahim, S.E.; Abdullah, S.M.; Al-Musawi, T.J. Removal of lead, cadmium, and mercury ions using biosorption. *Desalination Water Treat.* **2010**, *24*, 344–352. [CrossRef]
52. Aksu, Z. Application of biosorption for the removal of organic pollutants: A review. *Process Biochem.* **2005**, *40*, 997–1026. [CrossRef]
53. Guedes, A.C.; Malcata, F.X. Nutritional value and uses of microalgae in aquaculture. *Aquaculture* **2012**, *10*, 59–78.
54. Rajeswari, M.V.; Balasubramanian, T. Comparative study on growth of *Skeletonema costatum*: A microalga as live feed for aquaculture importance. *Int. J. Res. Fish. Aquac.* **2014**, *4*, 117–121.
55. Azmi, K.; Arsad, S.; Sari, L. The effect of commercial nutrients to increase the population of *Skeletonema costatum* on laboratory and mass scales. In Proceedings of the IOP Conference Series: Earth and Environmental Science, Surabaya, Indonesia, 26 September 2019; p. 12039.
56. Pratama, N.; Rahardja, B.; Sari, L. The effect of density as *Skeletonema costatum* bioremediation agent of copper (Cu) heavy metal concentration. In Proceedings of the IOP Conference Series: Earth and Environmental Science, Surabaya, Indonesia, 26 September 2019; p. 12039. [CrossRef]
57. Zhang, C.; Chen, X.; Wang, J.; Tan, L. Toxic effects of microplastic on marine microalgae *Skeletonema costatum*: Interactions between microplastic and algae. *Environ. Pollut.* **2017**, *220*, 1282–1288. [CrossRef] [PubMed]
58. Mangla, D.; Sharma, A.; Ikram, S. Synthesis of ecological chitosan/PVP magnetic composite: Remediation of amoxicillin trihydrate from its aqueous solution, isotherm modelling, thermodynamic, and kinetic studies. *React. Funct. Polym.* **2022**, *175*, 105261. [CrossRef]
59. Nassiri, Y.; Mansot, J.; Wéry, J.; Ginsburger-Vogel, T.; Amiard, J. Ultrastructural and electron energy loss spectroscopy studies of sequestration mechanisms of Cd and Cu in the marine diatom *Skeletonema costatum*. *Arch. Environ. Contam. Toxicol.* **1997**, *33*, 147–155. [CrossRef]
60. Özer, A.; Akkaya, G.; Turabik, M. The removal of Acid Red 274 from wastewater: Combined biosorption and biocoagulation with *Spirogyra rhizopus*. *Dye. Pigment.* **2006**, *71*, 83–89. [CrossRef]
61. Mohan, S.V.; Rao, N.C.; Prasad, K.K.; Karthikeyan, J. Treatment of simulated Reactive Yellow 22 (Azo) dye effluents using Spirogyra species. *Waste Manag.* **2002**, *22*, 575–582. [CrossRef]
62. Kumar, P.S.; Ramalingam, S.; Senthamarai, C.; Niranjanaa, M.; Vijayalakshmi, P.; Sivanesan, S. Adsorption of dye from aqueous solution by cashew nut shell: Studies on equilibrium isotherm, kinetics and thermodynamics of interactions. *Desalination* **2010**, *261*, 52–60. [CrossRef]
63. Senthilkumaar, S.; Kalaamani, P.; Subburaam, C. Liquid phase adsorption of crystal violet onto activated carbons derived from male flowers of coconut tree. *J. Hazard. Mater.* **2006**, *136*, 800–808. [CrossRef]
64. Salama, R.S.; El-Bahy, S.M.; Mannaa, M.A. Sulfamic acid supported on mesoporous MCM-41 as a novel, efficient and reusable heterogenous solid acid catalyst for synthesis of xanthene, dihydropyrimidinone and coumarin derivatives. *Colloids Surf. A Physicochem. Eng. Asp.* **2021**, *628*, 127261. [CrossRef]
65. Soedarti, T.; Tini, S.; Sucipto, H.; Kuncoro, E.P. Bioremediation of mercury (II) contaminated seawater using the diatom *Skeletonema costatum*. *KnE Life Sci.* **2017**, *3*, 62–68. [CrossRef]
66. Soedarti, T.; Surtiningsih, T.; Hariyanto, S.; Kuncoro, E.P. Remediation of Pb (II) and Cd (II) in seawater by Skeletonema costatum. *Sustinere J. Environ. Sustain.* **2017**, *1*, 63–143. [CrossRef]
67. Guillard, R.R. Culture of phytoplankton for feeding marine invertebrates. In *Culture of Marine Invertebrate Animals*; Springer: Berlin/Heidelberg, Germany, 1975; pp. 29–60. [CrossRef]
68. Ghoneim, M.M.; El-Desoky, H.S.; El-Moselhy, K.M.; Amer, A.; Abou El-Naga, E.H.; Mohamedein, L.I.; Al-Prol, A.E. Removal of cadmium from aqueous solution using marine green algae, *Ulva lactuca*. *Egypt. J. Aquat. Res.* **2014**, *40*, 235–242. [CrossRef]
69. Zar, J. *Biostatistical Analysis*, 2nd ed.; Prentice-Hall: Englewood Cliffs, NJ, USA, 1984.
70. Hashem, F. Removal of methylene blue by magnetitecovered bentonite nano-composite. *Eur. Chem. Bull.* **2013**, *2*, 524–529.
71. Eddebbagh, M.; Abourriche, A.; Berrada, M.; Zina, M.B.; Bennamara, A. Adsorbent material from pomegranate (*Punica granatum*) leaves: Optimization on removal of methylene blue using response surface methodology. *J. Mater. Environ. Sci.* **2016**, *7*, 2021–2033.
72. Zhao, D.; Zhang, W.; Chen, C.; Wang, X. Adsorption of methyl orange dye onto multiwalled carbon nanotubes. *Procedia Environ. Sci.* **2013**, *18*, 890–895. [CrossRef]
73. Aksu, Z.; Tezer, S. Equilibrium and kinetic modelling of biosorption of Remazol Black B by *Rhizopus arrhizus* in a batch system: Effect of temperature. *Process Biochem.* **2000**, *36*, 431–439. [CrossRef]
74. Yao, Y.; Bing, H.; Feifei, X.; Xiaofeng, C. Equilibrium and kinetic studies of methyl orange adsorption on multiwalled carbon nanotubes. *Chem. Eng. J.* **2011**, *170*, 82–89. [CrossRef]
75. Wu, K.; Yu, J.; Jiang, X. Multi-walled carbon nanotubes modified by polyaniline for the removal of alizarin yellow R from aqueous solutions. *Adsorpt. Sci. Technol.* **2018**, *36*, 198–214. [CrossRef]
76. Shirmardi, M.; Mahvi, A.H.; Hashemzadeh, B.; Naeimabadi, A.; Hassani, G.; Niri, M.V. The adsorption of malachite green (MG) as a cationic dye onto functionalized multi walled carbon nanotubes. *Korean J. Chem. Eng.* **2013**, *30*, 1603–1608. [CrossRef]

77. Gupta, A.; Das, A.; Majumder, K.; Arora, N.; Mayo, H.G.; Singh, P.P.; Beg, M.S.; Singh, S. Obesity is independently associated with increased risk of hepatocellular cancer-related mortality: A systematic review and meta-analysis. *Am. J. Clin. Oncol.* **2018**, *41*, 874. [CrossRef]
78. Doğar, Ç.; Gürses, A.; Açıkyıldız, M.; Özkan, E. Thermodynamics and kinetic studies of biosorption of a basic dye from aqueous solution using green algae *Ulothrix* sp. *Colloids Surf. B Biointerfaces* **2010**, *76*, 279–285. [CrossRef]
79. Amuda, O.S.; Olayiwola, A.O.; Alade, A.O.; Farombi, A.G.; Adebisi, S.A. Adsorption of methylene blue from aqueous solution using steam-activated carbon produced from *Lantana camara* Stem. *J. Environ. Prot.* **2014**, *5*, 1352. [CrossRef]
80. Alprol, A.E.; Heneash, A.M.M.; Ashour, M.; Abualnaja, K.M.; Alhashmialameer, D.; Mansour, A.T.; Sharawy, Z.Z.; Abu-Saied, M.A.; Abomohra, A.E. Potential Applications of *Arthrospira platensis* Lipid-Free Biomass in Bioremediation of Organic Dye from Industrial Textile Effluents and Its Influence on Marine Rotifer (*Brachionus plicatilis*). *Materials* **2021**, *14*, 4446. [CrossRef]
81. Inyinbor, A.; Adekola, F.; Olatunji, G.A. Kinetics, isotherms and thermodynamic modeling of liquid phase adsorption of Rhodamine B dye onto *Raphia hookerie* fruit epicarp. *Water Resour. Ind.* **2016**, *15*, 14–27. [CrossRef]
82. Taufik, A.; Saleh, R. Synthesis of iron (II, III) oxide/zinc oxide/copper (II) oxide (Fe3O4/ZnO/CuO) nanocomposites and their photosonocatalytic property for organic dye removal. *J. Colloid Interface Sci.* **2017**, *491*, 27–36. [CrossRef]
83. Hossain, M.; Ngo, H.H.; Guo, W.; Nguyen, T. Removal of copper from water by adsorption onto banana peel as bioadsorbent. *GEOMATE J.* **2012**, *2*, 227–234. [CrossRef]
84. Subbareddy, Y.; Jeseentharani, V.; Jayakumar, C.; Nagaraja, K.; Jeyaraj, B. Adsorptive removal of Malachite Green (oxalate) by low cost adsorbent. *J. Environ. Res. Dev.* **2012**, *7*, 275–284.
85. Achak, M.; Hafidi, A.; Ouazzani, N.; Sayadi, S.; Mandi, L. Low cost biosorbent "banana peel" for the removal of phenolic compounds from olive mill wastewater: Kinetic and equilibrium studies. *J. Hazard. Mater.* **2009**, *166*, 117–125. [CrossRef]
86. Memon, J.R.; Memon, S.Q.; Bhanger, M.; Memon, G.Z.; El-Turki, A.; Allen, G.C. Characterization of banana peel by scanning electron microscopy and FT-IR spectroscopy and its use for cadmium removal. *Colloids Surf. B Biointerfaces* **2008**, *66*, 260–265. [CrossRef]
87. Abdelwahab, O.; Amin, N. Adsorption of phenol from aqueous solutions by Luffa cylindrica fibers: Kinetics, isotherm and thermodynamic studies. *Egypt. J. Aquat. Res.* **2013**, *39*, 215–223. [CrossRef]
88. Langmuir, I. The constitution and fundamental properties of solids and liquids. II. Liquids. *J. Am. Chem. Soc.* **1917**, *39*, 1848–1906. [CrossRef]
89. Foo, K.Y.; Hameed, B.H. Insights into the modeling of adsorption isotherm systems. *Chem. Eng. J.* **2010**, *156*, 2–10. [CrossRef]
90. Tran, H.; You, S.-J.; Hosseini-Bandegharaei, A.; Chao, H.P. Mistakes and inconsistencies regarding adsorption of contaminants from aqueous solutions: A critical review. *Water Res.* **2017**, *120*, 88–116. [CrossRef]
91. Hall, K.R.; Eagleton, L.C.; Acrivos, A.; Vermeulen, T. Pore-and solid-diffusion kinetics in fixed-bed adsorption under constant-pattern conditions. *Ind. Eng. Chem. Fundam.* **1966**, *5*, 212–223. [CrossRef]
92. Ahalya, N.; Kanamadi, R.; Ramachandra, T. Biosorption of iron (III) from aqueous solutions using the husk of *Cicer arientinum*. *CSIR* **2006**, *13*, 122–127.
93. Adamson, A.W.; Gast, A.P. *Physical Chemistry of Surfaces*; Interscience publishers: New York, NY, USA, 1967; Volume 150.
94. Freundlich, H. *Über die Adsorption in Lösungen. Habilitationsschrift Durch Welche … zu Haltenden Probevorlesung" Kapillarchemie und Physiologie" Einladet Dr. Herbert Freundlich*; W. Engelmann: Leipzig, Germany, 1906.
95. Kargi, F.; Cikla, S. Biosorption of zinc (II) ions onto powdered waste sludge (PWS): Kinetics and isotherms. *Enzym. Microb. Technol.* **2006**, *38*, 705–710. [CrossRef]
96. Kavitha, D.; Namasivayam, C. Experimental and kinetic studies on methylene blue adsorption by coir pith carbon. *Bioresour. Technol.* **2007**, *98*, 14–21. [CrossRef]
97. Wang, X.; Guo, Y.; Yang, L.; Han, M.; Zhao, J.; Cheng, X. Nanomaterials as sorbents to remove heavy metal ions in wastewater treatment. *J. Environ. Anal. Toxicol.* **2012**, *2*, 154. [CrossRef]
98. Volesky, B. Detoxification of metal-bearing effluents: Biosorption for the next century. *Hydrometallurgy* **2001**, *59*, 203–216. [CrossRef]
99. Halsey, G. Physical adsorption on non-uniform surfaces. *J. Chem. Phys.* **1948**, *16*, 931–937. [CrossRef]
100. Kumar, Y.P.; King, P.; Prasad, V. Removal of copper from aqueous solution using *Ulva fasciata* sp.—A marine green algae. *J. Hazard. Mater.* **2006**, *137*, 367–373. [CrossRef] [PubMed]
101. Yang, S.I.; Seth, J.; Strachan, J.P.; Gentemann, S.; Kim, D.; Holten, D.; Lindsey, J.S.; Bocian, D.F. Ground and excited state electronic properties of halogenated tetraarylporphyrins. Tuning the building blocks for porphyrin-based photonic devices. *J. Porphyr. Phthalocyanines* **1999**, *3*, 117–147. [CrossRef]
102. Menkiti, M.; Aniagor, C.; Agu, C.; Ugonabo, V. Effective adsorption of crystal violet dye from an aqueous solution using lignin-rich isolate from elephant grass. *Water Conserv. Sci. Eng.* **2018**, *3*, 33–46. [CrossRef]
103. El Nemr, A.; Abdelwahab, O.; El-Sikaily, A.; Khaled, A. Removal of direct blue-86 from aqueous solution by new activated carbon developed from orange peel. *J. Hazard. Mater.* **2009**, *161*, 102–110. [CrossRef]
104. Kannan, N.; Meenakshisundaram, M. Adsorption of Congo Red on various activated carbons. A comparative study. *Water Air Soil Pollut.* **2002**, *138*, 289–305. [CrossRef]
105. Wright, E.; Partt, N. Current chemical Papars. *J. Chem. Soc. Faraday Trans.* **1961**, *70*, 1461. [CrossRef]

106. Volesky, B.; Weber, J.; Park, J. Continuous-flow metal biosorption in a regenerable Sargassum column. *Water Res.* **2003**, *37*, 297–306. [CrossRef]
107. Wu, Z.; Deng, W.; Tang, S.; Ruiz-Hitzky, E.; Luo, J.; Wang, X. Pod-inspired MXene/porous carbon microspheres with ultrahigh adsorption capacity towards crystal violet. *Chem. Eng. J.* **2021**, *426*, 130776. [CrossRef]
108. Mittal, H.; Al Alili, A.; Morajkar, P.P.; Alhassan, S.M. Graphene oxide crosslinked hydrogel nanocomposites of xanthan gum for the adsorption of crystal violet dye. *J. Mol. Liq.* **2021**, *323*, 115034. [CrossRef]
109. Satapathy, M.K.; Das, P. Optimization of crystal violet dye removal using novel soil-silver nanocomposite as nanoadsorbent using response surface methodology. *J. Environ. Chem. Eng.* **2014**, *2*, 708–714. [CrossRef]
110. Takabi, A.S.; Shirani, M.; Semnani, A. Apple stem as a high performance cellulose based biosorbent for low cost and eco-friendly adsorption of crystal violet from aqueous solutions using experimental design: Mechanism, kinetic and thermodynamics. *Environ. Technol. Innov.* **2021**, *24*, 101947. [CrossRef]
111. Putri, K.N.A.; Keereerak, A.; Chinpa, W. Novel cellulose-based biosorbent from lemongrass leaf combined with cellulose acetate for adsorption of crystal violet. *Int. J. Biol. Macromol.* **2020**, *156*, 762–772. [CrossRef] [PubMed]
112. Falaki, Z.; Bashiri, H. Preparing an adsorbent from the unused solid waste of Rosewater extraction for high efficient removal of Crystal Violet. *J. Iran. Chem. Soc.* **2021**, *18*, 2689–2702. [CrossRef]
113. Ji, Q.; Li, H. High surface area activated carbon derived from chitin for efficient adsorption of Crystal Violet. *Diam. Relat. Mater.* **2021**, *118*, 108516. [CrossRef]
114. Bertolini, T.C.R.; Izidoro, J.C.; Magdalena, C.P.; Fungaro, D.A. Adsorption of crystal violet dye from aqueous solution onto zeolites from coal fly and bottom ashes. *Orbital Electron. J. Chem.* **2013**, *5*, 179–191.
115. Sarabadan, M.; Bashiri, H.; Mousavi, S.M. Removal of crystal violet dye by an efficient and low cost adsorbent: Modeling, kinetic, equilibrium and thermodynamic studies. *Korean J. Chem. Eng.* **2019**, *36*, 1575–1586. [CrossRef]
116. Amodu, O.S.; Ojumu, T.V.; Ntwampe, S.K.; Ayanda, O.S. Rapid adsorption of crystal violet onto magnetic zeolite synthesized from fly ash and magnetite nanoparticles. *J. Encapsulation Adsorpt. Sci.* **2015**, *5*, 191. [CrossRef]
117. Ali, H.; Muhammad, S.K. Biosorption of crystal violet from water on leaf biomass of *Calotropis procera*. *J. Environ. Sci. Technol.* **2008**, *1*, 143–150. [CrossRef]
118. Guler, U.A.; Ersan, M.; Tuncel, E.; Dügenci, F. Mono and simultaneous removal of crystal violet and safranin dyes from aqueous solutions by HDTMA-modified *Spirulina* sp. *Process Saf. Environ. Prot.* **2016**, *99*, 194–206. [CrossRef]
119. Mondal, S. Methods of dye removal from dye house effluent—An overview. *Environ. Eng. Sci.* **2008**, *25*, 383–396. [CrossRef]
120. Knauer, K.; Behra, R.; Sigg, L. Adsorption and uptake of copper by the green alga scenedesmus subspicatus (chlorophyta) 1. *J. Phycol.* **1997**, *33*, 596–601. [CrossRef]
121. Perales-Vela, H.V.; Pena-Castro, J.M.; Canizares-Villanueva, R.O. Heavy metal detoxification in eukaryotic microalgae. *Chemosphere* **2006**, *64*, 1–10. [CrossRef] [PubMed]
122. Shaw, J. *Heavy Metal Tolerance in Plants: Evolutionary Aspects*; CRC Press: Boca Raton, FL, USA, 1989.

materials MDPI

Article

A Facile One-Pot Approach to the Fabrication of Nanocellulose–Titanium Dioxide Nanocomposites with Promising Photocatalytic and Antimicrobial Activity

Roberta G. Toro [1,*], Abeer M. Adel [2], Tilde de Caro [1], Bruno Brunetti [3], Mona T. Al-Shemy [2] and Daniela Caschera [1]

[1] Institute for the Study of Nanostructured Materials, National Research Council, Via Salaria Km 29,300, Monterotondo Stazione, 00015 Rome, Italy
[2] National Research Centre, Cellulose and Paper Department, 33El-Bohouth St. (Former El-Tahrir St.), Dokki, Giza P.O. Box 12622, Egypt
[3] Dipartimento di Chimica, Università degli Studi di Roma "La Sapienza" Institute for the Study of Nanostructured Materials, National Research Council c/o, Piazzale Aldo Moro 5, 00185 Rome, Italy
* Correspondence: robertagrazia.toro@cnr.it

Citation: Toro, R.G.; Adel, A.M.; de Caro, T.; Brunetti, B.; Al-Shemy, M.T.; Caschera, D. A Facile One-Pot Approach to the Fabrication of Nanocellulose–Titanium Dioxide Nanocomposites with Promising Photocatalytic and Antimicrobial Activity. *Materials* **2022**, *15*, 5789. https://doi.org/10.3390/ma15165789

Academic Editor: Lucjan Chmielarz

Received: 27 July 2022
Accepted: 19 August 2022
Published: 22 August 2022

Abstract: The combination of cellulosic materials and metal oxide semiconductors can provide composites with superior functional properties compared to cellulose. By using nanocellulose derived from agricultural waste, we propose a one-pot and environmentally friendly approach to the synthesis of nanocellulose–TiO_2 (NC–TiO_2) nanocomposites with peculiar photocatalytic activity and antibacterial effects. The as-prepared NC–TiO_2 composites were fully characterized by different techniques, such as X-ray diffraction (XRD), μ-Raman, Fourier transform infrared spectroscopy (FTIR), thermogravimetry analysis (TGA), scanning electron microscopy (SEM), transmission electron microscopy (TEM), and diffuse reflectance spectroscopy (DRS). The results showed that well crystalline anatase TiO_2 nanoparticles of about 5–6 nm were obtained. The photocatalytic activity in particular was evaluated by using methyl orange (MO) solution as a target pollutant at different pH values. It was found that all the tested NC–TiO_2 nanocomposites showed stable photocatalytic activity, even after consecutive photocatalytic runs. In addition, NCT nanocomposites with higher TiO_2 content showed degradation efficiency of almost 99% towards MO after 180 min of UV illumination. Finally, NC–TiO_2 nanocomposites also showed intriguing antimicrobial properties, demonstrating to be effective against Gram-positive (*Staphylococcus aureus*, *Bacillus subtilis*) with 20–25 mm of inhibition zone and Gram-negative bacteria (*Escherichia coli*, *Pseudomonas aeuroginosa*) with 21–24 mm of inhibition zone, and fungi (*Candida albicans*) with 9–10 mm of inhibition zone.

Keywords: nanotechnology; titanium dioxide; nanocellulose; hybrid nanocomposite; photocatalytic activity; photodegradation kinetic; antimicrobial effects

1. Introduction

Hybrid organic–inorganic materials have been the focus of intense research in the last few decades for the numerous advantages they potentially offer in many technological fields, such as optics, electronics, ionics, mechanics, and biology [1]. The particular interest around this class of materials relies on the enhanced optical, mechanical, electrical, and thermal properties the hybrid composites show with respect to the organic and inorganic counterparts. The new properties, in most cases, can be further tuned by simply varying the precursors ratio, their chemical functionalities, and, hence, the mutual chemical and/or physical interactions between the inorganic and organic components [2,3].

In this context, cellulose, the most abundant biopolymer in nature, is considered an excellent natural material with remarkable advantages in terms of good biodegradability, biocompatibility, and low cost, and can represent an ideal platform for the design and fabrication of hybrid organic–inorganic nanocomposites. Cellulose can be obtained from a variety of biomass sources, including wood, cotton, linen, bamboo, and agriculture straw plant cells [4]. The repeat unit of cellulose consists of a cyclic backbone of sugar rings with three hydroxyls on each ring. The presence of such high-density oxygenated functional groups as hydroxyl and ether groups provides peculiar chemical functionalities and makes nanocellulose an ideal matrix to promote the synthesis of metallic and metal oxide nanoparticles [5,6]. Many studies have indicated that nanocellulose can act either as templates or precursors [7,8]. Luo et al. [9], for example, reported the hydrothermal synthesis of hierarchical fibrous anatase–cellulose composites and demonstrated that, although no additional functional groups or binders were introduced into the cellulose, the anatase nanoparticles were homogeneously anchored to each cellulose nanofiber. Caschera et al. [10] fabricated silver–carboxylate nanocellulose (Ag–ONCs) nanocomposites by a facile two-step green method where the ONCs acted as a template and reducing agent for silver nanoparticles formation due to the specific hydroxyl and carboxyl groups on the cellulose surface. Recently, we demonstrated that paper sheets fabricated from nanocellulose recycled by agriculture wastes could be used as stable and robust matrices for the ex situ loading of pre-formed TiO_2 nanoparticles [11] since the hydroxyl groups on the surface of nanocellulose fibers offered stable anchoring sites for the stabilization of the TiO_2 nanoparticles. The TiO_2-modified paper sheets showed not only new photocatalytic activity towards methylene blue but also improved mechanical performances in terms of better tensile index, breaking length, tear and burst index, and air permeability. Methyl orange (MO) is an azo dye commonly known as a pH indicator. It is very soluble in aqueous solution, and it is extensively used in several industries, including the textile, paper, printing, and food industries, and mostly discharged in industrial wastewater [12]. Furthermore, MO is toxic and brings about serious water pollution, causing anomalous coloration to the surface water, blocking activity of photosynthetic bacteria and aquatic plants, and further representing a hazard to all mankind [13]. In this context, the removal of MO from effluent is strongly required.

TiO_2 is certainly the most studied inorganic semiconductor as it is the key component in many relevant applications, including aspects related to the environment and clean energy, antimicrobial agent, food, and biomedical [14]. In particular, as a photocatalyst, TiO_2 has many advantages in terms of strong oxidation activity towards a wide range of organic pollutants, environmental stability, easy synthesis, and low cost [15–19]. However, its practical application as a photocatalyst is still limited due to some issues regarding stability under real operating conditions (e.g., spontaneous aggregation of TiO_2 nanoparticles under the complex physical and chemical wastewater conditions), recovery, and reuse of the catalyst. In this perspective, the development of TiO_2 bio-composite photocatalysts can be an effective approach to overcome these limits as it can promote a more efficacious dispersion of catalyst nanoparticles and enhance the functional properties of the as-synthesized composite. Liu et al. [20] reported the preparation of anatase TiO_2/cellulose composites at low temperature, and the prepared composite displayed outstanding photocatalytic performance for the degradation of methyl orange. However, they used highly corrosive nitric acid to catalyze the hydrolysis of titanium precursor, and, in addition, the final stage of lyophilization was time-consuming and required quite expensive equipment. Xiao et al. [21] instead synthesized TiO_2 nanoparticles by using nanofibrillated cellulose as a template and the hazardous $TiCl_4$ as a titanium source. The cellulose was then removed by the final annealing step carried out at different temperatures in order to obtain crystalline TiO_2 nanoparticles.

In this study, we propose the one-pot synthesis of hybrid nanocellulose–TiO_2 nanocomposites (NC–TiO_2) for the decolorization of MO under UV light. The NC–TiO_2 nanocomposites were obtained by a simple, cost effective, and environmentally friendly sol–gel method using non-toxic chemicals (water and acetic acid) and nanocellulose (NC) extracted from agriculture residues [22]. The NC–TiO_2 nanocomposites with different amounts of NC were prepared at room temperature using a green solvent, such as distilled water, and mild acetic acid as a catalyst for the hydrolysis of the titanium butoxide precursor. A cost-effective aging process at room temperature and ambient pressure was carried out instead of traditional heat treatment to favor the complete crystallization of TiO_2. Furthermore, the nanocellulose (NC) obtained by recycling agricultural waste in the present study had high crystallinity and active carboxylate groups, in addition to the intrinsic hydroxyl groups of cellulose, which are expected to stabilize TiO_2 nanoparticles during the hydrolysis step of the titanium precursor. A complete physicochemical characterization of the NC–TiO_2 composites was carried out using X-ray diffraction (XRD), attenuated total reflectance-Fourier transform infrared (ATR-FTIR), μ-Raman spectroscopy, UV–visible spectroscopy, field emission scanning electron microscopy (FESEM), and thermogravimetric analysis (TGA). The NC–TiO_2 nanocomposites were tested in the photodegradation of the reactive azo dye methyl orange at different pHs, and their reusability in the degradation process was evaluated as well. Furthermore, the antibacterial properties were also investigated by monitoring the inhibition growth of NCT nanocomposites towards Gram-positive and Gram-negative bacteria and fungi.

2. Materials and Methods

2.1. Materials and Reagents

The chemicals used for sol−gel synthesis include titanium (IV)-n-butoxide (≥99% Alfa Aesar, Harverhill, MA, USA) and glacial acetic acid (Alfa Aesar, Harverhill, MA, USA). The bagasse fibers were obtained from the Egyptian pulp and paper industry (Quena Company of Cairo, Egypt). Methyl orange was purchased at Sigma-Aldrich (St. Louis, MO, USA). Deionized water was used in all the experiments.

2.2. Synthesis of NC–TiO_2 Nanocomposite

Nanocellulose (NC) was prepared from bagasse raw material using ammonium persulphate hydrolysis (APS) method as previously reported [10,23]. Briefly, 1 g of raw bagasse fibers is hydrolyzed with 1.5 M APS solution at liquor ratio 1:100 and 60 °C for 24 h under continuous stirring. The resulting white suspension is dialyzed against distilled water till the solution reached pH $\approx$ 5. NC with pH 7 is prepared by adding 1 M NaOH.

The NC–TiO_2 nanocomposites were prepared by a simple and low-cost sol–gel method. Titanium (IV)-n-butoxide ($Ti(OBut)_4$)was used as Ti precursor. First, the proper amount of NC was dispersed in 100 mL of distilled water and then sonicated in an ice bath at 0 °C for 10 min to promote high dispersion. $Ti(OBut)_4$ and glacial acetic acid were carefully mixed in a fixed volume ratio of 1:2, under vigorous stirring at room temperature. After 10 min, the acidic dispersion containing the titanium precursor was added dropwise to the prepared NC dispersion under constant and vigorous stirring. Different amounts of NC were used to prepare the aqueous dispersion, and the resulting NC–TiO_2 nanocomposites were denoted by their NC content ranging from 10 to 80 wt% with respect to TiO_2. The mixture was vigorously stirred at room temperature for 24 h, and then the resulted aqueous white dispersion was stored in dark conditions without stirring at room temperature and atmospheric pressure for five days. The NC–TiO_2 powders were precipitated by adding a 10% *w*/*w* sodium carbonate solution until a neutral pH was reached and then collected by centrifugation at 10,000 rpm (Thermo Scientific IEC CL31R Multispeed centrifuge, Waltham, MA, USA), washed three times with deionized water, and finally dried at 45 °C under vacuum. The final product exhibited a light-yellow color, which was indicative of the formation of the NC–TiO_2 nanocomposite. The same procedure without nanocellulose was applied for the synthesis of pure TiO_2. The NC–TiO_2 nanocomposites with different amount of NC were coded as NC10–TiO_2, NC20–TiO_2, and NC80–TiO_2 depending on the % wt of NC.

2.3. Characterization of NC–TiO_2 Nanocomposite

Structural characterization was performed by *X-ray Diffraction (XRD)* using a Siemen D5000 X-ray diffractometer, equipped with a CuKα (λ = 1.5406 Å) radiation operating at 40 kV and 30 nA. Raman spectra were collected with a *μ-Raman spectrometer* (Renishaw

RM 2000, Gloucestershire, UK) operating in the backscattering configuration, and equipped with a Peltier cooled charge-coupled device (CCD) camera in combination with a Leica optical microscope with a 50× objective. Measurements were performed using the 785 nm emission line of a laser diode. *Fourier transform infrared (FTIR) spectroscopy* measurements were carried out using an Alpha FT-IR spectrometer (Brucker Optics, Ettingen, Germany) equipped with an external reflection exchangeable sampling module. The spectra were collected as the average of at least 200 scans at a resolution of 4 cm^{-1} in the frequency range 4000–500 cm^{-1}. *Thermogravimetric analysis (TGA)* was performed on Ugine-Eyraud Model B60 Setaram thermobalance under static air, at constant heating rate of 5 °C/min, using a platinum crucible. UV−vis diffuse reflectance spectra were recorded using a Jasco double-beam V660 UV−vis spectrophotometer (Tokyo, Japan) in reflectance mode, equipped with a 60 mm integrating sphere and $BaSO_4$ as standard diffuse reflectance material. Surface morphology was investigated by *SEM TESCAN VEGA 3* equipped with a LaB_6 filament. *Transmission Electron Microscope (TEM)* analysis was carried out using a high-resolution JEOL-JEM 2100 (Tokyo, Japan).

2.4. Photocatalytic Activity of NC–TiO_2 Nanocomposite

Photocatalytic degradation experiments were carried out at room temperature using MO as a model dye. 10 mg of NC–TiO_2 nanocomposites was added to a beaker containing 50 mL of 25 mg/L MO aqueous solution at different pH (pH = 2, 7, 10). The pH of solution was adjusted by adding HCl or NaOH. A 39 W lamp (λ = 365 nm) was used as the light source. Before irradiation, the samples were kept in the dark for 1 h to ensure the adsorption–desorption equilibrium of MO on the surface of the nanocomposites. The beaker was set at a distance of 9 cm from the light source and irradiated under UV light. The absorbance was taken at given time intervals and the degradation rate was calculated by using the following equation, *Degradation rate (%)* = $(C_0 - C_t)/C_0 \times 100$, where C_0 is the initial concentration and C_t is the concentration of MO at time t. For comparison, an extra test was carried out in the presence of pure TiO_2 under the same experimental conditions.

2.5. Antimicrobial Activity of NC–TiO_2 Nanocomposites

The antimicrobial examinations of prepared NC–TiO_2 nanocomposites were carried out by using qualitative process. The qualitative test is accomplished through solid media (LB agar) using the agar diffusion disk method (AATCC Test Method 147-1988). The qualitative antimicrobial activity of the NC–TiO_2 nanocomposites was examined on four different bacterial species G(+) bacteria as *Staphylococcus aureus, Bacillus subtilis;* G(−) bacteria as *Escherichia coli, Pseudomonas aeuroginosa*, and a fungi *(Candida albicans).* Blank paper disks (Schleicher & Schuell, Spain) with a diameter of 10 mm are soaked by 10 µL of the stock NC–TiO_2 solutions (nominal NC–$TiO_{2\,conc}$ = 0.3 mg/mL) and put on LB Agar previously treated with the microorganism. After 48 h of incubation, the diameters of the resulted growth inhibition zones were measured and the averaged and mean values recorded in millimeters were tabulated. The antimicrobial agents Ampicillin and Amphotericin B were used as standard reference for G(+) & G(−) bacteria and fungi, respectively. The activity index (%) was calculated as [24]:

Activity index (%) = 100× [zone of inhibition by tested sample (mm)/zone of inhibition by standard (mm)].

3. Results

3.1. Structural and Morphological Characterization of NC–TiO_2 Nanocomposites

The X-ray powder diffraction (XRD) analysis performed on the NC–TiO_2 nanocomposites pointed to the formation of TiO_2 in the anatase crystalline phase. In Figure 1a, the XRD patterns of the NC–TiO_2 nanocomposites are shown; the corresponding patterns of NC and neat TiO_2 were also provided for comparison. The XRD patterns of the NC–TiO_2 nanocomposites showed diffraction peaks at 2θ = 25.23°, 37.71°, 47.72°, 54.16°, and 55.32°, which correspond to the tetragonal crystal planes (101), (004), (200), (105), and (211), respectively, of the anatase TiO_2 (JCPDS file 21-1272). The crystalline peaks characteristic of NC can hardly be seen, probably because of the low amount of NC in the nanocomposite. However, a closer look at the XRD patterns of the NC–TiO_2 samples revealed the presence of a shoulder at lower 2θ values for the (101) diffraction peak (inset Figure 1) due to the contribution of the (200) peak of the nanocellulose. The broad XRD peaks further suggested that the NC–TiO_2 nanocomposites had nanometric sizes. The crystalline sizes were determined using the Debye–Scherrer equation, $D = K\lambda/\beta\cos\theta$, where D is the crystal size, K is dimensionless constant (0.9), λ is the wavelength of X-ray, β is the full width at half-maximum (FWHM) of the diffraction peak, and θ is the diffraction angle. The average crystalline sizes (D) of NC–TiO_2 nanocomposites were found to be very similar, 5.73 and 5.81 nm for NC10–TiO_2 and NC20–TiO_2, respectively. There was no obvious difference in the particle size of the TiO_2 nanoparticles with the nanocellulose as the template. Moreover, the calculated D values were less than the size of pure TiO_2 (D_{TiO2} = 7.10 nm). Biomolecules, such as cellulose, can control the particle size of the TiO_2, acting as a limiting spacer, and, therefore, the TiO_2 crystal growth was limited by interactions between NC and TiO_2 [25]. In the perspective of photocatalytic applications, small crystals can provide more available active sites that absorb more photon and produce more electrons-holes, thus enhancing the photocatalytic activity of the system.

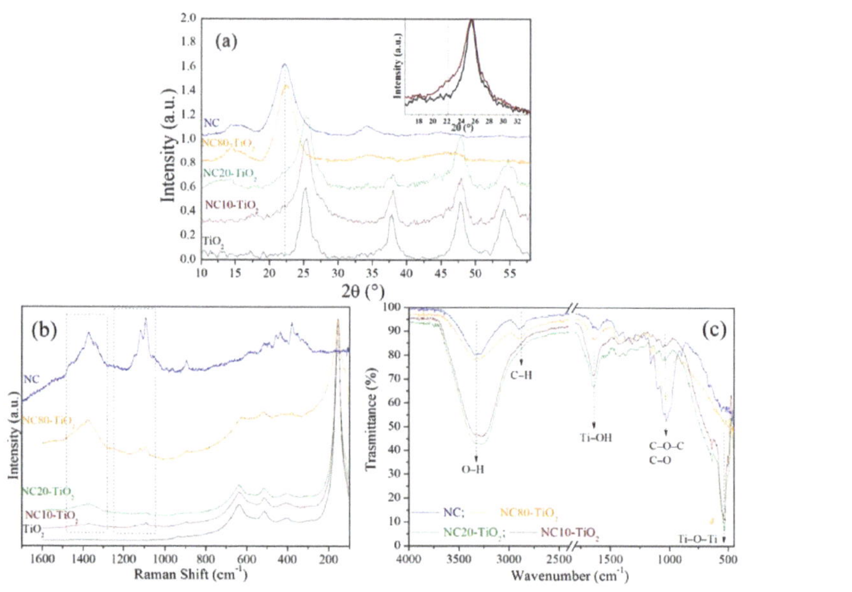

Figure 1. (**a**) XRD patterns of (–) NC, (–) TiO_2, (–) NC80-TiO_2, (–) NC20-TiO_2, and (–) NC10–TiO_2 nanocomposites. The inset shows a comparison between the (101) peak of TiO_2 and NC20–TiO_2; (**b**) μ- Raman spectra and (**c**) FTIR spectra for nanocellulose (NC) and NC–TiO_2 nanocomposites.

The NC–TiO_2 nanocomposites were further investigated by Raman spectroscopy, and the results were reported in Figure 1b. The Raman spectrum of neat NC confirmed the presence of peaks related to cellulose I structure [26]. In particular, the spectrum reported at 1093 and 1117 cm^{-1} the bands characteristic of the β-1,4-glucosidic linkages of the glucopyranose units of cellulose, the bands related to H-C-H and to H-O-C bending in the 1300–1500 cm^{-1}, and, at 329, 378, and 435 cm^{-1}, the bands related to the cellulosic ring deformation [27]. These bands were also visible in the Raman spectra of NC–TiO_2 nanocomposites in addition to the high intensity bands at 155, 400, 514, and 638 cm^{-1}, typical of TiO_2 anatase, thus confirming that anatase was the predominant phase. No rutile or brookite phase bands were detected in the Raman spectra, as already evidenced by XRD analyses.

The chemical nature of surface functional groups in NC–TiO_2 nanocomposites was further investigated by FTIR spectroscopy (Figure 1c). The FTIR spectra of nanocomposites showed an intense absorption below 800 cm^{-1}, which derived from the contribution of the Ti–O bond in TiO_2. In addition, the characteristic peaks of cellulose were also visible: the broad band in the 3600–3000 cm^{-1} region due to stretching vibrations of intra and inter O–H bonds, and the large absorptions in the 1200–860 cm^{-1} spectral region assigned to the glucosidic ring vibrations of C−H, C−O, and C−O−C [28,29]. These NC–TiO_2 bands represented a unique fingerprint of cellulose, and their presence, although less intense, in the spectra of the NC–TiO_2 nanocomposites, suggested that the cellulose retained its chemical structure. Nevertheless, the FTIR of the NC–TiO_2 nanocomposites showed intense absorption peaks associated with the O–H stretching at 3300 cm^{-1} and the O–H bending at 1636 cm^{-1}, which were shifted to lower wavenumbers with respect to those observed in nanocellulose. The observed shift was correlated to the involvement of –OH groups in the formation of a strong intermolecular hydrogen bonding interaction between the hydroxyl groups in the cellulose chain and Ti–OH groups on the surface of TiO_2 nanoparticles [30,31]. Furthermore, the increase in the –OH surface groups is particularly interesting because it may affect positively the NC–TiO_2 nanocomposites' photodegradation activity as they can promote the formation of $^{\bullet}$OH radicals during the photocatalytic process [32].

Thermogravimetric analysis (Figure 2) was carried out to obtain information on the thermal stability of the synthesized NC–TiO_2 nanocomposites. TG curves of TiO_2 and NC were also reported for comparison. The NC–TiO_2 nanocomposites decomposed at a relatively higher temperature than NC (Figure 2). As expected, the TG curve of TiO_2 showed a small weight loss of about 2% in the range 105–250 °C, related to the total content of residual water, acetic acid, and some organic groups deriving from the synthetic procedure. In the case of the NC10–TiO_2, the TG curve showed an initial weight loss of about 5% up to 130 °C caused by the removal of physically absorbed water, followed by a second weight loss step of 16.8% in the 130–520 °C range, attributed to the burning of the carbon species in the nanocomposites. Further, there was no significant weight loss observed above 520 °C, and an RC_{650} of about 84% remained. A similar trend was observed also for the NC20–TiO_2 except that for the lower RC_{650} of about 76% at 650 °C. Contrarily, the NC80–TiO_2 followed a thermal degradation path similar to that of the NC, with two distinct steps of decomposition. The first step between 215° and 400 °C with a weight loss of 39% could be attributed to the reactions of dehydration and decarboxylation of the NC that produced the combustion gases, such as levoglucosan [33]. The second step in the range of 400–580 °C with $RC_{650} \approx 21\%$ was the oxidative decomposition of the char formed from the first step [23,34]. In all cases, the residue was the inorganic component of TiO_2, so we can infer the TiO_2 content was about 84, 76, and 21 by wt. % for NC10–TiO_2, NC20–TiO_2, and NC80–TiO_2, respectively.

Assuming pseudo-first-order kinetics, the kinetic parameters of each step in the pyrolysis range were estimated by employing Coats and Redfern's theory [35]. The kinetic and thermodynamic parameters, such as activation energies (Ea; kJ mol^{-1}), pre-exponential factors (A; s^{-1}), enthalpy (ΔH; kJ mol^{-1}), entropy (ΔS; JK^{-1} mol^{-1}), and Gibbs free energy of activation (ΔGa; kJ mol^{-1}), based on the weight loss data obtained from TG analyses, are given in Table S1. The Ea and A values were lower for NC10–TiO_2 and NC20–TiO_2 compared to NC80–TiO_2, and this may be attributed to the presence of a higher content of non-flammable TiO_2 covering NC nanocrystals, which requires lower energies to activate the pyrolysis reactions [36]. This means that the presence of TiO_2 bonded to NC affected the thermal reaction pathway responsible for the different activation energies. From the overall results obtained in Table S1, we can observe the positive effect of TiO_2 on the thermal stability of the corresponding NC–TiO_2 nanocomposites, which follows the order NC10–TiO_2 > NC20–TiO_2 > NC80–TiO_2.

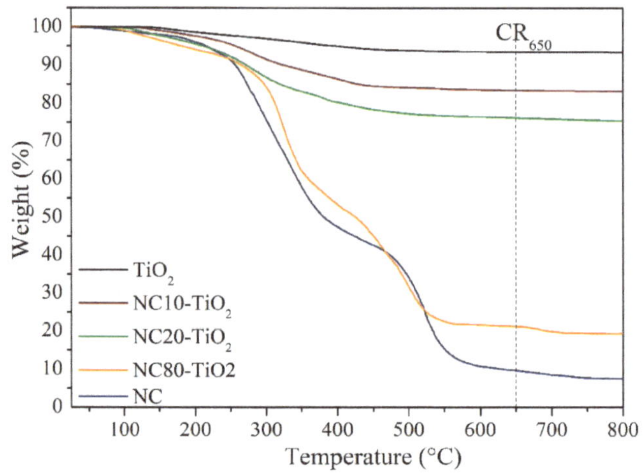

Figure 2. TGA of (–) TiO_2, (–) NC, (–) NC10-TiO_2, (–) NC20-TiO_2, and (–) NC80-TiO_2.

The morphology of pure TiO_2 and NC–TiO_2 was studied by SEM, and the corresponding images were reported in Figure 3a,b. TiO_2 is composed of large sheet-like aggregates, heterogeneous in size (Figure 3a). The NC–TiO_2 composites had very similar morphologies as a result of the aggregation of several thousands of nanoparticles whose final size decreased slightly upon increasing the amount of nanocellulose (Figure 3b). The TEM image (Figure 3c) showed that the NC had nanosized rod-shaped whiskers, bunched together into a tendon-like structure, with an average NC width and length of about 6.2 ± 0.96 nm and 39.4 ± 8.68 nm, respectively. The NC–TiO_2 nanocomposites had instead a granular shape (Figure 3d), with the TiO_2 covering the surface of the NC. Moreover, the TiO_2 particles not only covered the surface but also deposited on the pores of NC.

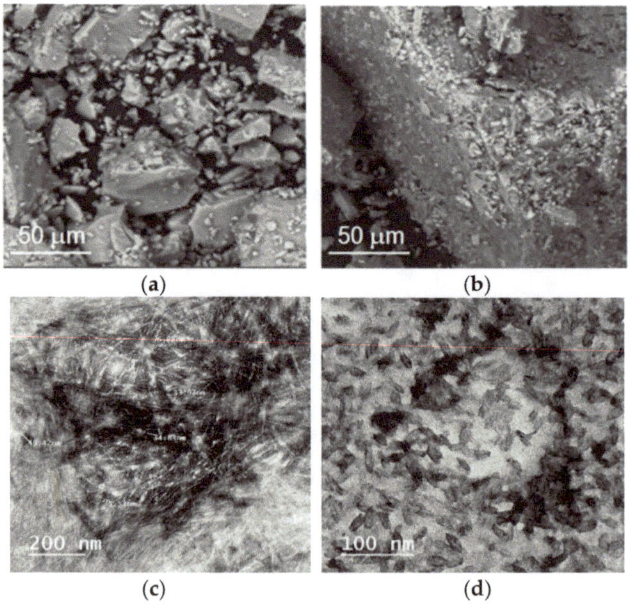

(a) (b) (c) (d)

Figure 3. FE-SEM images of (**a**) TiO_2, (**b**) NC20–TiO_2, and TEM images of (**c**) NC and (**d**) NC20–TiO_2.

3.2. Photocatalytic and Antimicrobial Activity of NC–TiO_2 Nanocomposites

The optical properties of NC–TiO_2 nanocomposites were investigated by UV–vis diffuse reflectance spectroscopy (Figure 4). The UV absorption spectra of the NC–TiO_2 nanocomposites resembled that of pure TiO_2, with the maximum UV absorption redshifted from 330 nm to 360 nm. The observed shift for the nanocomposites could be indicative of the chemical interactions of TiO_2 NPs with the large molecular framework of the nanocellulose. The NCT nanocomposites also showed a certain level of visible light absorption, which, in general, could be associated with a decrease in the band gap following the formation of the NC–TiO_2 hybrid nanocomposites [13,37]. The corresponding band gap values were obtained by plotting the Kubelka–Munk function against the photon energy (inset in Figure 4). Pure TiO_2 showed a band gap of 3.37 eV, very close to the reported value for bulk TiO_2 (3.40 eV). The band gap values of NCT nanocomposites were very similar and sensibly lower than that found for the pureTiO_2 (E_g = 3.00 and E_g = 3.03 eV for NC10–TiO_2 and NC20–TiO_2, respectively). This result confirmed the bonding interactions between NC and TiO_2 after the formation of the NC–TiO_2 nanocomposites, and they were directly correlated to the observed absorption of visible radiation.

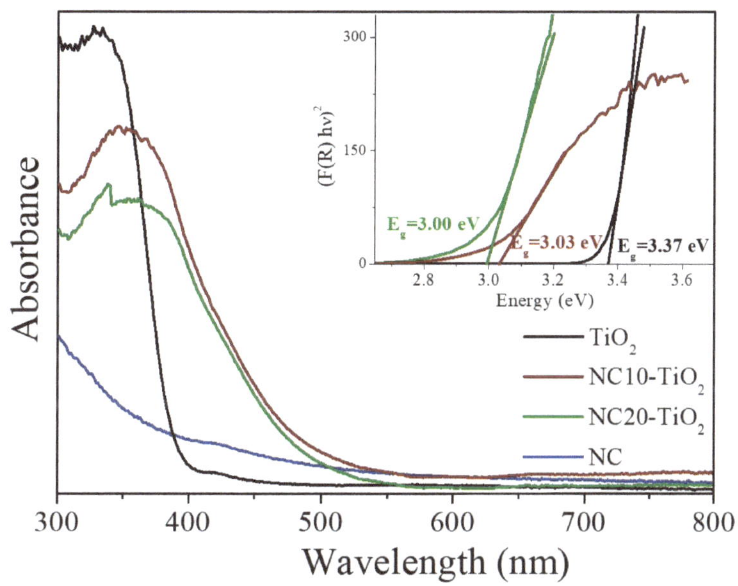

Figure 4. UV–vis spectra for (–) TiO_2, (–) NC10–TiO_2, (–) NC20–TiO_2 nanocomposites, and (–) NC. The inset shows the Kubelka–Munk plot for (–) TiO_2, (–) NC10–TiO_2, and (–) NC20–TiO_2 nanocomposites.

Photodegradation tests were carried out in the presence of NC–TiO_2 nanocomposite powders dispersed in 50 mL of aqueous solution with an initial MO concentration of 25 ppm at pH = 2. The pH of the solution was adjusted before irradiation, and it was not controlled during the course of the reaction. The residual concentration of the organic dye was followed by UV–vis spectroscopy monitoring the variation of absorbance at its λ_{max}. Figure 5a showed the evolution of MO absorbance during the photocatalytic experiment. The MO absorption spectrum had a maximum at λ = 507 nm corresponding to the $\pi \rightarrow \pi^*$ transitions of the dimethylamino electron donors. A progressive decrease in the $\pi \rightarrow \pi^*$ MO absorption peak with the increasing irradiation time and a corresponding blue-shift during the kinetic experiments were observed, which confirmed the degradation of the dye associated with the removal of the N-methyl groups [38].

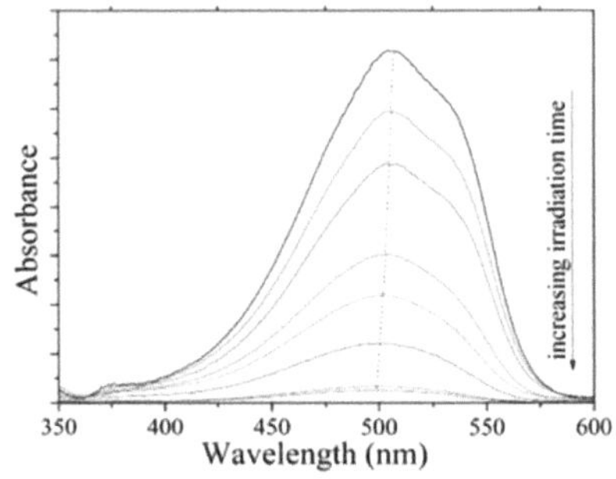

(**a**)

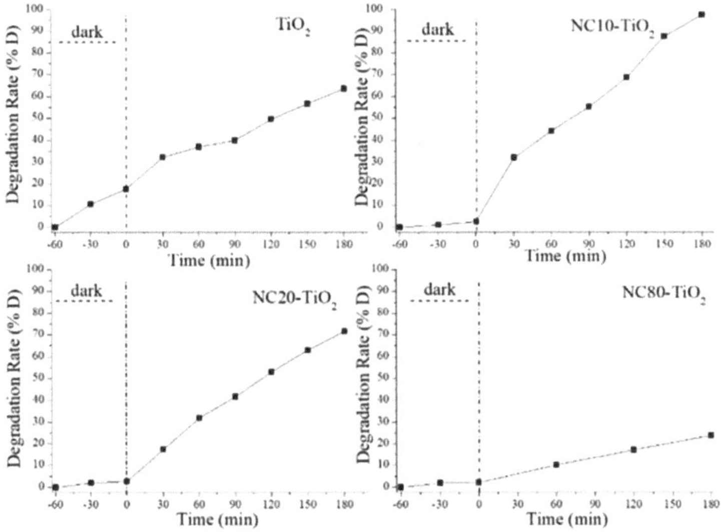

(**b**)

Figure 5. (**a**) Time–dependent UV–vis spectra of methyl orange and its degradation by NC10–TiO_2 under UV irradiation; (**b**) quantitative evaluation of photocatalytic degradation rate of methyl orange by TiO_2, NC10–TiO_2, NC20–TiO_2, and NC80–TiO_2 under UV illumination.

Figure 5b showed the degradation rates of MO by NC–TiO_2 nanocomposites under UV light. Before UV irradiation, solutions containing the nanocomposites were stirred for 60 min in the dark at room temperature to establish the adsorption–desorption equilibrium. Apart from pure TiO_2, none of the nanocomposites showed noticeable absorption phenomena. During the first stage, indeed, it was observed that 17.5%, 2%, 2.5%, and 1.3% MO were absorbed for pure TiO_2, NC10–TiO_2, NC20–TiO_2, and NC80–TiO_2, respectively, indicating that the absorption had negligible effects on the removal of MO in the case of the NCT composites.

Under UV light and in the presence of the raw NC alone, MO did not display any significant photodegradation after 120 min, indicating that nanocellulose had no photocatalytic activity (data not shown). When NC–TiO_2 samples were irradiated with UV light, the situation changed drastically. The highest degradation rate was reached by the NC10–TiO_2, which degraded almost 99% of MO after 180 min of UV illumination, while NC20–TiO_2 reached 72%. For NC80–TiO_2, the photocatalytic activity was estimated at about 23% after 180 min. These findings confirmed that NC10–TiO_2 composite showed excellent photocatalytic performance, similar and, in some cases, higher with respect to analogous systems, as reported in Table 1.

Table 1. Comparison of the photocatalytic efficiency NC–TiO_2 with other reported works for the catalytic degradation of various organic pollutants.

Catalyst	Pollutant	Wavelength	Efficiency	Reference
TiO_2/CNF	MO	365 nm	99% within 30 min	[20]
TiO_2/NFC	MO	365 nm	95% within 60 min	[21]
CNF/TiO_2	MB	UV	97% within 30 min	[30]
TiO_2/RC	phenol	365 nm	96% within 360 min	[15]
TiO_2/Agar	MB	UV	62% within 105 min	[25]
NC/TiO_2	Mefenamic acid	UV	89% within 150 min	[39]
NC/TiO_2	MO	365 nm	99% within 180 min	This work

The kinetics of the MO photodegradation was described by using the pseudo-first-order model $\ln(C_0/C) = k_{app}t$, where C_0 is the initial concentration of MO and k_{app} is the apparent rate constant. The kinetic and regression constants are reported in Figure 6. The NC10–TiO_2 had the highest reaction rate constant (k_{app} = 0.016 min^{-1}), about four times higher than that of bare TiO_2, thus confirming the best degradation efficiency, suggesting that low amounts of nanocellulose had beneficial effects on the degradation efficiency of the corresponding NC–TiO_2 nanocomposites. This result was in contrast with those reported in some cases, as in Ref. [39], according to which the efficiency in mefenamic acid (MEF) degradation decreased by increasing the TiO_2 content in the NC–TiO_2 nanocomposites. These differences suggest that not only the amount of NC in the nanocomposites affected the photodegradation activity but also the NC–TiO_2 intrinsic chemical structure. In the present study, the observed behavior could be explained by considering the intrinsically different amounts of hydroxyl groups available at the surface of the catalyst particles. As already known, the species responsible for photodegradation of dyes in water were hydroxyl radicals, $^\bullet OH$, and anion radical superoxides, $O_2^{-\bullet}$. Hydroxyl radicals were produced by the photogenerated holes (h+) in the conduction band from the direct oxidation of H_2O, OH^- ions, or terminal –OH groups on the surface of the catalyst. In this respect, NC–TiO_2 nanocomposites with low amounts of NC provided a large number of hydroxyl groups on the catalyst surface, as evidenced by the FTIR results, which could promptly react with h^+, locally producing $^\bullet OH$ radicals, which degraded the organic dyes in solution and reduced the electron-hole recombination responsible for the decrease in photocatalytic activity.

It is well known that pH plays an important role in determining the efficiency of the photocatalytic degradation of sulfonated dyes, such as methyl orange, so we evaluated the activity of NC–TiO_2 nanocomposites at different pHs, and the results are summarized in Figure 7. For all the nanocomposites, the best photocatalytic efficiency was obtained at pH = 2 under continuous UV irradiation. At basic pH, however, NC–TiO_2 nanocomposites encountered intrinsic chemical instability, and it was not possible to carry out photocatalytic measurements. Such an issue was also reported by Comparelli et al. [40] for TiO_2 nanoparticles covered by oleic acid, who assumed that such instability was owed to the removal of the oleic acid ligand from the surface of TiO_2 nanoparticles caused by the enhanced solubility of the ligand at such an extreme pH. Moreover, the Raman characterization (Figure S2) performed on our NC–TiO_2 nanocomposites after photocatalytic experiments at different pHs confirmed the hypothesis about the occurrence of some degradations at a basic pH. In particular, the Raman spectrum of NC20–TiO_2 after photocatalysis at pH 10 showed no longer peaks typical of cellulose, while only Raman peaks of TiO_2 anatase were visible.

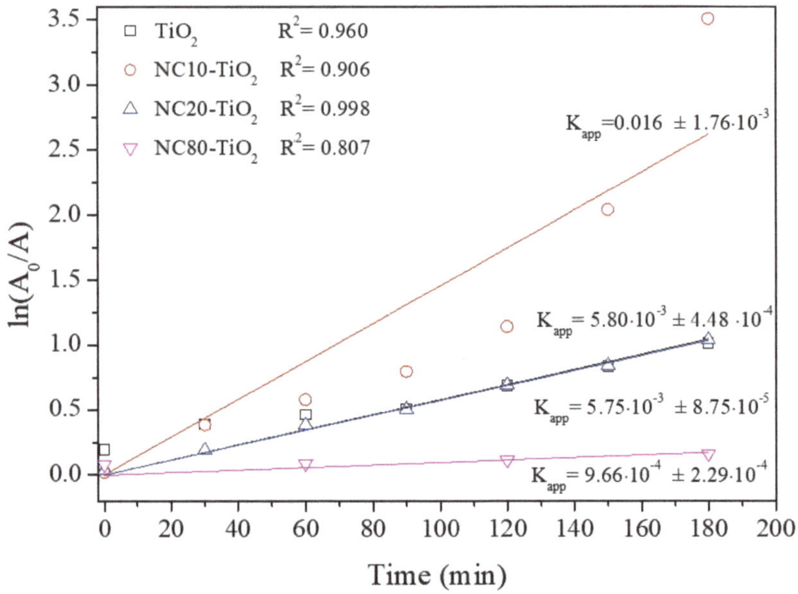

Figure 6. Kinetics of methyl orange photodegradation using NC–TiO_2 nanocomposites.

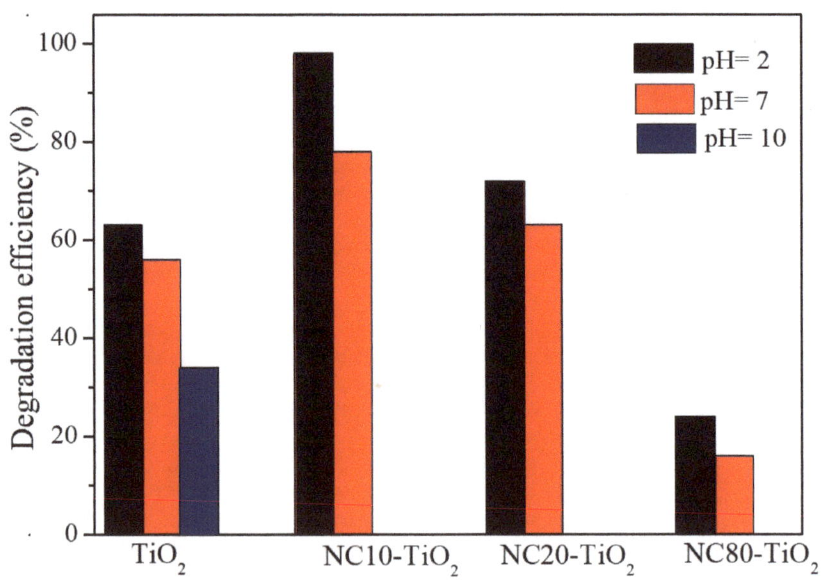

Figure 7. Percentages of the degradation efficiency for methyl orange obtained with different NC–TiO_2 nanocomposites and TiO_2 at different pH values.

Finally, the recyclability of the photocatalyst was evaluated for the NC10–TiO_2 sample, being the most efficient in MO degradation. After each photocatalytic run, the sample was recovered from solution by centrifugation, washed with ethanol, and dried. The composite was then reused for a new MO photodegradation test at pH 2.

The results were reported in Figure 8: after three times of repeated use, the degradation rate decreased from 98% to 75%, which could be attributed mainly to the inevitable loss of small amounts of photo-active material after every run (about 20% mass loss for NC10–TiO_2 at the end of the collection processes).

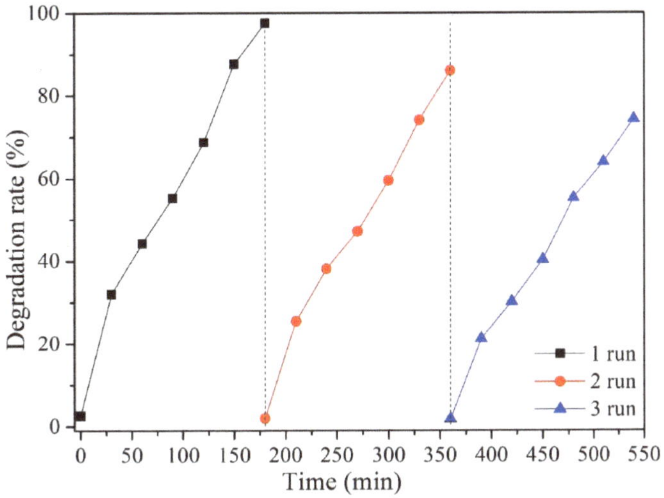

Figure 8. Evaluation of the degradation rate for methyl orange solution over UV irradiation NC10–TiO_2 nanocomposites for three consecutive cycles.

The antimicrobial activity of NC–TiO_2 nanocomposites was qualitatively assessed by applying an agar well diffusion disk test against four bacterial species, *Staphylococcus aureus, Bacillus subtilis* as Gram (+) bacteria, *Escherichia coli, Pseudomonas aeuroginosa* as Gram (–) bacteria, and a yeast pathogen *Candida albicans*. According to this method, the nanocomposites impregnated a paper filter placed on an agar plate. After that, the bacteria were allowed to grow on the agar plate and the evaluation of the diameter of the inhibitory zones allowed to determine the efficiency of bacterial growth inhibition [41]. The pure TiO_2 and two nanocomposites having different NC contents (NC20–TiO_2 and NC80–TiO_2) were selected for this study. The antibacterial activity of the selected samples was represented in Figure 9.

The results, expressed by measuring the size of the inhibition zone, were summarized in Table S2 and highlighted the influence of the different TiO_2:NC ratios on the antimicrobial properties of the nanocomposites. Comparing the results, indeed, the formation of a clear inhibition zone around all the samples was evident, indicating that the NC–TiO_2 nanocomposites possessed peculiar bactericide effects and were effective in inhibiting the growth of all the four bacteria species and *Candida albicans* fungi, as well as the neat TiO_2. Furthermore, the results clearly indicated that the antimicrobial activity of the nanocomposites was induced by TiO_2 nanoparticles, and it was strictly dependent on their concentration: the NC20–TiO_2 nanocomposite sample showed much better antibacterial properties against *Bacillus subtilis* and *Staphylococcus aureus* than G(+) and *Pseudomonas aeruginosa as* G(−) bacteria. Our results were in accordance with those reported by Zong [31], which observed an improvement in the antibacterial properties of cellulose after its functionalization with TiO_2 nanoparticles: in particular, when the contents of Ti increased, obvious inhibition zones in the Petri dishes were observed (0.4–2.8 mm, depending on Ti content). The exact mechanism involved in the inactivation of microorganisms by TiO_2-based materials has not been clear until now, and it is usually associated with different phenomena involving, for example, the photocatalytic generation, under band-gap irradiation, of reactive oxygen species (ROS) with high oxidative potentials, oxidative stress caused by ROS, interruption of microorganisms' DNA replication, and protein synthesis processes caused by reaction with ions released by nanoparticles [42]. Moreover, our results demonstrated that NC–TiO_2

nanocomposites were more active against the Gram-positive *Staphylococcus aureus* and displayed very high activity compared to the standard drug (Figure 10). This behavior has already been observed [43,44] and could be explained considering the differences in the structure of the cell walls of these bacteria: Gram-negative bacteria (e.g., *Pseudomonas aeruginosa*) usually have an additional outer membrane consisting of lipopolysaccharide and peptidoglycan, which helps them to resist to the oxidative damage caused by the interaction with the TiO_2 nanoparticles.

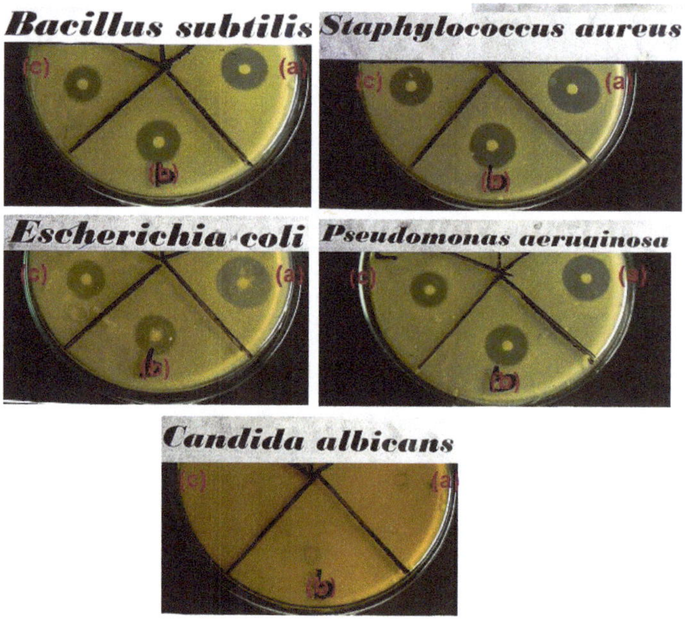

Figure 9. Antimicrobial activity for TiO_2 (**a**), NC20–TiO_2 (**b**), and NC80–TiO_2 (**c**) nanocomposites.

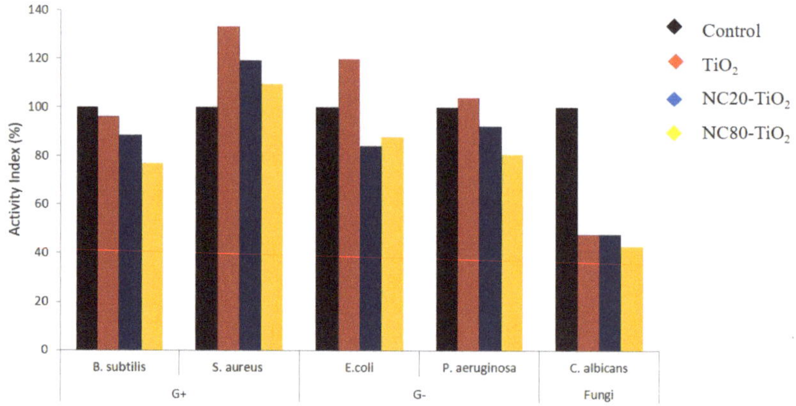

Figure 10. Antimicrobial activities (activity index %) for TiO_2 and NC–TiO_2 nanocomposites. Blank paper is added as control.

The good antimicrobial performance and the photocatalytic activity of the nanocellulose–TiO_2 nanocomposites make these materials the key candidates in many important industrial sectors, such as food manufacturing and packaging, water remediation, pharmaceuticals, and medical treatments.

4. Conclusions

We demonstrated the successful one-pot synthetic procedure for the fabrication of nanocellulose–TiO_2 nanocomposites. With regard to hydrothermal methods and other sol–gel synthesis methods that normally require high-temperature calcination steps or harmful solvents, the proposed approach is simple, environmentally friendly, low-cost, and energy-saving. Structural and morphological characterization revealed the formation of well crystalline anatase nanoparticles with a final size of about 5–6 nm. The chemical interactions between the surface chemical groups on the nanocellulose and the titania precursor are responsible for the significant narrowing of the energy band gap with respect to the pure TiO_2, thus leading to a more efficient photocatalyst towards the photodegradation of methyl orange (almost 99% degradation after 180 min of UV illumination). In particular, the NC–TiO_2 nanocomposites with a low amount of nanocellulose revealed particular efficiency in increasing the density of surface hydroxyl groups, which can easily react with holes to produce hydroxyl radicals that degrade dye molecules. Furthermore, the NC–TiO_2 nanocomposites also possessed considerable antimicrobial activity against Gram-negative and Gram-positive bacterial strains. Therefore, this study provides new insight for fabrication of efficient NC–TiO_2 photocatalysts and enables their application in different sectors, including water remediation and packaging.

Supplementary Materials: The following supporting information can be downloaded at: https://www.mdpi.com/article/10.3390/ma15165789/s1, Figure S1: DTG analyses of NC–TiO_2 nanocomposites; Table S1: TG, DTG, kinetic, and thermodynamic data from the thermal decomposition steps of prepared samples; Figure S2: Raman spectra of NC20–TiO_2 sample after photocatalytic tests at different pH; Table S2: Diameter size of the inhibition zone for TiO_2, NC20–TiO_2, and NC80–TiO_2 nanocomposites through different bacteria.

Author Contributions: R.G.T.: Conceptualization, Methodology, Investigation, Writing—original draft, Validation, Writing—review & editing. D.C.: Conceptualization, Methodology, Validation, Writing—review & editing, Funding acquisition. A.M.A.: Conceptualization, Methodology, Validation, Formal analysis, Investigation, Writing—review & editing, Funding acquisition. T.d.C.: Investigation, Validation. B.B.: Investigation, Validation. M.T.A.-S.: Methodology, Investigation, Validation. All authors have read and agreed to the published version of the manuscript.

Funding: The activities have been performed in the framework of the Joint Bilateral Agreement CNR/NRC (Italy/Egypt), Biennial Programme 2018–2019, for the project "Improvement of mechanical and barrier properties of biopolymer nano-composites for packaging applications".

Institutional Review Board Statement: Not applicable.

Informed Consent Statement: Not applicable.

Data Availability Statement: Data presented in this study are available on request from the authors.

Acknowledgments: The authors thank Fulvio Federici for his technical support during laboratory procedure, and Claudio Veroli for his technical support during XRD measurements. The activities have been performed in the framework of the Joint Bilateral Agreement CNR/NRC (Italy/Egypt), Biennial Programme 2018–2019, for the project "Improvement of mechanical and barrier properties of biopolymer nano-composites for packaging applications".

Conflicts of Interest: The authors declare no conflict of interest. The funders had no role in the design of the study; in the collection, analyses, or interpretation of data; in the writing of the manuscript; or in the decision to publish the results.

References

1. Saveleva, M.S.; Eftekhari, K.; Abalymov, A.; Douglas, T.E.L.; Volodkin, D.; Parakhonskiy, B.V.; Skirtach, A.G. Hierarchy of Hybrid Materials—The Place of Inorganics-*in*-Organics in it, Their Composition and Applications. *Front. Chem.* **2019**, *7*, 179. [CrossRef] [PubMed]

2. Singh, S.; Chen, H.; Shahrokhi, S.; Wang, L.P.; Lin, C.-H.; Hu, L.; Guan, X.; Tricoli, A.; Xu, Z.J.; Wu, T. Hybrid Organic–Inorganic Materials and Composites for Photoelectrochemical Water Splitting. *ACS Energy Lett.* **2020**, *5*, 1487–1497. [CrossRef]
3. Mir, S.H.; Nagahara, L.A.; Thundat, T.; Mokarian-Tabari, P.; Furukawa, H.; Khosla, A. Organic-Inorganic Hybrid Functional Materials: An Integrated Platform for Applied Technologies. *J. Electrochem. Soc.* **2018**, *165*, B3137–B31556. [CrossRef]
4. Adel, A.M.; Ibrahim, A.A.; El-Shafei, A.M.; Al-Shemy, M.T. Inclusion complex of clove oil with chitosan/β-cyclodextrin citrate/oxidized nanocellulose biocomposite for active food packaging. *Food Packag. Shelf Life* **2019**, *20*, 100307. [CrossRef]
5. Adel, A.M.; Al-Shemy, M.T.; Diab, M.A.; El-Sakhawy, M.; Toro, R.G.; Montanari, R.; de Caro, T.; Caschera, D. Fabrication of packaging paper sheets decorated with alginate/oxidized nanocellulose-silver nanoparticles bio-nanocomposite. *Int. J. Biol. Macromol.* **2021**, *181*, 612–620. [CrossRef]
6. Adel, A.M.; El-Shafei, A.M.; Ibrahim, A.A.; Al-Shemy, M.T. Chitosan/Nanocrystalline Cellulose Biocomposites Based on Date Palm (*Phoenix dactylifera* L.) Sheath Fibers. *J. Renew. Mater.* **2019**, *7*, 567–582. [CrossRef]
7. Shopsowitz, K.E.; Stahl, A.; Hamad, W.Y.; MacLachlan, M.J. Hard templating of nanocrystalline titanium dioxide with chiral nematic ordering. *Angew. Chem.* **2012**, *51*, 6886–6890. [CrossRef]
8. Zuo, H.-F.; Guo, Y.-R.; Li, S.-J.; Pan, Q.-J. Application of microcrystalline cellulose to fabricate ZnO with enhanced photocatalytic activity. *J. Alloy. Compd.* **2014**, *617*, 823–827. [CrossRef]
9. Luo, Y.; Xu, J.; Huang, J. Hierarchical nanofibrous anatase-titania-cellulose composite and its photocatalytic property. *Cryst. Eng. Comm.* **2014**, *16*, 464–471. [CrossRef]
10. Caschera, D.; Toro, R.G.; Federici, F.; Montanari, R.; de Caro, T.; Al-Shemy, M.T.; Adel, A.M. Green approach for the fabrication of silver-oxidized cellulose nanocomposite with antibacterial properties. *Cellulose* **2020**, *27*, 8059–8073. [CrossRef]
11. Toro, R.G.; Diab, M.; de Caro, T.; Al-Shemy, M.; Adel, A.; Caschera, D. Study of the Effect of Titanium Dioxide Hydrosol on the Photocatalytic and Mechanical Properties of Paper Sheets. *Materials* **2020**, *13*, 1326. [CrossRef]
12. Malviya, A.; Jaspal, D.; Sharma, P.; Dubey, A. Isothermal mathematical modeling for decolorizing water—A comparative approach. *Sustain. Environ. Res.* **2015**, *25*, 53–58.
13. Nyangiwe, N.N.; Baatjie, B.; Greyling, C.; Khenfouch, M.; Maaza, M. The decolourisation of Methyl Orange and textile effluent under UV using commercial and synthesized nano-TiO_2. *J. Phys. Conf. Ser.* **2018**, *984*, 012009. [CrossRef]
14. Anaya-Esparza, L.M.; Ruvalcaba-Gómez, J.M.; Maytorena-Verdugo, C.I.; González-Silva, N.; Romero-Toledo, R.; Aguilera-Aguirre, S.; Pérez-Larios, A.; Montalvo-González, E. Chitosan-TiO_2: A Versatile Hybrid Composite. *Materials* **2020**, *13*, 811. [CrossRef]
15. Mohamed, M.A.; Salleh, W.N.W.; Jaafar, J.; Ismail, A.F.; Mutalib, M.A.; Sani, N.A.A.; Asri, S.; Ong, C.S. Physiochemical characteristic of regenerated cellulose/N-doped TiO_2 nanocomposite membrane fabricated from recycled newspaper with photocatalytic activity under UV and visible light irradiation. *Chem Engin. J.* **2016**, *284*, 202–215. [CrossRef]
16. Zhan, C.; Li, Y.; Sharma, P.R.; He, H.; Sharma, S.K.; Wang, R.; Hsiao, B.S. A study of TiO_2 nanocrystal growth and environmental remediation capability of TiO_2/CNC nanocomposites. *RSC Adv.* **2019**, *9*, 40565–40576. [CrossRef]
17. Zhao, X.; Zhang, G.; Zhang, Z. TiO_2-based catalysts for photocatalytic reduction of aqueous oxyanions: State-of-the-art and future prospects. *Environ. Int.* **2020**, *136*, 105453. [CrossRef]
18. Li, G.; Park, S.; Rittmann, B.E. Developing an efficient TiO_2-coated biofilm carrier for intimate coupling of photocatalysis and biodegradation. *Water Res.* **2012**, *46*, 6489–6496. [CrossRef]
19. Toro, R.G.; Calandra, P.; Federici, F.; de Caro, T.; Mezzi, A.; Cortese, B.; Pellegrino, A.L.; Malandrino, G.; Caschera, D. Development of superhydrophobic, self-cleaning, and flame-resistant DLC/TiO_2 melamine sponge for application in oil–water separation. *J. Mater. Sci.* **2020**, *55*, 2846–2859. [CrossRef]
20. Liu, G.-Q.; Pan, X.-J.; Li, J.; Li, C.; Ji, C.-L. Facile preparation and characterization of anatase TiO_2/nanocellulose composite for photocatalytic degradation of methyl orange. *J. Saudi Chem. Soc.* **2021**, *25*, 101383. [CrossRef]
21. Xiao, H.; Li, J.; He, B. Anatase-Titania Templated by Nanofibrillated Celluloseand Photocatalytic Degradation for Methyl Orange. *J. Inorg. Organomet Polym.* **2017**, *27*, 1022–1027. [CrossRef]
22. Adel, M.A.; El-Gendy, A.A.; Mohamed, D.A.; Abou-Zeid, R.E.; El-Zawawy, W.K.; Dufresne, A. Microfibrillated cellulose from agricultural residues. Part I: Papermaking application. *Ind. Crops Prod.* **2016**, *93*, 161–174. [CrossRef]
23. Adel, A.M.; El-Shafei, A.; Ibrahim, A.; Al-Shemy, M.T. Extraction of oxidized nanocellulose from date palm (*Phoenix dactylifera* L.) sheath fibers: Influence of CI and CII polymorphs on the properties of chitosan/bionanocomposite films. *Ind. Crops Prod.* **2018**, *124*, 155–165. [CrossRef]
24. Ibrahim, A.A.; Adel, A.M.; AbdEl–Wahab, Z.H.; Al–Shemy, M.T. Utilization of carboxymethyl cellulose based on bean hulls as chelating agent. Synthesis, characterization and biological activity. *Carbohydr. Polym.* **2011**, *83*, 94–115. [CrossRef]
25. Najafidoust, A.; Allahyari, S.; Rahemi, N.; Tasbihi, M. Uniform coating of TiO_2 nanoparticles using biotemplates for photocatalytic wastewater treatment. *Ceram. Int.* **2020**, *46*, 4707–4719. [CrossRef]
26. Schenzel, K.; Fischer, S. NIR FT Raman Spectroscopy—A Rapid Analytical Tool for Detecting the Transformation of Cellulose Polymorphs. *Cellulose* **2001**, *8*, 49–57. [CrossRef]
27. Schenzel, K.; Almlof, H.; Germgard, U. Quantitative analysis of the transformation process of cellulose I→cellulose II using NIR FT Raman spectroscopy and chemometric methods. *Cellulose* **2009**, *16*, 407–415. [CrossRef]
28. Adel, A.M.; El-Wahab, Z.H.A.; Ibrahim, A.A.; Al-Shemy, M.T. Characterization of microcrystalline cellulose prepared from lignocellulosic materials. Part I: Acid catalyzed hydrolysis. *Bioresour. Technol.* **2010**, *101*, 124446–124455. [CrossRef]

29. Shankar, S.; Rhim, J.W. Preparation of nanocellulose from micro-crystalline cellulose: The effect on the performance and properties of agar-based composite films. *Carbohydr. Polym.* **2016**, *135*, 18–26. [CrossRef]
30. Li, S.; Hao, X.; Dai, X.; Tao, T. Rapid Photocatalytic Degradation of Pollutant from Water under UV and Sunlight via Cellulose Nanofiber Aerogel Wrapped by TiO_2. *J. Nanomater.* **2018**, *2018*, 8752015. [CrossRef]
31. Zong, E.; Wang, C.; Yang, J.; Zhu, H.; Jiang, S.; Liu, X.; Song, P. Preparation of TiO_2/cellulose nanocomposites as antibacterial bio-adsorbents for effective phosphate removal from aqueous medium. *Int. J. Biolog. Macromol.* **2021**, *182*, 434–444. [CrossRef]
32. De Almeida, J.C.; Correâ, M.T.; Koga, R.H.; Del Duque, D.M.S.; Lopes, O.F.; Gelson, T.S.; da Silva, T.; Ribeiro, C.; de Mendonça, V.R. Crystallization time in ZnO: The role of surface OH groups in its photoactivity. *New J. Chem.* **2020**, *44*, 18216–18224. [CrossRef]
33. Cao, W.; Li, J.; Martí-Rossellò, T.; Zhang, X. Experimental study on the ignition characteristics of cellulose, hemicellulose, lignin and their mixtures. *J. Energy Inst.* **2019**, *92*, 1303–1312. [CrossRef]
34. Ceylan, Ö.; Van Landuyt, L.; Rahier, H.; De Clerck, K. The effect of water immersion on the thermal degradation of cotton fibers. *Cellulose* **2013**, *20*, 1603–1612. [CrossRef]
35. Coats, A.W.; Redfern, J.P. Kinetic Parameters from Thermogravimetric Data. *Nature* **1964**, *201*, 68–69. [CrossRef]
36. El-Sabour, M.A.; Mohamed, A.L.; El-Meligy, M.G.; Al-Shemy, M.T. Characterization of recycled waste papers treated with starch/organophosphorus-silane biocomposite flame retardant. *Nord. Pulp Pap. Res. J.* **2021**, *36*, 108–124. [CrossRef]
37. Li, Y.; Zhang, J.; Zhan, C.; Kong, F.; Li, W.; Yang, C.; Hsiao, B.S. Facile synthesis of TiO_2/CNC nanocomposites for enhanced Cr(VI) photoreduction: Synergistic roles of cellulose nanocrystals. *Carbohydr. Polym.* **2020**, *233*, 115838. [CrossRef]
38. Petrella, A.; Spasiano, D.; Cosma, P.; Rizzi, V.; Race, M.; Mascolo, M.C.; Ranieri, E. Methyl Orange Photo-Degradation by TiO_2 in a Pilot Unit under Different Chemical, Physical, and Hydraulic Conditions. *Processes* **2021**, *9*, 205. [CrossRef]
39. Rathod, M.; Moradeeya, P.G.; Haldar, S.; Basha, S. Nanocellulose/TiO_2 composites: Preparation, characterization and application in the photocatalytic degradation of a potential endocrine disruptor, mefenamic acid, in aqueous media. *Photochem. Photobiol. Sci.* **2018**, *17*, 1301–1309. [CrossRef]
40. Comparelli, R.; Fanizza, E.; Curri, M.L.; Cozzoli, P.D.; Mascolo, G.; Passino, R.; Agostiano, A. Photocatalytic degradation of azo dyes by organic-capped anatase TiO_2 nanocrystals immobilized onto substrates. *Appl. Catal. B Envir.* **2005**, *55*, 81–91. [CrossRef]
41. Matar, M.J.; Ostrosky-Zeichner, L.; Paetznick, V.L.; Rodriguez, J.R.; Chen, E.; Rex, J.H. Correlation between E-test, disk diffusion, and microdilution methods for antifungal susceptibility testing of fluconazole and voriconazole. Antimicrob. *Agents Chemother.* **2003**, *47*, 1647–1651. [CrossRef] [PubMed]
42. Bhuiyan, M.S.H.; Miah, M.Y.; Pail, S.C.; Aka, T.D.; Saha, O.; Rahaman, M.M.; Sharif, M.J.I.; Habiba, O.; Ashaduzzaman, M. Green synthesis of iron oxide nanoparticle using Carica papaya leaf extract: Application for photocatalytic degradation of Remazol yellow RR dye and antibacterial activity. *Helyon* **2020**, *6*, e04603. [CrossRef] [PubMed]
43. Slavin, Y.N.; Asnis, J.; Hafeli, U.O.; Bach, H. Metal nanoparticles; understanding the mechanism behind antibacterial activity. *Int. J. Nanobiotech.* **2017**, *15*, 65–85. [CrossRef] [PubMed]
44. Maslana, K.; Zywicka, A.; Wenelska, K.; Mijowska, E. Boosting of Antibacterial Performance of Cellulose Based Paper Sheet via TiO_2 nanoparticles. *Int. J. Molec. Sci.* **2021**, *22*, 1451. [CrossRef]

materials

MDPI

Article

Green Synthesis of Zinc Oxide Nanoparticles Using Red Seaweed for the Elimination of Organic Toxic Dye from an Aqueous Solution

Abdallah Tageldein Mansour [1,2,*], Ahmed E. Alprol [3,*], Mohamed Khedawy [3], Khamael M. Abualnaja [4], Tarek A. Shalaby [5,6], Gamal Rayan [1], Khaled M. A. Ramadan [7,8] and Mohamed Ashour [3,*]

1 Animal and Fish Production Department, College of Agricultural and Food Sciences, King Faisal University, P.O. Box 420, Al-Ahsa 31982, Saudi Arabia; gahmed@kfu.edu.sa
2 Fish and Animal Production Department, Faculty of Agriculture (Saba Basha), Alexandria University, Alexandria 21531, Egypt
3 National Institute of Oceanography and Fisheries (NIOF), Cairo 11516, Egypt; dr_khedawy@yahoo.com
4 Department of Chemistry, College of Science, Taif University, P.O. Box 11099, Taif 21944, Saudi Arabia; k.ala@tu.edu.sa
5 Department of Arid Land Agriculture, College of Agricultural and Food Science, King Faisal University, P.O. Box 400, Al-Ahsa 31982, Saudi Arabia; tshalaby@kfu.edu.sa
6 Horticulture Department, Faculty of Agriculture, Kafrelsheikh University, Kafr El-Sheikh 33516, Egypt
7 Central Laboratories, Department of Chemistry, King Faisal University, P.O. Box 420, Al-Ahsa 31982, Saudi Arabia; kramadan@kfu.edu.sa
8 Department of Biochemistry, Faculty of Agriculture, Ain Shams University, Cairo 11566, Egypt
* Correspondence: amansour@kfu.edu.sa (A.T.M.); ah831992@gmail.com (A.E.A.); microalgae_egypt@yahoo.com (M.A.)

Citation: Mansour, A.T.; Alprol, A.E.; Khedawy, M.; Abualnaja, K.M.; Shalaby, T.A.; Rayan, G.; Ramadan, K.M.A.; Ashour, M. Green Synthesis of Zinc Oxide Nanoparticles Using Red Seaweed for the Elimination of Organic Toxic Dye from an Aqueous Solution. *Materials* **2022**, *15*, 5169. https://doi.org/10.3390/ma15155169

Academic Editors: Roberta G. Toro, Daniela Caschera and Abeer M. Adel

Received: 14 June 2022
Accepted: 20 July 2022
Published: 26 July 2022

Abstract: This study aims to produce green zinc oxide nanoparticles (ZnO-NPs) derived from red seaweed (*Pterocladia Capillacea*) and evaluate their potential to absorb Ismate violet 2R (IV2R) ions from an aqueous solution. UV-vis spectrophotometry, Fourier-transform infrared spectroscopy (FTIR), scanning electron microscopy (SEM), X-ray diffraction (XRD), and a Brunauer–Emmett–Teller surface area analysis (BET) were used to analyze the structural, morphological, and optical features of the synthesized nanoparticles. The change in color of the chemical solution revealed the formation of zinc oxide nanoparticles. The FTIR examination confirmed the synthesis of both Zn and ZnO nanoparticle powder, with a BET surface area of 113.751 $m^2\ g^{-1}$ and an average pore size of 2.527 nm for the synthesized adsorbent. Furthermore, the maximum removal effectiveness of IV2R was 99% when 0.08 g ZnO-NPs was applied at a pH of 6, a temperature of 55 °C, and a contact time of 120 min. The dye adsorption capacity of the ZnO-NPs was 72.24 mg g^{-1}. The adsorption process was also controlled by the Freundlich adsorption model and pseudo-second-order reaction kinetics. The adsorption of IV2R ions onto the ZnO-NPs could be represented as a nonideal and reversible sorption process of a nonuniform surface, according to Freundlich adsorption isotherms. In addition, the constant values of the model parameters were determined using various nonlinear regression error functions. Moreover, thermodynamic parameters such as entropy change, enthalpy change, and free energy change were investigated; the adsorption process was spontaneous and endothermic. The high capacity of the ZnO-NPs synthesized by red seaweed promotes them as promising substances for applications in water treatment for the removal of IV2R dye from aqueous systems.

Keywords: green ZnO-NPs; IV2R; equilibrium modeling; thermodynamic parameters; seaweed extract; adsorption process

1. Introduction

Green nanotechnology is a methodology for creating nanomaterials that is devoid of or employs fewer toxic substances throughout the generation process [1]. Nanomaterial synthesis is one of the most challenging and rapidly growing areas of nanotechnology [2]. Metal nanoparticles have several applications in environmental pollution treatment, including heavy metals, degradation of toxic dyes, and antimicrobial substances [3]. Several researchers have presented numerous mechanistic ways to explain the hidden process behind the green synthesis of ZnO-NPs. As reported by Frattini et al. [4], NPs are generated both inside living algae and within sun-dried biomass. Nanomaterials are produced by reducing the amount of energy used throughout the production process [5]. In water or organic solvents, zinc ions are stable and dispersed colloidally. Nevertheless, zinc is a highly toxic chemical element, and zinc oxide has a more pronounced effect on microorganisms than on eukaryotic cells, as shown in the work of [6]. Nanoparticle synthesis can be achieved via physical, chemical, or biological means [7]. Although all chemical techniques are capable of producing clean, well-defined nanoparticles, they are costly and potentially hazardous to the environment. The various utilizations of biological biomasses, including yeast, bacteria, fungi, several plant extracts, and different algae extracts, are low-cost, environmentally friendly, and safe alternatives to existing nanoparticle production technologies [8]. The green synthesis of nanoparticles is used in the creation of highly stable nanoparticles.

Aquatic plants (microalgae and seaweeds) have the highest photosynthetic efficiency, are the highest biomass producers, are resistant to several pollutants, and have the ability to grow on land that is often inappropriate for other uses [9]. The cells of aquatic plants, such as seaweeds, have bioactive compounds and contain functional groups, including carboxyl, hydroxyl, amino, and sulphate, that can act as metal-binding sites between the adsorbent and adsorbate [10].

Algae nanoparticles are quite stable in solutions, eco-friendly, and safe to a large extent in several fields [9]. Furthermore, algal cells are sustainable resources that are extensively used as biomaterials for food [11], feed [5,12,13], fertilizers [14,15], bioenergy [16], and wastewater treatments [14,17]. Seaweeds have significant metal-binding capacities due to the presence of proteins, polysaccharides, and lipids on their cell wall surfaces that include several functional groups [18].

One of the most important environmental challenges, particularly in emerging countries, has been identified as irregular industrial development. The textile, plastic, and paper sectors all produce large amounts of wastewater containing synthetic dyes [19]. Synthetic dyes are utilized in a variety of industries, including textiles, leather, plastics, photography, medicine, cosmetics, and food [20,21]. Direct, reactive, acidic, and basic dyes are the most common types of dyes used in these industries [22]. Dye-containing wastewater has negative consequences for both the aquatic environment [23] and human health [24,25]. It is poisonous and can harm human health by causing damage to the neurological system, brain, eyes, liver, and skin [26]. These poisonous substances can endanger animals, plants, and humans. Dye removal from wastewater is accomplished using physical, chemical, and biological techniques. To remove colors from wastewater, a variety of processes are applied, including adsorption, advanced oxidation, membrane separation, and biological and electrochemical degradation [27].

Many companies utilize the adsorption process to eliminate harmful pollutants because it is one of the most important methods for water purification [1,28,29]. Adsorption is a viable approach because of its simple design, low cost, availability, and desire to eliminate color at high concentrations [30,31]. Adsorption is also a favored way of treating aqueous effluent in industrial operations for a variety of separation and purification applications. The sorption process is without a doubt the most essential physicochemical process for several organic and inorganic materials' uptake in aquatic environments [32]. The extent of adsorption is affected by factors including pH, contact time, doses of substrates, adsorbing ion, temperature, and the presence of competing and complex ions [33].

Nanosized zinc oxides have a number of advantages, including strong surface reactivity and ease of synthesis from abundant natural materials, and they are a destructive sorbent compared to their commercial analogues. They are also confirmed to be excellent materials as adsorbents [34]. Moreover, ZnO-NPs have been selected as an adsorbent material because they have greater surface areas, more active surface sites than bulk materials, and excellent pollutant removal capacities [35–37]. In addition, ZnO-NPs as semiconductors have gained attention for their wide range of applications, including optoelectronics, optics, electronics, and dye removal employing environmentally benign synthesis components including fungi, bacteria, and marine macroalgae [38,39].

In the present study, a green synthesis of zinc oxide (ZnO) nanoparticles is created as an adsorbent biomaterial for the removal of Ismate violet 2R dye from an aqueous solution under varied circumstances by applying a batch adsorption technique. Additionally, ZnO is discovered to be more effective and, possibly, more beneficial than other metals for the biosynthesis of nanoparticles (NPs) for medical applications. Furthermore, the green synthesis of zinc oxide nanoparticles is characterized using UV-vis, SEM, FTIR, X-ray diffraction, and a PET surface area analysis. Additionally, the kinetics, thermodynamics, and equilibrium of IV2R removal from aqueous solutions applying this adsorbent are examined.

2. Materials and Methods

2.1. Collection and Preparation of Seaweeds

Seaweed (*Pterocladia Capillacea*) samples were collected from the Alexandria coast of Egypt. The harvested *P. Capillacea* biomass was washed once with both seawater and tap water and twice with distilled water to remove the exterior grit and salt [40]. The seaweed biomass was dried for 72 h at 70 °C, ground with a blender, and sieved to obtain particles that passed through a laboratory test sieve with a mesh size of 125 µm. The homogenous suspension of *P. Capillacea* biomass was filtered and dried at 70 °C for 72 h or until a consistent weight was achieved and then stored at room temperature.

2.2. ZnO Nanoparticle Synthesis

The dried *P. Capillacea* powder (2 g) was mixed with 100 mL double-distilled water, heated to 100 °C, and then filtered through filter paper. The production of green ZnO-NPs was applied according to a procedure described previously [41,42]. An amount of 50 mL aqueous extract of *P. Capillacea* was added to a 2 mM solution of Zn (Ac) $2H_2O$, followed by the drop-wise addition of 2 M NaOH until the pH reached 12. The mixture was stirred continuously at 65 °C for 2–3 h. The pale white solid product was collected using centrifugation at 5000 rpm for 10 min, followed by meticulous washing with distilled water and drying overnight at 80 °C.

2.3. Reagents and Chemicals

Table 1 shows the chemical structure and certain features of Ismate violet 2R (IV2R) dye. All the used reagents and chemicals were of analytical quality (Merck Company, Darmstadt, Germany). The IV2R dye was dissolved in 1 L double-distilled water to make a stock solution (1000 mg L^{-1}). Dye solutions with concentrations in the range of 10–80 mg L^{-1} were generated for treatment studies by diluting the stock solution with double-distilled water.

Table 1. Physical characteristics of the dyes.

Characteristics	Value
Dye name	Ismate violet 2R
Wavelength (λ max)	550 nm
Mol. wt.	700
Molecular formula	$C_{22}H_{14}N_4O_{11}S_3CuCl$
C.I. name	IV2R
Molecular structure	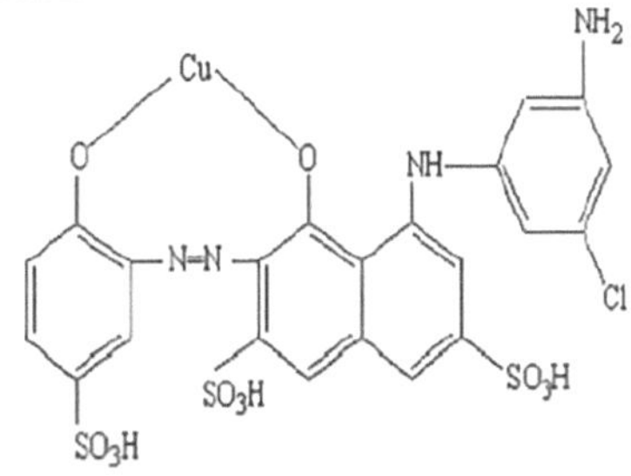

2.4. *Characterization*

SEM was used to examine the morphology of the ZnO-NPs (SEM, Vega Tescan TS5136MM). Using KBr pellets, the FTIR spectra were recorded with a Nicolet 6700 FTIR spectrometric analyzer. Furthermore, the BET technique was used to compute the specific surface area in the relative pressure range of 0.05–0.3. The Barrett–Joyner–Halenda (BJH) method was used to compute the mesopore size and distribution from the desorption curves, whereas the t-plot approach was used to calculate the micropore area values. The porosity (using the DFT approach) and surface area were measured using a N_2 adsorption–desorption isotherm analysis (Tristar 3000 apparatus, Micrometrics Instrument Corp., Norcross, GA, USA) (BET method). A UV-vis spectrophotometer was used to obtain the spectrophotometric readings (UV 4000, MRI, Stuttgart, Germany).

An XRD (D/Max 2550PC, Rigaku, Japan) diffractometer was used for X-ray diffractometry. The reflection-scanning mode was used to record the radical scan while changing 2 θ from 0° to 100°. A pH meter was used to determine the pH of the solution (Metrohm Herisau Digital E 532, Herisau, Switzerland).

2.5. *Adsorption Equilibrium Procedure*

2.5.1. Batch Adsorption Studies

In a 100 mL conical flask, batch experimental trials for adsorption equilibrium studies were performed by mixing 50 mL IV2R solution with several concentrations of 10, 20, 40, 60, and 80 mg L^{-1}. In the flasks, the adsorbent concentrations were 0.005, 0.01, 0.02, 0.04, and 0.8 g. The mixtures were shaken for 180 min at a steady speed of 200 rpm at an appropriate temperature. The temperatures used in the studies were 25, 30, 35, 45, and 55 °C. With 0.1 M HCl or 0.1 M NaOH, the pH values were altered to 2, 4, 6, 8, and 10. The mixing time was also examined (15, 30, 60, 120, and 180 min). The mixtures were filtered after the solutions reached equilibrium, and thus, the adsorbents were removed from the solutions. A UV-visible spectrophotometer set to maximum adsorption wavelength was used to determine the residual IV2R concentrations in the solutions (550 nm). Equation (1) was used to calculate the equilibrium value of the adsorbed dye per unit amount of adsorbent (q_e mg g^{-1}). In addition, Equation (2) was used to compute the IV2R removal percentage (%).

$$q_e = \frac{(C_i - C_f) \times V}{W} \quad (1)$$

$$\text{Percentage removal } (\%) = \frac{(C_i - C_f)}{C_i} \times 100 \quad (2)$$

where the concentrations of IV2R (mg g^{-1}) at the initial ions of adsorption and equilibrium, respectively, are C_i and C_f; m is the adsorbent mass (g); and V is the volume of the solution (L).

2.5.2. Adsorption Kinetics, Equilibrium, and Thermodynamic Studies

The adsorption kinetics analysis showed the adsorption rate, the performance of the adsorbent, and the mass transfer mechanisms. Adsorption dynamics must be understood to design effective adsorption systems.

Pseudo-first-order, pseudo-second-order, intraparticle diffusional, and Elovich models were used to evaluate the adequacy of different models for predicting adsorption kinetics. The IV2R adsorption kinetics of ZnO-NP samples were investigated at initial concentrations of 10 mg L^{-1}, and 0.02 g dry adsorbent was mixed with 50 mL IV2R solution in batch studies carried out in a shaker for 200 rpm at 25 °C for 15, 30, 60, 120, and 180 min.

Each adsorption system (adsorbent material and adsorbate) had its isotherm, and the amount of the adsorbed adsorbate on an adsorbent was determined by ZnO-NPs and the characteristics of the solution. The most applicable models used to fit the kinetic sorption experiments were Lagergren's pseudo-first-order model (Equation (3)) [43], a pseudo-second-order model (Equation (4)) [44], intraparticle diffusion (Equation (5)) [45], and the Elovich model (Equation (6)) [46]:

$$\text{Log } (q_e - q_t) = \log q_e - k_1 t/2.303 \quad (3)$$

$$t/q_t = 1/K_2\, q_e^{\,2} + t/q_e \quad (4)$$

$$q_t = K_{dif}\, t^{1/2} + C \quad (5)$$

$$q_t = \frac{1}{\beta} \ln (\alpha\beta) + \ln(t)\frac{1}{\beta} \quad (6)$$

where q_e (mg g^{-1}) is the quantity of dye adsorbed at equilibrium, and q_t (mg g^{-1}) is the amount of dye adsorbed at a time (t). The pseudo-first-order and pseudo-second-order adsorption rate constants are k_1 (min^{-1}) and k_2 (g mg^{-1} min), respectively. The intraparticle diffusion rate constant is determined using the slope of the regression line, and K_{dif} (mg g^{-1} $min^{0.5}$) is the intercept. The Elovich constants are α (mg g^{-1} min^{-1}) for the initial sorption rate and β (g mg^{-1}) for the surface coverage and chemisorption activation energy. Nonetheless, the isotherm studies were performed using 0.02 g of dry adsorbent at initial IV2R concentrations of 10, 20, 40, 60, and 80 mg L^{-1}, with enough time for adsorption equilibrium (180 min). Furthermore, thermodynamic studies of the adsorption experiments were carried out using the same approach at 25, 30, 35, 45, and 55 °C.

2.5.3. Analysis of Errors

Traditional linear regression approaches for determining isotherm parameters appeared to provide a satisfactory fit for the data of the experiments. Different error functions of nonlinear regression were employed to calculate the constant model parameters, and they were compared to those determined from less accurate, linearized data-fitting techniques to evaluate the fit of the isotherm equations to the experimental data. The chi-square test and the residual root mean square error (RMSE) were utilized. *RMSE* is a term that can be defined as:

$$\text{RMS} = 100\sqrt{\frac{1}{N}} \times \sqrt{\sum_{I=1}^{N}(1 - \frac{q_e, \text{calc}}{q_e, \text{isotherm}})2} \quad (7)$$

The subscripts "exp" and "cal" denote the experimental and calculated values, respectively, while n denotes the number of experimental isotherm observations and N is the number of the experimental data.

Moreover, X^2 is a small number of the data from the model calculation that is closest to the experimental data; the X^2 value can be used to gauge how well the model fits. The better the curve fitting, the lower the RMSE value. The formula for the chi-square test (8) is as follows:

$$X^2 = [(q_{e,\,\text{isotherm}} - q_{e,\,\text{calc}})^2]/q_{e,\,\text{isotherm}} \quad (8)$$

3. Results and Discussion

3.1. ZnO-NP Characterization

3.1.1. FTIR

The FTIR spectra of synthesized zinc oxide nanoparticles in the range of 400–4000 cm^{-1} are shown in Figure 1. A peak at about 3465 cm^{-1} that was assigned to N–H overlapped with a stretching vibration of hydroxyl groups (–OH) in the sample spectra. The weak absorption bands seen between 2344 and 2365 cm^{-1} corresponded to C–O stretching, which is one of the carboxylic group's distinctive peaks [47]. The bands detected at 1638 cm^{-1} could be attributed to the samples' C=O. The C=C stretching of aromatics was represented by the intensity peaks at 1553 and 1517 cm^{-1}. The C–C stretching vibration of aromatic ring compounds caused the bands to be around 1416 cm^{-1}. C–H asymmetric vibration was related to the bands at 1373 cm^{-1}. C–O stretching vibration was assigned to the peaks of 1124 and 1038 cm^{-1} in the synthesis spectra. A band at 419 cm^{-1} confirmed the production of ZnO-NPs [48]. C–H bending vibration was associated with peaks at 832, 708, and 734 cm^{-1}. Finally, the peak in the 400–600 cm^{-1} range was due to stretching vibrations of Zn–O [45,49]. The metallic group was responsible for the peak at 571 cm^{-1} in the FTIR diagram of the synthesized adsorbent. The presence of ZnO in the adsorbent was responsible for this peak [50]. In fact, electrons from either the C=O or C=C groups could transfer to the Zn^{2+} ions' free orbital. Some functional groups in this sample, such as N–H, O–H, C=C, and C–O, were favorable to the adsorption process [48].

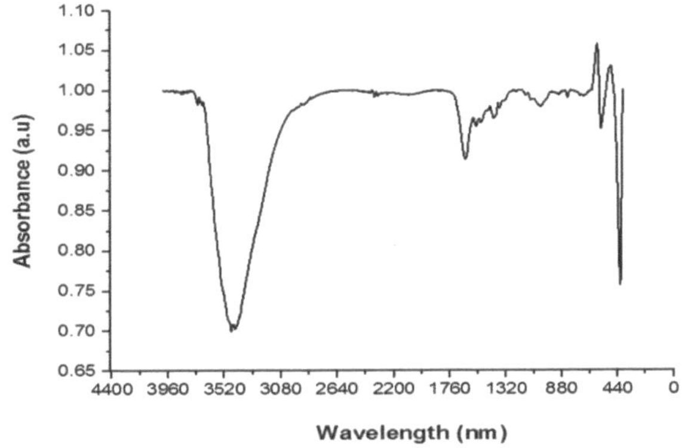

Figure 1. FTIR spectra of green ZnO-NPs.

3.1.2. SEM

Scanning electron microscopy (SEM) is a vital technique for examining the textural structure, surface morphology, and particle size of an adsorbent. Figure 2 shows SEM pictures of the synthesized adsorbent at a magnification of ×25,000. The photographs demonstrate the likeness of microparticles with an average particle size of 1 μm and show that the outer surface of the ZnO-NP powder was composed of rod-like particles and small flower forms that were interconnected, producing holes and, therefore, yielding a large surface area. In addition, the microparticles on the sample surface had an uneven and porous texture, so there was a chance that IV2R molecules would be trapped. Additionally, the presence of these small particles increased the porosity and surface area of the synthetic adsorbent. This was supported by the measured surface area obtained for the synthetic adsorbent (113.751 m^2g^{-1}).

Figure 2. SEM images of synthesized adsorbent.

3.1.3. BET

For BET, when the ratio is close to 1, the curve may reach a limiting value or climb if larger pores are present in micropore materials, which absorb more N_2 at low relative pressures. The produced adsorbent's N_2 adsorption–desorption isotherm could be classified as a type I isotherm, as shown in Figure 3. The adsorption isotherm directly touched the line of P/Po = 1 in this diagram. The volumes adsorbed at a low relative pressure (P/Po 0.1), indicating that nitrogen molecules were mainly adsorbed on the micropores. The synthesized adsorbent had a BET surface area of 113.751 m^2 g^{-1}, whereas the BJH desorption was 82.23 m^2 g^{-1}. It had a large surface area, which gave this low-cost material a valuable property. Furthermore, the overall pore volume was 0.143 C^2 g^{-1}, and the average pore size was 2.527 nm, as presented in Table 2. The adsorbent had a high volume of broad micropores and a high proportion of micropore volume.

Table 2. BET surface area and porosity of ZnO-NPs.

Surface Characteristics	Value
BET surface area	113.751 m^2 g^{-1}
Langmuir method	215.295 m^2 g^{-1}
Average pore size	2.52797 nm
Total pore volume	0.14378 C^2 g^{-1}
DH desorption	84.0658 m^2 g^{-1}
BJH desorption	82.2329 m^2 g^{-1}
Average particle radius	1.1988 nm
Polarizability	1.46 (mL mol^{-1})
Surface atom density	13.1 (mol cm^2)

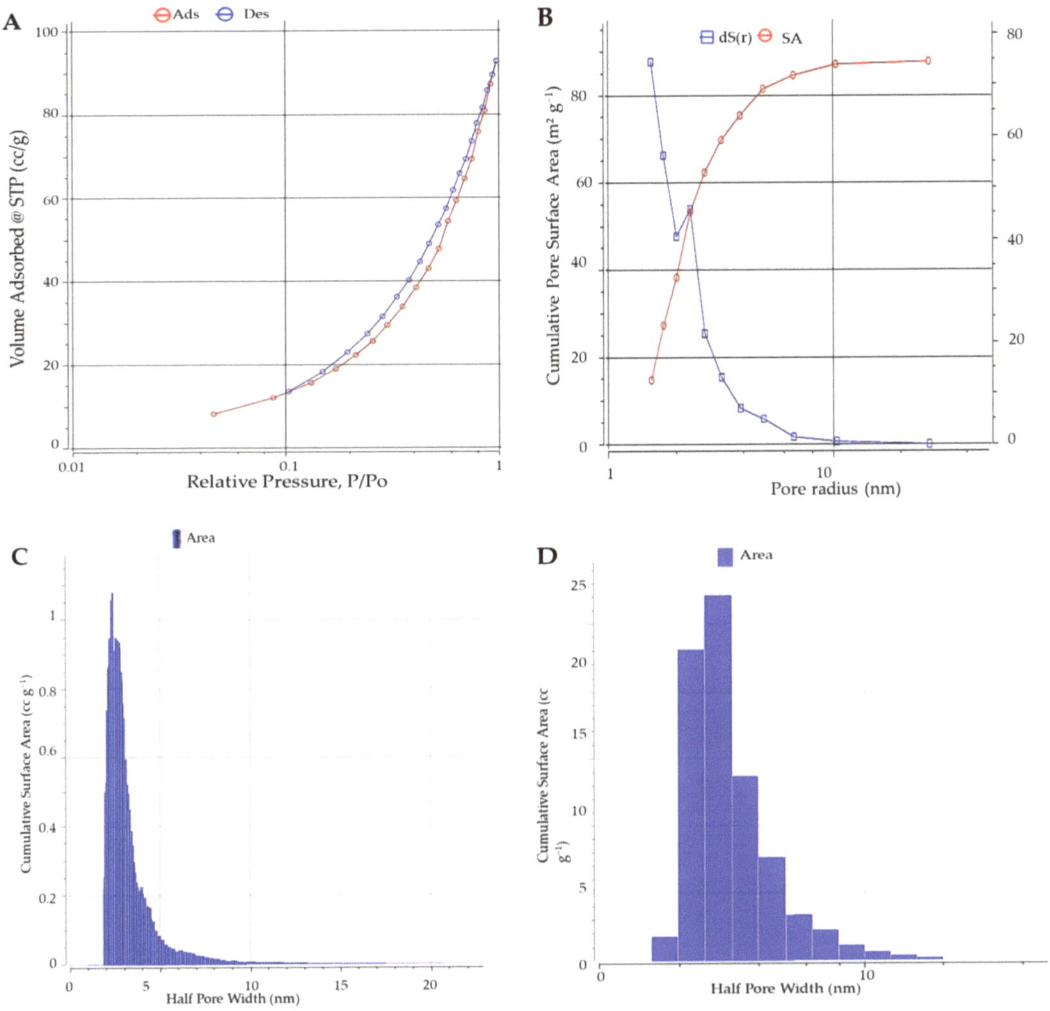

Figure 3. BET analyses and N_2 adsorption–desorption isotherms of green ZnO-NPs: (**A**) isotherm–isotherm, (**B**) BJH adsorption results, (**C**) DFT method histogram of surface area (linear), and (**D**) DFT method histogram of surface area (log).

3.2. XRD Study

The crystal structure of the produced material was investigated using X-ray diffraction [42]. ZnO-NPs' X-ray diffraction pattern was discovered utilizing an XRD system with a range of 0 to 100° A. The XRD patterns of green synthetic ZnO-NPs are displayed in Figure 4. The ZnO powder was extremely crystalline, according to the XRD peak, and all of the peaks were arranged in a way that corresponded to either the cubic shape of ZnO, which is the reference pattern for synthetic ZnO-NPs (65-2800), or the hexagonal structure of ZnO-NPs (05-20664), which is the reference pattern for green ZnO-NPs [41]. Additionally, the sharp and intense peaks showed the crystal structure of the produced ZnO nanoparticles [51]. Additionally, a mixed hexagonal and cubic phase was visible in the results when compared to the previous reference patterns.

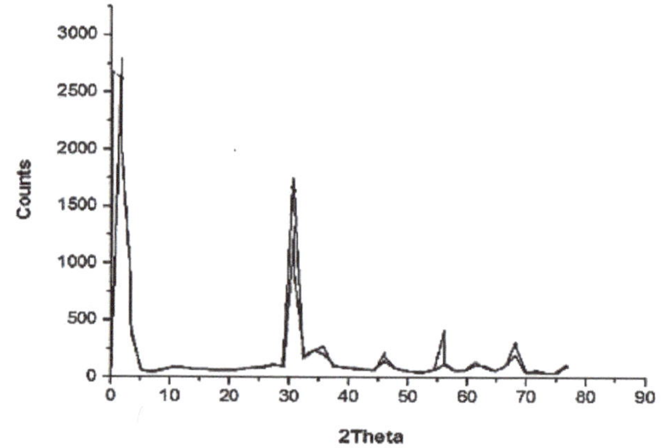

Figure 4. Pattern of XRD of synthesized ZnO.

3.3. UV

The key benefits of this method are that it is readily available, affordable, and simple to use. Since the sample was nontoxic, thus there was no pollution. The wavelength range of the metal nanoparticles was determined with UV-vis, as shown in Figure 5. The maximal absorbance was shown by the highest peak. The production of zinc oxide nanoparticles was confirmed by the UV spectrometer absorption peaks in the 300–400 nm range.

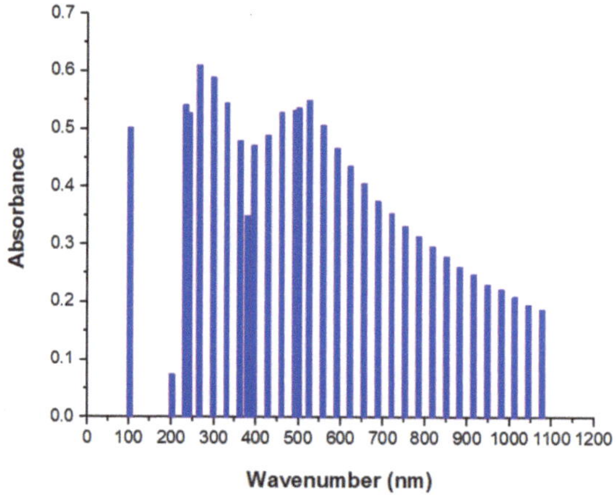

Figure 5. UV-vis spectra of green ZnO-NPs.

These peaks likewise reflected the particles' narrow particle size distribution and nanoscale size. The absorption peak at 243 nm, which is associated with $\pi \rightarrow \pi^*$ transitions in the sesquiterpene system, could hardly be seen in the UV spectra of the ZnO-NPs, which could be due to a change in the sesquiterpene structure and absence of this $\pi \rightarrow \pi^*$ transition [52].

3.4. Influence of Experimental Parameters on the Adsorption Mechanism

3.4.1. pH

The pH of the solution had an impact on the surface charge of the adsorbent, the levels of ionization of various pollutants, the dissociation of functional groups on the active sites of the adsorbent, and the structure of the dye molecule [53]. As a result, during the dye adsorption process, the pH of the solution was a key parameter. The IV2R molecules' ability to ionize and the surface properties of the adsorbent were both influenced by pH, hence the effectiveness of the adsorption was tested at varied solution pH levels. The effect of the initial pH of the solution on IV2R adsorption is shown in Figure 6. IV2R adsorption increased from pH 2 to 6 and then decreased. It was found from the results that the IV2R dye elimination was maximal at 98.78% for pH 2. This could be because, at low pH values, the external surface of the adsorbent material was positively charged via absorbing H^+ ions [54]. Furthermore, the change in surface charge of the adsorbent could cause a high percentage of removal at pH 2 and 6. As the pH of the IV2R solution increased, the adsorbent's surface became negatively charged, aiding the electrostatic-attraction-based adsorption of the positively charged cationic dye [27].

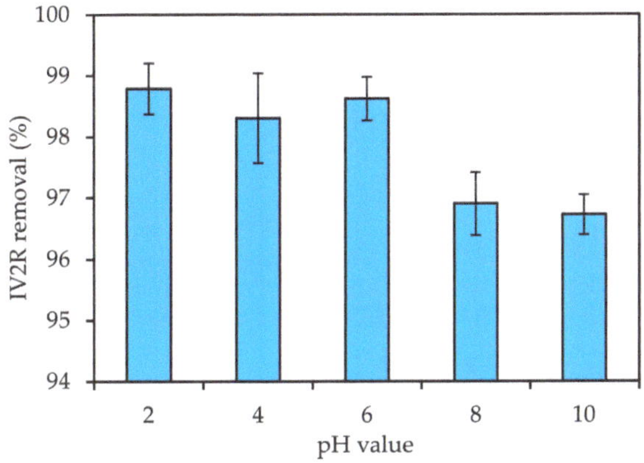

Figure 6. Influence of pH on removal of IV2R using green ZnO-NPs (adsorbent amount: 0.02 g; temperature: 25 ± 2 °C; initial volume: 50 mL; adsorbent dosage: 0.02 g; initial dye concentration: 10 mg L^{-1}; and contact time: 180 min).

3.4.2. ZnO-NP Dosages

The dose of the adsorbent determined its capacity for a specific initial concentration of the adsorbate under the operating conditions. Figure 7 depicts the effect of adsorbent dosage on IV2R removal. For this study, adsorbent dosages ranging from 0.005 to 0.8 g were added to 50 mL of IV2R solution. Because of the increase in the number of adsorption sites, the percentage of the removal of IV2R increased as the adsorbent dosage increased. As expected, the percentage of elimination increased from 96.47% to 99% when the adsorbent dosage increased from 0.005 to 0.8 g. As a result, only 0.8 g of adsorbent were required to remove 99% of the IV2R dye when the quantity of the adsorbent material was augmented. It is evident that, as the adsorbent dosage was raised, the IV2R dye ions' capacity for adsorption grew quickly due to the increased availability of exchangeable sites and surface area, and that, after equilibrium was established, increasing these sites had little effect [55], as the adsorption process is a surface phenomenon. According to the results, as the adsorbent dosage was increased, the removal efficiency increased, but the adsorption level capacity decreased. This was caused by the polysaccharide and fiber-like elements found in the algae cell walls, the adsorbent active surface, and dynamic elements such as an increased level of contact and an increased number of free bonds on the adsorbent surface, which could be specified using adsorption isotherms and synthetic equations, according to Aziz et al. [52].

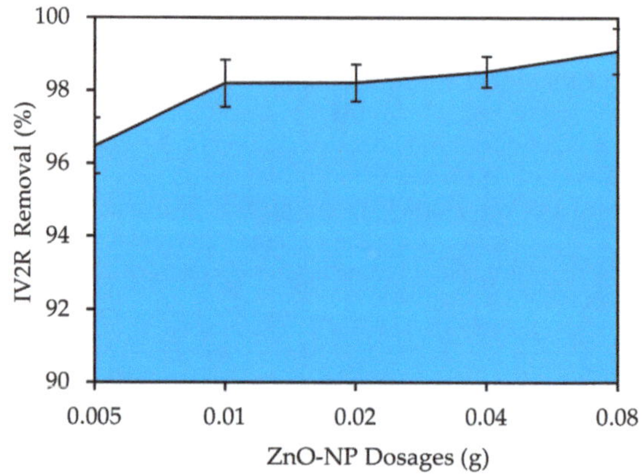

Figure 7. Influence of ZnO-NP adsorbent dosage on removal of IV2R (pH: 6; temperature: 25 ± 2 °C; initial volume: 50 mL; initial dye concentration: 10 mg L^{-1}; and contact time: 180 min).

3.4.3. Contact Times

The effect of contact time on IV2R adsorption in solutions with 10 mg L^{-1} initial concentrations is shown in Figure 8. It reveals that the adsorption rate rose significantly with increasing contact time to reach 98.4% after 120 min, and then the IV2R dye reached adsorption equilibrium. Furthermore, once equilibrium was attained, the adsorption capacity remained constant. This was most likely caused by the IV2R ion saturation of nano-adsorbent surfaces, which was followed by adsorption and desorption processes [52]. Furthermore, it occurred due to the adsorbent's huge surface area, as well as strong attraction forces between the adsorbent and dye molecules. The rate of removal increased at first and then lowered to a near-equilibrium condition after 30–120 min. According to Abate et al. [56], the maximum number of active sites is initially available but becomes saturated over time. The remaining vacant surface sites are challenging to occupy as a result of the formation of a repulsive force between the adsorbate on the solid surface of the adsorbent and the adsorbate on the bulk phase (solutions), as well as the saturation of the active site of the adsorbent.

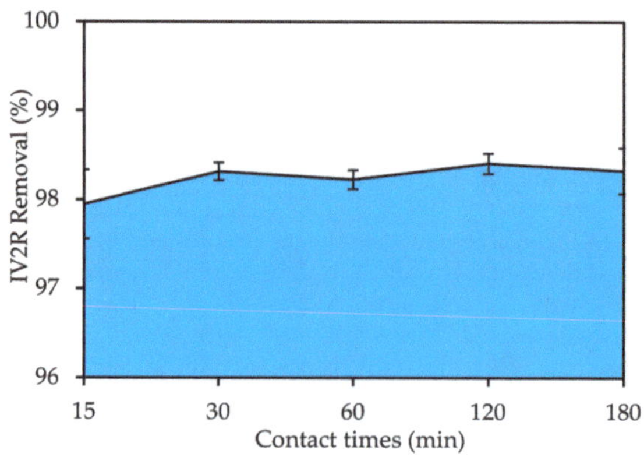

Figure 8. Influence of contact time on IV2R removal by green ZnO-NPs (pH: 6; temperature: 25 ± 2 °C; initial volume: 50 mL; initial dye concentration: 10 mg L^{-1}; and adsorbent dosage: 0.02 g).

3.4.4. Initial IV2R Concentration

Figure 9 shows the effects of initial dye concentrations ranging from 10 to 80 mg L^{-1}. With increasing initial concentrations, the IV2R adsorption increased progressively because the initial dye concentration was crucial to the adsorption process, as it served as a driving force for transferring dyes from the aqueous solution to the solid surface. The ratio of IV2R molecules to the accessible surface area increased as concentrations rose. As a result, the adsorption process was influenced by the dye concentration at the start. The available active sites of the adsorbent were degraded at high dye concentrations, and the amount of adsorbed IV2R decreased. However, when the initial concentration of IV2R dyes increased, the adsorption capacity (q_e) significantly increased. The adsorption capacity (q_e) increased from 1 to 42.1 mg g^{-1} as the dye concentration was increased from 10 to 80 mg L^{-1}. Therefore, the interaction between the dye molecules and the surface of the adsorbent was increased with increasing initial IV2R dye concentrations [57].

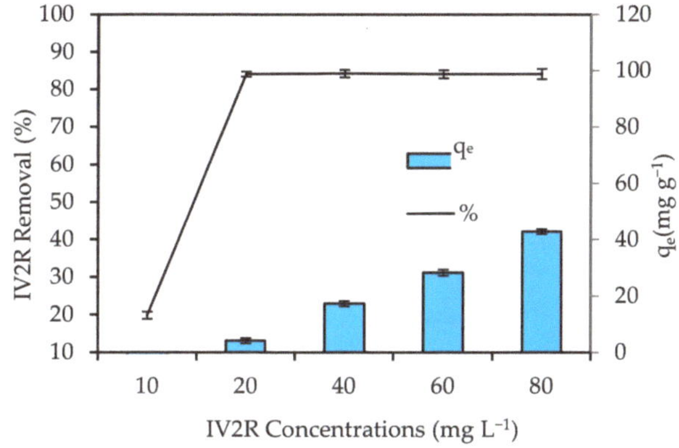

Figure 9. Influence of initial dye concentrations on removal of IV2R by green ZnO-NPs (pH: 6; temperature: 25 ± 2 °C; initial volume: 50 mL; adsorbent dosage: 0.05 g; and contact time: 180 min).

3.4.5. Temperature

The influence of temperature on IV2R removal by the synthesized adsorbent was investigated in the range of 25–55 °C, as shown in Figure 10. The IV2R removal increased as the temperature rose from 30 to 55 °C, and it was nearly constant at higher temperatures. At 55 °C, the highest adsorption of IV2R was observed at higher temperatures, and the removal was practically constant, with a percentage of clearance of 99.61%. The fact that adsorption increased with the temperature increase from 30 to 55 °C implied that IV2R adsorption is an endothermic process. Changing the dye adsorption by altering the temperature revealed that, as the temperature rose, the adsorbed surface activity rose and that IV2R adsorption is mostly controlled by chemical forces [58].

3.5. Adsorption Isotherms

Adsorption isotherms were used to determine the interaction between the adsorbent and the dye, as well as the dye's adsorption capacity. The first empirical technique used to investigate the nature of adsorption was the isotherm shape. The IV2R adsorption isotherms of ZnO-NPs are shown in Figures 11–16. The fitting of isotherm data to various isotherm models aided in the development of an appropriate model for the adsorption process, as well as the estimation of the adsorbent's adsorption capacity. The equilibrium results were compared with ZnO-NPs, which have a large surface area, to better understand the synthesized adsorbent adsorption capacity. Langmuir, Freundlich, Tempkin, Harkins–Jura, Halsey, and Sips isotherms were used to assess the experimental equilibrium data. The four equilibrium isotherm models utilized for IV2R adsorption on the generated ZnO-NPs presented in Table 3 yielded all the correlation coefficients, R^2 values, and adsorption parameters at equilibrium.

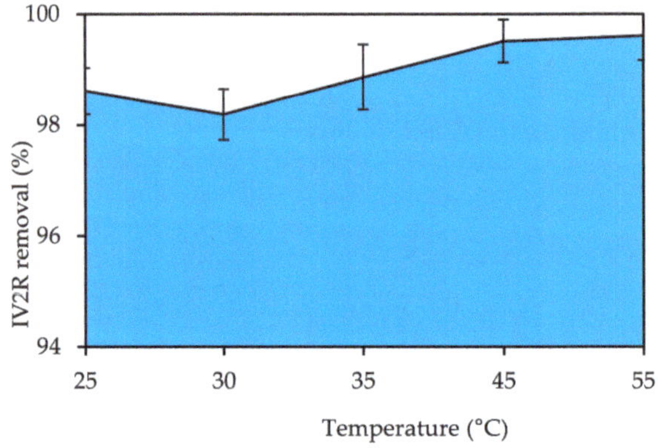

Figure 10. Influence of temperature on IV2R removal by green ZnO-NPs (pH: 6; initial volume: 50 mL; initial dye concentration: 10 mg L^{-1}; adsorbent dosage: 0.02g; and contact time: 180 min).

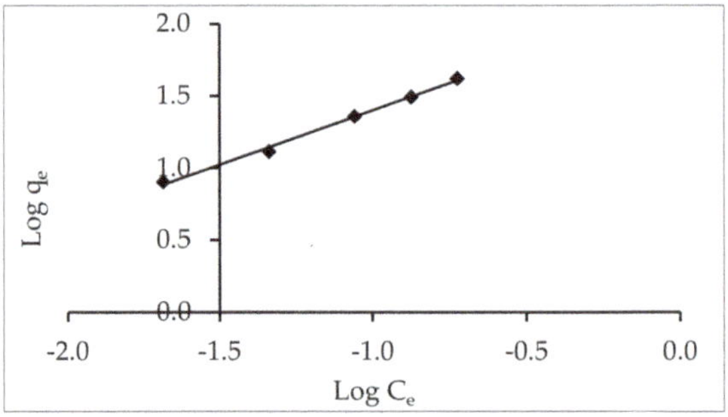

Figure 11. Freundlich equation for the sorption of IV2R onto Zn-ONPs.

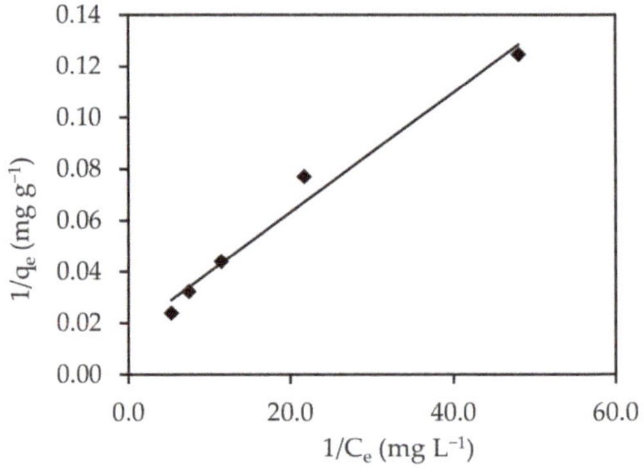

Figure 12. Langmuir model for the adsorption of IV2R onto ZnO-NPs.

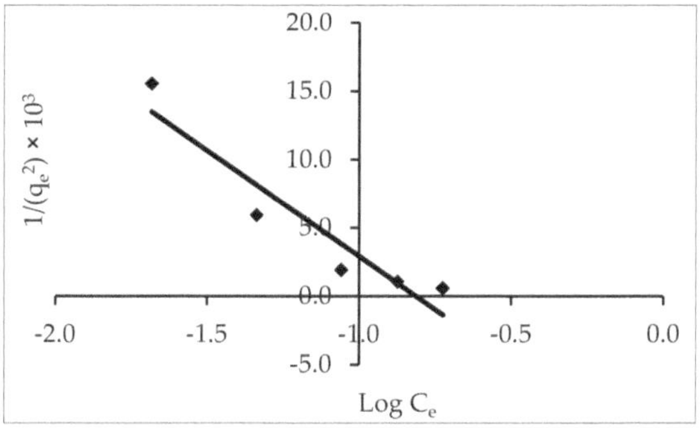

Figure 13. Harkins–Jura adsorption isotherm for the adsorption of IV2R onto ZnO-NPs.

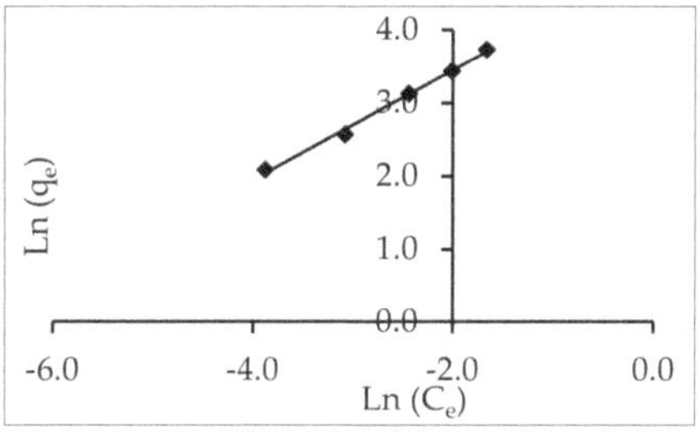

Figure 14. Halsay adsorption isotherm for the adsorption of IV2R onto ZnO-NPs.

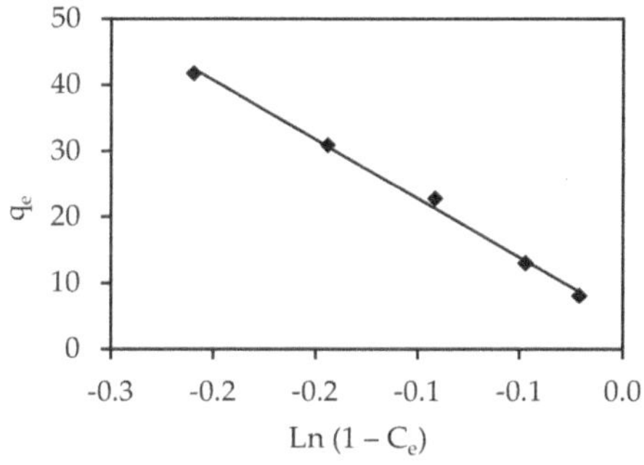

Figure 15. Smith isotherms for the adsorption of IV2R onto ZnO-NPs.

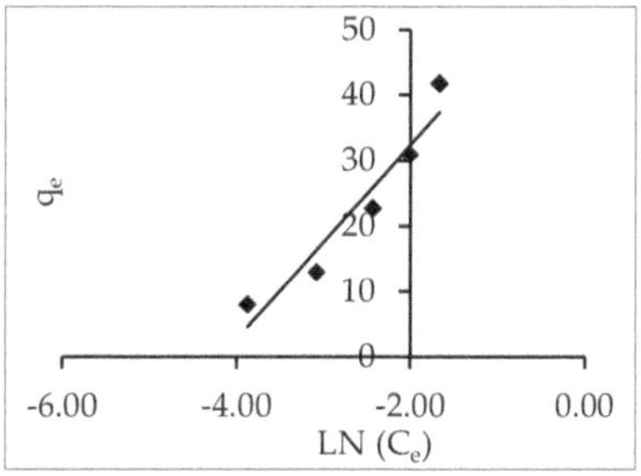

Figure 16. Tempkin isotherms for the adsorption of IV2R onto ZnO-NPs.

Table 3. Correlation coefficients and isotherm model constants of adsorption of IV2R on ZnO-NPs.

RMSE	X^2	Value	Isotherm Parameter	Isotherm Model
0.046	0.001	0.754 34.30 0.995	$1/n$ K_F ($mg^{-1-1/n}$ $L^{1/n}$ g^{-1}) R^2	Freundlich
0.047	0.002	59.88 7.26 0.979 0.038	Q_{max} ($mg\ g^{-1}$) K_a R^2 R_L	Langmuir
14.995	65.439	0.06 0.81 0.894	A_{HJ} B_{HJ} R^2	Harkins–Jura
0.054	0.001	0.753 721 0.995	$1/n_H$ K_H R^2	Halsey
0.004	0.000	4.940 178.730 0.995	W_{bs} W_s R^2	Smith
16.192	76.305	14.65 65.36 592.46 0.926	A_T B_T b_T R^2	Tempkin

3.5.1. Freundlich Isotherm

Adsorption varies in a manner that is directly proportional to pressure, according to the Freundlich isotherm. This empirical equation, which assumes that different locations have various adsorption energies involved, describes the multilayer adsorption of heterogeneous systems [44]. The isotherm's linear model can be written logarithmically, as demonstrated in the following equation:

$$\text{Log } q_e = \text{Log } K_f + 1/n \text{ Log } C_e \quad (9)$$

The intercept and slope of the plot of log C_e against log q_e (Figure 11) can be used to determine the values of parameters K_f and n (listed in Table 3), where K_f (1 g^{-1}) is the Freundlich adsorption capacity constant, and 1/n is the adsorption intensity constant that varies with the heterogeneity of the adsorbate. The Freundlich isotherm, with R^2 = 0.995, was determined to be the best fit for the adsorption experimental data; the high correlation coefficient suggested a high affinity between the ZnO-NP adsorbent surface and the IV2R dye, which played a large role in the adsorption mechanism. This could be due to the nonhomogeneous nature of the adsorbent surface, which could include multiple types of sorption sites [37]. The adsorption favorability is determined by the magnitude of the (n) parameter. A favorable adsorption condition is defined as $n > 1$ in the range of 1–10 [38], (in the present study, it was 1.32 for IV2R dye). Moreover, Table 3 shows that the Freundlich isotherm fit the experimental data well, as evidenced by the RMSE and chi-square values. This result also indicates that IV2R adsorption on ZnO nanoparticles was based on multilayer adsorption.

3.5.2. Langmuir

The pressure dependence of molecule (adsorbate) adsorption is demonstrated by the Langmuir adsorption isotherm model. When the energy of adsorption is constant, a saturated monolayer of adsorbate molecules is present on the adsorbent surface, there is no migration or contact between the adsorbate molecules on the surface plane, and maximum adsorption occurs [59]. The Langmuir isotherm model's linear expression is defined using the following Equation (10).

$$1/q_e = 1/(Ka\ q_m\ C_e) + 1/q_m \tag{10}$$

The values of Ka and qm can be determined from the slope and intercept of the linear graph of $1/C_e$ against $1/q_e$ (Figure 12), where qm (mg g^{-1}) is the maximum adsorption capacity of dye per unit mass of sorbent to form a complete monolayer on the surface-bound adsorption, and Ka (L mg^{-1}) is the Langmuir energy constant, which is related to the heat of adsorption. The Langmuir constants are listed in Table 3 (q_m = 59.88 mg g^{-1} and Ka = 7.26 L mg^{-1}). The Langmuir isotherm's essential properties can be stated in terms of a dimensionless constant separation factor R_L, which is given in Equation (11):

$$R_L = \frac{1}{1 + bC_i} \tag{11}$$

where C_i (mg L^{-1}) denotes the greatest initial adsorbate concentration.

The value of R_L shows the type of isotherm to be either linear (R_L = 1), favorable ($0 < R_L < 1$), unfavorable ($R_L > 1$), or irreversible (R_L = 0). In this study, the value of R_L was determined to be 0.038, and a value between 0 and 1 denoted favorable adsorption, which indicated that IV2R adsorption on the surface of ZnO-NPs was advantageous. Because the Langmuir equation assumes that the adsorbent surface is energetically homogeneous with high root mean square error (RMSE) and chi-square (X^2) values, the fact that the Langmuir isotherm fit the experimental data well could be due to the uniform dispersion of active sites on the ZnO-NPs.

3.5.3. Harkins–Jura

Multilayer adsorption is explained by the Harkins–Jura adsorption isotherm, which is explained by the presence of heterogeneous pore distribution. The Harkins–Jura adsorption isotherm can be expressed as:

$$\frac{1}{qe^2} = \left(\frac{B_2}{A}\right) - \left(\frac{1}{A}\right) \log C_e \tag{12}$$

where the isotherm constants are A and B.

The value of $1/q_e$ is plotted against log C_e; the isotherm constants are B (intercept/slope; $mg^2\ L^{-1}$) and A (1/slope; $g^2\ L^{-1}$) (Figure 13).

For ZnO-NPs, this model exhibited the best fit for the results, with R^2 equal to (0.900). The obtained results showed that multilayers of adsorption formed and that the ZnO-NP adsorbent was porous. In addition to considering the existence of a heterogeneous pore distribution, the Harkins–Jura isotherm is related to the Freundlich model. Because of large RMSE and X^2 values, the Harkins–Jura model was less suitable to describe IV2R adsorption onto ZnO-NPs.

3.5.4. Halsay

The experimental data that fit this equation attest to the adsorbent's heterosporous nature, and the equation is appropriate for multilayer adsorption. The slope and intercept of a plot of ln q_e vs. ln C_e, as illustrated in Figure 14, can be used to determine n and K with the following equation:

$$\mathrm{Ln}[-\mathrm{Ln}(1 - C_e)] = \mathrm{Ln}K + \left(\frac{1}{n}\right)\mathrm{Ln}q_e \tag{13}$$

Table 3 lists the model parameters. Because the Halsay model assumes a multilayer behavior for the adsorption of an adsorbate onto an adsorbent, it may also be used to represent the adsorption of IV2R onto ZnO-NPs. This was supported by high RMSE and X^2 values (Table 4).

Table 4. Adsorption kinetics of sorption of IV2R onto ZnO-NPs.

Model	Parameter	Value
First-order kinetic	q_e (calc.)	82.41
	K_1 (min^{-1})	4.66×10^{-3}
	R^2	0.018
Second-order kinetic	q_e (calc.) ($mg\ g^{-1}$)	4.00
	K_2 ($mg\ g^{-1}\ min^{-1}$)	5.69
	R^2	1
	q_e (exp.)	3.99
Interparticle diffusion	K_{dif} ($min^{1/2}$)	0.0014
	C	0.60
	R^2	0.615
Elovich	B ($mg\ g^{-1}\ min^{-1}$)	1.31
	α ($g\ mg^{-1}$)	25,510.02
	R^2	0.771

3.5.5. Smith

The Smith model is suitable for multilayer adsorption and heteroporous solids. This model is frequently derived using the equation shown below:

$$q_e = W_{bS} - W_S \text{ Ln } (1 - C_e) \quad (14)$$

where W_{bS} and W_s are the Smith model parameters.

The Smith model is applicable for heteroporous solids and multilayer adsorption, as well as explaining the adsorption isotherms of biological substances such as cellulose and starch. A plot of q_e vs. Ln (1− C_e) can be used to solve the Smith model, as shown in Figure 15. The sorption isotherms were properly represented by the Smith model across the entire range of water activity. The Smith equation, on the other hand, was adsorption. The correlation coefficients and isotherm constants varied from R^2 = 0.995.

3.5.6. Tempkin

Lastly, the Tempkin isotherm considers the adsorption process's indirect adsorbent–adsorbate interactions [60]. Due to adsorbent–adsorbate interactions, the heat of adsorption of all the molecules in a layer drops linearly with coverage. It can be written as q in the linear form [61]:

$$q_e = B \text{ Ln } A + B \text{ Ln } C_e \quad (15)$$

where B = (RT)/b is the universal gas constant of 8.314 J (mol K^{-1}), and T is the absolute temperature in Kelvin. The heat of adsorption is proportional to the constant b. The isotherm constants A and B can be calculated from the slope and intercept of a plot of q_e vs. ln C_e (Figure 16).

The computed A was 14.65 L g^{-1}, which was the equilibrium binding constant corresponding to the maximal binding energy, as shown in Figure 14. The heat of adsorption for the ZnO nanoparticles was related to constant B, which was equal to 65.36 J mol^{-1}. Moreover, the Tempkin model was shown to be unsuitable for IV2R adsorption onto ZnO-NPs by comparing the values of the error functions.

3.6. Models of Sorption Kinetics

Adsorption is a physiochemical process in which a solute (adsorbate) is transferred from the liquid phase to the adsorbent surface through mass transfer. The study of adsorption kinetics is desirable because it gives information about the adsorption mechanism, which is critical for process efficiency.

3.6.1. Pseudo-First-Order Model

The first model of the adsorption rate based on adsorption capacity was the pseudo-first-order kinetic model. Log ($q_e - q_t$) against (t) plots showed a linear relationship between k_1 and q_e, as measured by the slope and intercept, respectively. $R^2 = 0.018$ indicated that the correlation coefficients were not good enough. Furthermore, the calculated value of q_e from the equation, 82.41 mg g^{-1}, differed with the experimental value of 3.99 mg g^{-1}) Table 4 and Figure 17).

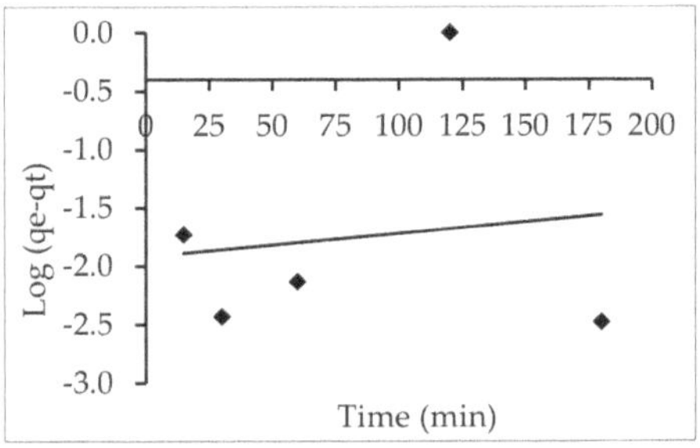

Figure 17. The pseudo-first-order equation for the elimination of IV2R onto ZnO-NPs.

3.6.2. Pseudo-Second-Order Model

A pseudo-second-order model was used to further examine the experimental kinetic data. By graphing t/q_t against t for IV2R, a straight line was obtained. The slope and intercept of the curve in Figure 18 were used to calculate the second-order rate constant and q_e values, which are reported in Table 4. The correlation coefficients (R^2) for IV2R adsorption onto ZnO-NPs were determined to be equal to one. The kinetics of IV2R adsorption onto ZnO-NPs could be well-represented by the second-order equation. This showed that chemisorption, which involves valent forces and the sharing or exchange of electrons between the adsorbent and the adsorbate, could be the rate-limiting phase in this sorption process [44].

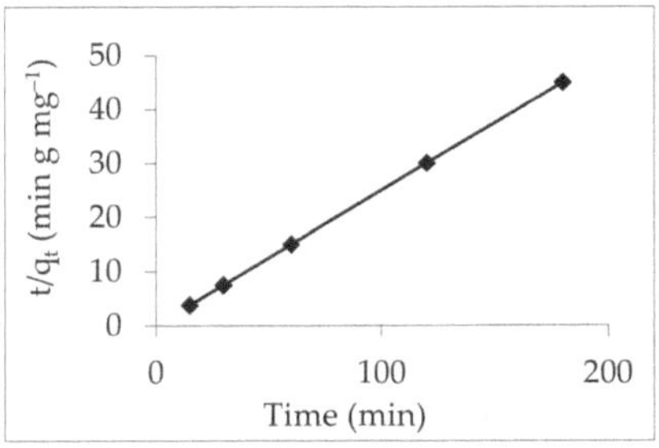

Figure 18. The pseudo-second-order equation for the elimination of IV2R onto ZnO-NPs.

3.6.3. Model of Interparticle Diffusion

The overall adsorption rate in a porous adsorbent, according to the literature, must account for the three processes of external mass transfer, intraparticle diffusion, and adsorption on active sites inside the pores. Film or intraparticle diffusion, or a mixture of both mechanisms, affects the overall rate of adsorption. Plotting (q_t) versus ($T^{0.5}$) yields a linear relationship, and the slope and intercept can be used to calculate the K_{dif} parameter [62]. Table 4 lists the values of K_{dif} and the R^2 correlation coefficients determined from the intraparticle diffusion plots. The fact that the linear curve in Figure 19 did not pass through the origin suggests that boundary layer diffusion had some influence on adsorption and that intraparticle diffusion was not the primary factor dictating the rate [63]. The discrepancy in mass transfer rates between the starting and final stages of adsorption could be the cause of this deviation [64].

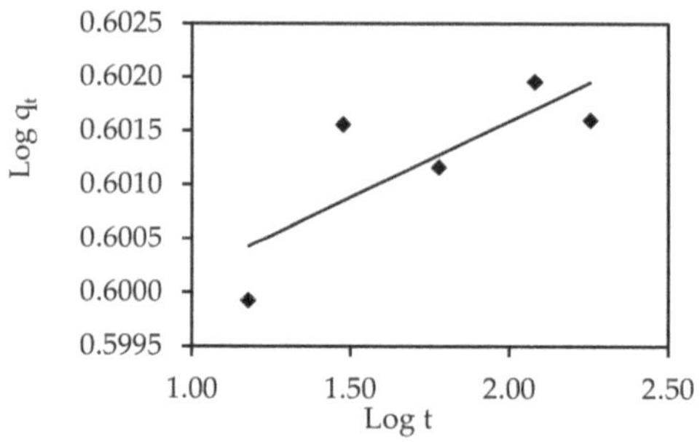

Figure 19. Interparticle diffusion equation for the elimination of IV2R onto ZnO-NPs.

3.6.4. Elovich Model

One of the most effective models for explaining the kinetics of gas chemisorption onto solid systems is the simple Elovich model. However, it has recently been used to describe the process of contaminants adsorbing from aqueous solutions. The plot of q_t against Ln (time) for IV2R sorption onto ZnO-NPs is shown in Figure 20. The estimated Elovich equation parameters were derived from the slope and intercept of the linearization of the simple Elovich equation. The number of adsorption sites available was indicated by the value of β. Table 4 shows the adsorption quantity when Ln t was equal to zero, i.e., the adsorption quantity when t equaled 1 h. This value added knowledge of the first step's adsorption behavior. In addition, it is obvious from this diagram that the Elovich equation did not fit the experimental data well. For the Elovich model, a small β (mg g^{-1} min^{-1}) (desorption constant) value (<1) indicates an irreversible adsorption method [65]. Nevertheless, the average β parameter of 1.31 (g mg^{-1}) achieved in this study indicated that the adsorption process was readily reversible.

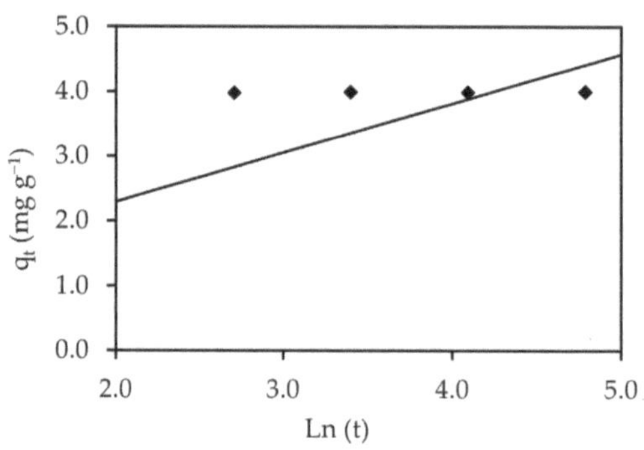

Figure 20. The Elovich equation for the elimination of IV2R onto ZnO-NPs.

3.7. Thermodynamic Studies

The influence of temperature on adsorption capacity was investigated using thermodynamic analyses. The following Equations (16)–(19) were used to determine the Gibb's free energy change (ΔG°), the enthalpy change (ΔH°), and the change in entropy (ΔS°):

$$Kd = qe/Ce \tag{16}$$

$$\Delta G^\circ = -RT \, Ln \, K_d \tag{17}$$

$$\Delta G^\circ = \Delta H^\circ - T\,\Delta S^\circ, \text{ or} \tag{18}$$

$$\Delta G^\circ = T\,(\Delta S^\circ) + \Delta H^\circ \tag{19}$$

where q_e is the solid phase equilibrium concentration in milligrams per liter (mg L^{-1}), C_e is the equilibrium concentration in milliliters per liter (mg L^{-1}), T is the temperature in degrees Celsius, and R is the gas constant (8.314 J mol^{-1} K). In addition to ΔG°, which may be obtained from Equation (17) as shown in Figure 21, the values of ΔH° and ΔS° were determined using the intercept and slope of the displayed curve of T vs. ΔG° from Equations (18) or (19). Table 5 lists the thermodynamic parameters. The fact that the change in free energy (ΔG°) was negative showed that the adsorption process was feasible and that IV2R adsorption onto ZnO-NPs occurred spontaneously at the temperature range investigated. Furthermore, a positive enthalpy change (ΔH°) suggested that the adsorption was chemical and endothermic. The entropy value (ΔS°) was negative. This showed that, during the adsorption process, the unpredictability between the solid–solution contacts was reduced, as presented in Table 5.

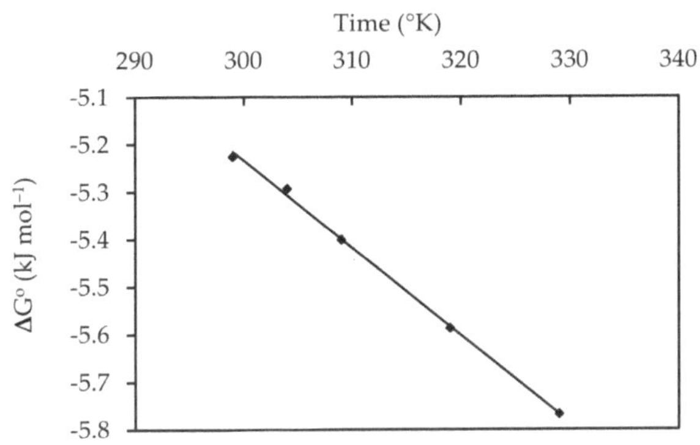

Figure 21. Thermodynamic behaviors of the removal of IV2R onto ZnO-NPs.

Table 5. Thermodynamic manners of adsorption of IV2R onto ZnO-NPs.

Temperature (°C)	ΔG° (kJ mol^{-1})	ΔH° (kJ mol^{-1})	ΔS° (J mol^{-1})
25	−5.22548		
30	−5.29248		
35	−5.3995	0.287	−0.018
45	−5.58672		
55	−5.76682		

3.8. Dye Removal Mechanism

The mechanism of adsorption is one of the most important tools to know when developing a system. The most prevalent adsorption mechanisms are chemical adsorption, physical adsorption, and ion exchange. The adsorption mechanism can be examined through the modelling of the data from the adsorption equilibrium and an analysis of the empirical results. It is feasible to indicate the basic mechanisms of adsorption for dye on an adsorbent by taking into consideration data such as pore size distribution, thermodynamics, and equilibrium results. Electrostatic interaction was the main force at work between the adsorbent and the adsorbed dye. Hydrogen bonding, hydrophobic interactions, and interactions of the aromatic rings of the ZnO-NPs with the aromatic rings of the dye all played significant roles in the adsorption process between the surfaces of the ZnO-NPs and IV2R. Furthermore, the aromatic rings of IV2R molecules interacted with the ZnO-NPs' C=O, OH, NH, and phenyl groups, which acted as adsorption sites. The evaluation of the generated adsorbent extraction also revealed the presence of ZnO. In aqueous solutions, zinc oxide is virtually insoluble and acts as a weak base. In this experiment, the surface synthesis of ZnO nanoparticles was primarily Zn $(OH)^+$ at pH = 6.0 As a result, ionic bonding between the negatively charged functional groups of IV2R dye ($-SO_3^-$) and the positively charged core of ZnO ($Zn(OH)^+$) may be the fundamental mechanism of interaction between IV2R dye and ZnO. As a result, IV2R adsorption by the synthesized adsorbent could be a chemisorption process.

3.9. Synthetic Dye Effluent Treatment

According to the adsorption data (kinetic and equilibrium), the ZnO-NP material was very effective at removing IV2R from aqueous solutions, suggesting that they might be used to treat actual effluents. As a result, ZnO-NPs were put to the test for treating IV2R actual and synthetic dye effluents. UV-vis spectra of the effluents were used to calculate the ZnO-NP removal percentage of the dye mixture in the effluents. It was explored whether widely a available water treatment could be used as a low-cost material for removing color from real textile wastewater. In the batch test, dried ZnO-NPs were utilized as an adsorbent. The results showed that the actual effluents were removed with the highest amount of color at 95.6% when compared to synthetic IV2R dye with distilled water at almost 99.6%. Industrial effluents may contain toxic compounds and compounds that contain a variety of organic and inorganic materials that are possibly mutagenic or carcinogenic, as well as acids, alkalis, and radioactive elements.

3.10. Comparison with Other Adsorbents

Table 6 compares the maximal IV2R adsorption capabilities of the synthesized adsorbent, ZnO-NPs, with those of several other adsorbents, such as modified biomasses, activated carbons, and zeolites. The synthesized adsorbent had a greater maximum adsorption capacity (59.88 mg g^{-1}) than the other adsorbents listed in Table 6. The Cu(II) ions in organic dye could be combined with the hydroxyl groups and amino groups of the adsorbent through a strong electrostatic attraction process. Consequently, the solid green ZnO-NPs could be used as favorable and cost-effective sorption material for eliminating IV2R ions from industrial wastewater.

Table 6. Comparison of the adsorption of organic dyes with other adsorbent materials.

Ref.	q_e (mg g^{-1})	Contact Time (min)	pH	Organic Dyes	Adsorptions
[66]	116.29	4		Ay 199	ZnO-NPs-AC
[19]	66.66	120	9	methylene blue	AC–ZnO
[67]	9.7	180	7	indigo carmine	Charcoal from rice bran
[68]	30	1200	-	indigo carmine	Charcoal from extracted residue of coffee beans
[69]	55.25	5	4	Titan yellow	Aloe vera
[70]	312.5	60–120	7.5	methylene blue	*Arthrospira platensis* biomass
[71]	2.13	480	4.9	tartrazine	Cellulose extracted from wheat residue
Current study	59.88	60–120	2	Ismate violet 2R (IV2R)	Green ZnO-NPs (prepared from red seaweed (*Pterocladia Capillacea*))

4. Conclusions

The current study suggested that macro-algae's ability to synthesize ZnO-NPs, which could be used to cure IV2R dye, makes them a strong contender. FTIR spectroscopy, SEM, UV-vis, and BET measurements were used to characterize the synthesis. The presence of ZnO-NPs was indicated by a 419 cm^{-1} FTIR band with a sharp, significant band at 571 cm^{-1}. The synthetic adsorbent had a BET surface area of 113.751 $m^2\ g^{-1}$. The starting dye concentration, pH, contact time, solution temperature, and adsorbent dosage were found to be the optimal process parameters for the saturation of accessible binding sites. The adsorption of IV2R on the synthetic adsorbent was aided by a temperature of 55 °C, a pH of 6, and a high adsorbent dosage. According to Langmuir, the adsorbent was a low-cost adsorbent with an IV2R adsorption capability of 59.88 mg g^{-1}. Furthermore, ZnO-NPs had a 95% absorption rate for the treatment of the genuine effluents tested. For linearly and nonlinearly regressed data, the resulting isotherm data fit best with the Freundlich, Smith, Halsey, and Tempkin isotherm models, while the kinetic data fit best with a pseudo-first-order model. The computed thermodynamics parameters revealed that the adsorption system in concern was favorable, endothermic, and spontaneous. As a result, ZnO-NPs may be thought of as a possible adsorbent for removing IV2R dye from polluted water.

Author Contributions: Conceptualization, A.E.A., M.K. and M.A.; methodology, A.E.A., M.A. and M.K.; software, A.T.M., A.E.A. and M.A.; validation, A.E.A. and M.A.; formal analysis, A.E.A. and M.A.; investigation, A.E.A., A.T.M. and M.A.; resources, A.T.M., K.M.A., G.R., T.A.S., K.M.A.R., A.E.A. and M.A.; data curation, A.E.A. and M.A.; writing—original draft preparation, A.E.A., A.T.M. and M.A.; writing—review and editing, A.T.M., M.A. and A.E.A.; visualization, A.E.A., A.T.M. and M.A.; supervision, A.T.M., A.E.A. and M.A.; project administration, A.T.M., A.E.A. and M.A.; funding acquisition, K.M.A., G.R., T.A.S., K.M.A.R. and A.T.M. All authors have read and agreed to the published version of the manuscript.

Funding: The open access publication fee of this work was supported by the Deanship of Scientific Research and the Vice-Presidency for Graduate Studies and Scientific Research of King Faisal University, Saudi Arabia (Project number GRANT719), as well as the Taif University Researcher-Supporting (Project number TURSP-2020/267) of Taif University, Taif, Saudi Arabia.

Institutional Review Board Statement: Not applicable.

Informed Consent Statement: Not applicable.

Data Availability Statement: The data that support the findings of this study are available from the authors upon reasonable request.

Acknowledgments: The authors gratefully appreciate the Deanship of Scientific Research and the Vice-Presidency for Graduate Studies and Scientific Research of King Faisal University, Saudi Arabia (Project No. GRANT719). In addition, the authors appreciate the Taif University Researcher-Supporting Project (number TURSP-2020/267) of Taif University, Taif, Saudi Arabia.

Conflicts of Interest: The authors declare no conflict of interest.

References

1. Mansour, A.T.; Alprol, A.E.; Abualnaja, K.M.; El-Beltagi, H.S.; Ramadan, K.; Ashour, M. The Using of Nanoparticles of Microalgae in Remediation of Toxic Dye from Industrial Wastewater: Kinetic and Isotherm Studies. *Materials* **2022**, *15*, 3922. [CrossRef]
2. Abualnaja, K.M.; Alprol, A.E.; Ashour, M.; Mansour, A.T. Influencing Multi-Walled Carbon Nanotubes for the Removal of Ismate Violet 2R Dye from Wastewater: Isotherm, Kinetics, and Thermodynamic Studies. *Appl. Sci.* **2021**, *11*, 4786. [CrossRef]
3. Barwal, I.; Ranjan, P.; Kateriya, S.; Yadav, S.C. Cellular oxido-reductive proteins of *Chlamydomonas reinhardtii* control the biosynthesis of silver nanoparticles. *J. Nanobiotechnol.* **2011**, *9*, 56. [CrossRef]
4. Frattini, A.; Pellegri, N.; Nicastro, D.; De Sanctis, O. Effect of amine groups in the synthesis of Ag nanoparticles using aminosilanes. *Mater. Chem. Phys.* **2005**, *94*, 148–152. [CrossRef]
5. Abualnaja, K.M.; Alprol, A.E.; Abu-Saied, M.A.; Mansour, A.T.; Ashour, M. Studying the Adsorptive Behavior of Poly(Acrylonitrile-co-Styrene) and Carbon Nanotubes (Nanocomposites) Impregnated with Adsorbent Materials towards Methyl Orange Dye. *Nanomaterials* **2021**, *11*, 1144. [CrossRef]

6. Burmistrov, D.E.; Yanykin, D.V.; Paskhin, M.O.; Nagaev, E.V.; Efimov, A.D.; Kaziev, A.V.; Ageychenkov, D.G.; Gudkov, S.V. Additive Production of a Material Based on an Acrylic Polymer with a Nanoscale Layer of Zno Nanorods Deposited Using a Direct Current Magnetron Discharge: Morphology, Photoconversion Properties, and Biosafety. *Materials* **2021**, *14*, 6586. [CrossRef]
7. Abdulwahid, K.E.; Dwaish, A.S.; Dakhil, O.A. Green synthesis and characterization of zinc oxide nanoparticles from Cladophora glomerata and its antifungal activity against some fungal isolates. *Plant Arch.* **2019**, *19*, 3527–3532.
8. Vilar, V.J.; Botelho, C.M.; Boaventura, R.A. Chromium and zinc uptake by algae Gelidium and agar extraction algal waste: Kinetics and equilibrium. *J. Hazard. Mater.* **2007**, *149*, 643–649. [CrossRef]
9. Sharawy, Z.Z.; Ashour, M.; Labena, A.; Alsaqufi, A.S.; Mansour, A.T.; Abbas, E. Effects of dietary *Arthrospira platensis* nanoparticles on growth performance, feed utilization, and growth-related gene expression of Pacific white shrimp, *Litopenaeus vannamei*. *Aquaculture* **2022**, *551*, 737905. [CrossRef]
10. Metwally, A.S.; El-Naggar, H.A.; El-Damhougy, K.A.; Bashar, M.A.E.; Ashour, M.; Abo-Taleb, H.A.H. GC-MS analysis of bioactive components in six different crude extracts from the Soft Coral (*Sinularia maxim*) collected from Ras Mohamed, Aqaba Gulf, Red Sea, Egypt. *Egypt. J. Aquat. Biol. Fish.* **2020**, *24*, 425–434. [CrossRef]
11. Kaur, P.J.; Kaushik, G.; Siddiqui, R.; Goyal, P.K. Biomedical applications of some green synthesized metal nanomaterials. In *Green Nanomaterials for Industrial Applications*; Elsevier: Amsterdam, The Netherlands, 2022; pp. 71–91. [CrossRef]
12. A Abo-Taleb, H.; F Zeina, A.; Ashour, M.; M Mabrouk, M.; E Sallam, A.; MM El-feky, M. Isolation and cultivation of the freshwater amphipod *Gammarus pulex* (Linnaeus, 1758), with an evaluation of its chemical and nutritional content. *Egypt. J. Aquat. Biol. Fish.* **2020**, *24*, 69–82. [CrossRef]
13. Ashour, M.; Mabrouk, M.M.; Abo-Taleb, H.A.; Sharawy, Z.Z.; Ayoub, H.F.; Van Doan, H.; Davies, S.J.; El-Haroun, E.; Goda, A.M.S.A. A liquid seaweed extract (TAM®) improves aqueous rearing environment, diversity of zooplankton community, whilst enhancing growth and immune response of Nile tilapia, *Oreochromis niloticus*, challenged by *Aeromonas hydrophila*. *Aquaculture* **2021**, *543*, 736915. [CrossRef]
14. Ashour, M.; Alprol, A.E.; Heneash, A.M.M.; Saleh, H.; Abualnaja, K.M.; Alhashmialameer, D.; Mansour, A.T. Ammonia Bioremediation from Aquaculture Wastewater Effluents Using *Arthrospira platensis* NIOF17/003: Impact of Biodiesel Residue and Potential of Ammonia-Loaded Biomass as Rotifer Feed. *Materials* **2021**, *14*, 5460. [CrossRef] [PubMed]
15. Ashour, M.; Hassan, S.M.; Elshobary, M.E.; Ammar, G.A.G.; Gaber, A.; Alsanie, W.F.; Mansour, A.T.; El-Shenody, R. Impact of Commercial Seaweed Liquid Extract (TAM®) Biostimulant and Its Bioactive Molecules on Growth and Antioxidant Activities of Hot Pepper (*Capsicum annuum*). *Plants* **2021**, *10*, 1045. [CrossRef] [PubMed]
16. Alprol, A.E.; Heneash, A.M.; Ashour, M.; Abualnaja, K.M.; Alhashmialameer, D.; Mansour, A.T.; Sharawy, Z.Z.; Abu-Saied, M.A.; Abomohra, A.E.-F. Potential Applications of *Arthrospira platensis* Lipid-Free Biomass in Bioremediation of Organic Dye from Industrial Textile Effluents and Its Influence on Marine Rotifer (*Brachionus plicatilis*). *Materials* **2021**, *14*, 4446. [CrossRef]
17. Khalaf, M.A. Biosorption of reactive dye from textile wastewater by non-viable biomass of Aspergillus niger and *Spirogyra* sp. *Bioresour. Technol.* **2008**, *99*, 6631–6634. [CrossRef]
18. Ghoneim, M.M.; El-Desoky, H.S.; El-Moselhy, K.M.; Amer, A.; Abou El-Naga, E.H.; Mohamedein, L.I.; Al-Prol, A.E. Removal of cadmium from aqueous solution using marine green algae, Ulva lactuca. *Egypt. J. Aquat. Res.* **2014**, *40*, 235–242. [CrossRef]
19. Nourmoradi, H.; Ghiasvand, A.; Noorimotlagh, Z. Removal of methylene blue and acid orange 7 from aqueous solutions by activated carbon coated with zinc oxide (ZnO) nanoparticles: Equilibrium, kinetic, and thermodynamic study. *Desalination Water Treat.* **2015**, *55*, 252–262. [CrossRef]
20. AlProl, A.E. Study of environmental concerns of dyes and recent textile effluents treatment technology: A review. *Asian J. Fish. Aquat. Res* **2019**, *3*, 1–18.
21. Clarke, E.; Anliker, R. Organic dyes and pigments. In *Anthropogenic Compounds*; Springer: Berlin/Heidelberg, Germany, 1980; pp. 181–215.
22. Amer, A.; Abdel-Moneim, M. Bioremediation of Reactive Blue 19 and Reactive Black 5 from Aqueous Solution by using Fungi Aspergillus niger. *Int. J. Curr. Microbiol. Appl. Sci.* **2017**, *6*, 1676–1686.
23. Alprol, A.E.; Heneash, A.M.M.; Soliman, A.M.; Ashour, M.; Alsanie, W.F.; Gaber, A.; Mansour, A.T. Assessment of Water Quality, Eutrophication, and Zooplankton Community in Lake Burullus, Egypt. *Diversity* **2021**, *13*, 268. [CrossRef]
24. Keharia, H.; Madamwar, D. Bioremediation concepts for treatment of dye containing wastewater: A review. *Indian J. Exp. Biol.* **2003**, *41*, 1068–1075. [PubMed]
25. Al-Tohamy, R.; Ali, S.S.; Li, F.; Okasha, K.M.; Mahmoud, Y.A.-G.; Elsamahy, T.; Jiao, H.; Fu, Y.; Sun, J. A critical review on the treatment of dye-containing wastewater: Ecotoxicological and health concerns of textile dyes and possible remediation approaches for environmental safety. *Ecotoxicol. Environ. Saf.* **2022**, *231*, 113160. [CrossRef] [PubMed]
26. Sun, D.; Zhang, Z.; Wang, M.; Wu, Y. Adsorption of reactive dyes on activated carbon developed from Enteromorpha prolifera. *Am. J. Anal. Chem.* **2013**, *4*, 33867. [CrossRef]
27. Falaki, Z.; Bashiri, H. Preparing an adsorbent from the unused solid waste of Rosewater extraction for high efficient removal of Crystal Violet. *J. Iran. Chem. Soc.* **2021**, *18*, 2689–2702. [CrossRef]
28. Babel, S.; Opiso, E.M. Removal of Cr from synthetic wastewater by sorption into volcanic ash soil. *Int. J. Environ. Sci. Technol. (Tehran)* **2007**, *4*, 99–107. [CrossRef]
29. El-Moselhy, K.; Abdel-Azzem, M.; Amer, A.; Al-Prol, A. Adsorption of Cu (II) and Cd (II) from aqueous solution by using rice husk adsorbent. *Phys. Chem. Indian J.* **2017**, *12*, 109–122.

30. Al Prol, A.E.; Abd El Azzem, M.; Amer, A.; El-Metwally, M.E.; Abd El-Hamid, H.T.; El-Moselhy, K.M. Adsorption of cadmium (II) ions from aqueous solution onto mango leaves. *Asian J. Phys. Chem. Sci.* **2017**, *2*, 1–11. [CrossRef]
31. Rajasulochana, P.; Preethy, V. Comparison on efficiency of various techniques in treatment of waste and sewage water–A comprehensive review. *Resour. Effic. Technol.* **2016**, *2*, 175–184. [CrossRef]
32. Abul, A.; Samad, S.; Huq, D.; Moniruzzaman, M.; Masum, M. Textile dye removal from wastewater effluents using chitosan-ZnO nanocomposite. *J. Text. Sci. Eng.* **2015**, *5*, 3.
33. Fomina, M.; Gadd, G.M. Biosorption: Current perspectives on concept, definition and application. *Bioresour. Technol.* **2014**, *160*, 3–14. [CrossRef]
34. Zafar, M.N.; Dar, Q.; Nawaz, F.; Zafar, M.N.; Iqbal, M.; Nazar, M.F. Effective adsorptive removal of azo dyes over spherical ZnO nanoparticles. *J. Mater. Res. Technol.* **2019**, *8*, 713–725. [CrossRef]
35. Wang, X.; Guo, Y.; Yang, L.; Han, M.; Zhao, J.; Cheng, X. Nanomaterials as sorbents to remove heavy metal ions in wastewater treatment. *J. Environ. Anal. Toxicol.* **2012**, *2*, 154–158. [CrossRef]
36. Özgür, Ü.; Alivov, Y.I.; Liu, C.; Teke, A.; Reshchikov, M.; Doğan, S.; Avrutin, V.; Cho, S.-J.; Morkoç, H. A comprehensive review of ZnO materials and devices. *J. Appl. Phys.* **2005**, *98*, 11. [CrossRef]
37. Borysiewicz, M.A. ZnO as a functional material, a review. *Crystals* **2019**, *9*, 505. [CrossRef]
38. Azizi, S.; Ahmad, M.B.; Namvar, F.; Mohamad, R. Green biosynthesis and characterization of zinc oxide nanoparticles using brown marine macroalga Sargassum muticum aqueous extract. *Mater. Lett.* **2014**, *116*, 275–277. [CrossRef]
39. Naseer, M.; Aslam, U.; Khalid, B.; Chen, B. Green route to synthesize Zinc Oxide Nanoparticles using leaf extracts of Cassia fistula and Melia azadarach and their antibacterial potential. *Sci. Rep.* **2020**, *10*, 1–10. [CrossRef]
40. Ashour, M.; El-Shafei, A.A.; Khairy, H.M.; Abd-Elkader, D.Y.; Mattar, M.A.; Alataway, A.; Hassan, S.M. Effect of Pterocladia capillacea Seaweed Extracts on Growth Parameters and Biochemical Constituents of Jew's Mallow. *Agronomy* **2020**, *10*, 420. [CrossRef]
41. Hassaan, M.A.; Hosny, S.; ElKatory, M.R.; Ali, R.M.; Rangreez, T.A.; El Nemr, A. Dual action of both green and chemically synthesized zinc oxide nanoparticles: Antibacterial activity and removal of Congo red dye. *Desalin. Water Treat* **2021**, *218*, 423–435. [CrossRef]
42. Doustkhah, E.; Esmat, M.; Fukata, N.; Ide, Y.; Hanaor, D.A.; Assadi, M.H.N. MOF-derived nanocrystalline ZnO with controlled orientation and photocatalytic activity. *Chemosphere* **2022**, *303*, 134932. [CrossRef]
43. Lagergren, S.K. About the theory of so-called adsorption of soluble substances. *Sven. Vetenskapsakad. Handingarl.* **1898**, *24*, 1–39.
44. Yang, S.I.; Seth, J.; STRACHAN, J.P.; Gentemann, S.; Kim, D.; Holten, D.; Lindsey, J.S.; Bocian, D.F. Ground and excited state electronic properties of halogenated tetraarylporphyrins. Tuning the building blocks for porphyrin-based photonic devices. *J. Porphyr. Phthalocyanines* **1999**, *3*, 117–147. [CrossRef]
45. Ng, J.; Cheung, W.; McKay, G. Equilibrium studies of the sorption of Cu (II) ions onto chitosan. *J. Colloid Interface Sci.* **2002**, *255*, 64–74. [CrossRef]
46. Sparks, D.L. *Soil Physical Chemistry*; CRC press: Boca Raton, FL, USA, 1998.
47. Anand, M.; Suresh, S. Marine seaweed Sargassum wightii extract as a low-cost sensitizer for ZnO photoanode based dye-sensitized solar cell. *Adv. Nat. Sci. Nanosci. Nanotechnol.* **2015**, *6*, 035008. [CrossRef]
48. Pandimurugan, R.; Thambidurai, S. Novel seaweed capped ZnO nanoparticles for effective dye photodegradation and antibacterial activity. *Adv. Powder Technol.* **2016**, *27*, 1062–1072. [CrossRef]
49. Sparks, D. Kinetics of reactions in pure and mixed systems. In *Soil Physical Chemistry*; Sparks, D.L., Ed.; CRC Press: Boca Raton, FL, USA, 1986; pp. 83–145.
50. Wahab, R.; Kim, Y.-S.; Shin, H.-S. Fabrication, characterization and growth mechanism of heterostructured zinc oxide nanostructures via solution method. *Curr. Appl. Phys.* **2011**, *11*, 334–340. [CrossRef]
51. Zandsalimi, Y.; Taymori, P.; Darvishi Cheshmeh Soltani, R.; Rezaee, R.; Abdullahi, N.; Safari, M. Photocatalytic removal of Acid Red 88 dye using zinc oxide nanoparticles fixed on glass plates. *J. Adv. Environ. Health Res.* **2015**, *3*, 102–110.
52. Azizi, S.; Mahdavi Shahri, M.; Mohamad, R. Green synthesis of zinc oxide nanoparticles for enhanced adsorption of lead ions from aqueous solutions: Equilibrium, kinetic and thermodynamic studies. *Molecules* **2017**, *22*, 831. [CrossRef] [PubMed]
53. Yu, J.-G.; Yu, L.-Y.; Yang, H.; Liu, Q.; Chen, X.-H.; Jiang, X.-Y.; Chen, X.-Q.; Jiao, F.-P. Graphene nanosheets as novel adsorbents in adsorption, preconcentration and removal of gases, organic compounds and metal ions. *Sci. Total Environ.* **2015**, *502*, 70–79. [CrossRef]
54. Gupta, V.; Agarwal, A.; Singh, M.; Singh, N. Removal of Red RB dye from aqueous solution by belpatra bark charcoal (BBC) adsorbent. *J. Mater. Environ. Sci.* **2017**, *8*, 3654–3665.
55. Meena, A.K.; Kadirvelu, K.; Mishra, G.; Rajagopal, C.; Nagar, P. Adsorptive removal of heavy metals from aqueous solution by treated sawdust (*Acacia arabica*). *J. Hazard. Mater.* **2008**, *150*, 604–611. [CrossRef]
56. Abate, G.Y.; Alene, A.N.; Habte, A.T.; Getahun, D.M. Adsorptive removal of malachite green dye from aqueous solution onto activated carbon of Catha edulis stem as a low cost bio-adsorbent. *Environ. Syst. Res.* **2020**, *9*, 1–13. [CrossRef]
57. Mondal, N.K.; Kar, S. Potentiality of banana peel for removal of Congo red dye from aqueous solution: Isotherm, kinetics and thermodynamics studies. *Appl. Water Sci.* **2018**, *8*, 1–12. [CrossRef]
58. Rafatullah, M.; Sulaiman, O.; Hashim, R.; Ahmad, A. Adsorption of methylene blue on low-cost adsorbents: A review. *J. Hazard. Mater.* **2010**, *177*, 70–80. [CrossRef] [PubMed]

59. Kumar, P.S.; Kirthika, K. Equilibrium and kinetic study of adsorption of nickel from aqueous solution onto bael tree leaf powder. *J. Eng. Sci. Technol.* **2009**, *4*, 351–363.
60. Hameed, B.; Chin, L.; Rengaraj, S. Adsorption of 4-chlorophenol onto activated carbon prepared from rattan sawdust. *Desalination* **2008**, *225*, 185–198. [CrossRef]
61. Tempkin, M.; Pyzhev, V. Kinetics of ammonia synthesis on promoted iron catalyst. *Acta Phys. Chim. USSR* **1940**, *12*, 327.
62. Su, J.; Lin, H.-f.; Wang, Q.-P.; Xie, Z.-M.; Chen, Z.-l. Adsorption of phenol from aqueous solutions by organomontmorillonite. *Desalination* **2011**, *269*, 163–169. [CrossRef]
63. Cheung, W.; Szeto, Y.; McKay, G. Intraparticle diffusion processes during acid dye adsorption onto chitosan. *Bioresour. Technol.* **2007**, *98*, 2897–2904. [CrossRef]
64. Liu, Q.-S.; Zheng, T.; Wang, P.; Jiang, J.-P.; Li, N. Adsorption isotherm, kinetic and mechanism studies of some substituted phenols on activated carbon fibers. *Chem. Eng. J.* **2010**, *157*, 348–356. [CrossRef]
65. Hadi, M.; Samarghandi, M.R.; McKay, G. Equilibrium two-parameter isotherms of acid dyes sorption by activated carbons: Study of residual errors. *Chem. Eng. J.* **2010**, *160*, 408–416. [CrossRef]
66. Jamshidi, H.; Ghaedi, M.; Sabzehmeidani, M.M.; Bagheri, A.R. Comparative study of acid yellow 119 adsorption onto activated carbon prepared from lemon wood and ZnO nanoparticles loaded on activated carbon. *Appl. Organomet. Chem.* **2018**, *32*, e4080. [CrossRef]
67. Nakamura, T.; Kawasaki, N.; Tanada, S.; Tamura, T.; Shimizu, Y. Indigo carmine removal by charcoal from rice bran as an agricultural by-product. *Toxicol. Environ. Chem.* **2005**, *87*, 321–327. [CrossRef]
68. Nakamura, T.; Hirata, M.; Kawasaki, N.; Tanada, S.; Tamura, T.; Nakahori, Y. Decolorization of indigo carmine by charcoal from extracted residue of coffee beans. *J. Environ. Sci. Health Part A* **2003**, *38*, 555–562. [CrossRef] [PubMed]
69. El-Azazy, M.; Dimassi, S.N.; El-Shafie, A.S.; Issa, A.A. Bio-waste Aloe vera leaves as an efficient adsorbent for titan yellow from wastewater: Structuring of a novel adsorbent using Plackett-Burman factorial design. *Appl. Sci.* **2019**, *9*, 4856. [CrossRef]
70. Mitrogiannis, D.; Markou, G.; Çelekli, A.; Bozkurt, H. Biosorption of methylene blue onto Arthrospira platensis biomass: Kinetic, equilibrium and thermodynamic studies. *J. Environ. Chem. Eng.* **2015**, *3*, 670–680. [CrossRef]
71. Villabona-Ortíz, Á.; Figueroa-Lopez, K.J.; Ortega-Toro, R. Kinetics and Adsorption Equilibrium in the Removal of Azo-Anionic Dyes by Modified Cellulose. *Sustainability* **2022**, *14*, 3640. [CrossRef]

materials

MDPI

Article

Synthesis of Mg-Al Hydrotalcite Clay with High Adsorption Capacity

Zhaoyi Li [1], Jie Zhang [1,2], Chengtun Qu [1,3], Ying Tang [1,2,*] and Michal Slaný [4,5,*]

1 State Key Laboratory of Petroleum Pollution Control, Xi'an Shiyou University, Xi'an 710065, China; 210104@xsyu.edu.cn (Z.L.); zhangjie@xsyu.edu.cn (J.Z.); xianquct@xsyu.edu.cn (C.Q.)
2 Shaanxi Key Laboratory of Carbon Dioxide Storage and Enhanced Oil Recovery, Xi'an 710060, China
3 Shaanxi Province Key Laboratory of Environmental Pollution Control and Reservoir Protection Technology of Oilfields, Xi'an Shiyou University, Xi'an 710065, China
4 Institute of Inorganic Chemistry, Slovak Academy of Sciences, Dúbravská cesta 9, 845 36 Bratislava, Slovakia
5 Institute of Construction and Architecture, Slovak Academy of Sciences, Dúbravská cesta 9, 845 03 Bratislava, Slovakia
* Correspondence: tangying78@xsyu.edu.cn (Y.T.); michal.slany@savba.sk (M.S.)

Abstract: A novel Mg-Al metal oxide has been successfully synthesized by the calcination of hierarchical porous Mg-Al hydrotalcite clay obtained by using filter paper as a template under hydrothermal conditions. Various characterizations of the obtained nanoscale oxide particles verified the uniform dispersion of Mg-Al metal oxides on the filter paper fiber, which had a size of 2–20 nm and a highest specific surface area (SSA) of 178.84 m^2/g. Structural characterization revealed that the as-prepared Mg-Al metal oxides preserved the tubular morphology of the filter paper fibers. Further experiments showed that the as-synthesized Mg-Al metal oxides, present at concentrations of 0.3 g/L, could efficiently remove sulfonated lignite from oilfield wastewater (initial concentration of 200 mg/L) in a neutral environment (pH = 7) at a temperature of 298 K. An investigation of the reaction kinetics found that the adsorption process of sulfonated lignite (SL) on biomorphic Mg-Al metal oxides fits a Langmuir adsorption model and pseudo-second-order rate equation. Thermodynamic calculations propose that the adsorption of sulfonated lignite was spontaneous, endothermic, and a thermodynamically feasible process.

Keywords: metal oxides; filter paper; adsorption; sulfonated lignite; hydrotalcite; clay

Citation: Li, Z.; Zhang, J.; Qu, C.; Tang, Y.; Slaný, M. Synthesis of Mg-Al Hydrotalcite Clay with High Adsorption Capacity. *Materials* **2021**, *14*, 7231. https://doi.org/10.3390/ma14237231

Academic Editors: Roberta G. Toro, Daniela Caschera and Abeer M. Adel

Received: 14 October 2021
Accepted: 25 November 2021
Published: 26 November 2021

Publisher's Note: MDPI stays neutral with regard to jurisdictional claims in published maps and institutional affiliations.

1. Introduction

Petroleum is the lifeblood of many industries; however, its extraction and processing—especially the development and construction of oil fields, due to drilling operations, pipeline laying, and other projects—will inevitably have an impact on the natural ecosystem. Sulfonated lignite is a good freshwater viscosity reducer and fluid loss reducer that can withstand high temperatures of 200–240 °C. SL is widely used in the application of drilling fluid, resulting in effects on the chemical oxygen demand (COD) and biochemical oxygen demand (BOD) of drilling wastewater [1,2], and millions of tons of high-concentration toxic organic drilling wastewater are produced during oilfield exploration in China every year. This often becomes a difficult problem for oilfield-produced water discharge, and it is very relevant to develop technologies to avoid further drilling fluid pollution. So far, a variety of methods, such as chemical precipitation, electrolysis, ion exchange, membrane separation, and adsorption, have been examined to remove pollutants from oilfield wastewater. Between these, adsorption technology has received the greatest attention due to the number of advantages, such as being relatively low-cost, highly efficient, simple, and eco-friendly. Various classes of adsorbents are utilized to remove pollutants from oilfield wastewater, including activated carbon, layered double hydroxides, metal oxides, and biosorbents. In recent years, metal oxide nanoparticles have become one of the most widely studied adsorption materials due to their large specific surface area with abundant active sites [3]. However, their larger specific area than traditional adsorbents, instability, single-pore structure, and their strong tendency for agglomeration limits their application range. Therefore, hierarchically porous materials with large surface area, low density, variable chemical compositions, large accessible space, and interconnected hierarchical porosity at different length scales have attracted considerable attention as important functional materials [4].

Layered double hydroxides (LDHs) are one of the attractive materials used to improve the adsorption activities of metal oxide materials [5]. Moreover, the calcined product of LDHs has a larger specific surface area and abundant pore structure. However, high calcination temperatures cause a collapse of the pore structure, and the particles agglomerate easily, which is not conducive to the improvement of the degree of activation of the metal oxides [6,7]. Although there are extensive studies on metal oxides, the preparation of a hierarchically porous LDO is still complicated and extremely challenging. Over billions of years, organisms have evolved and optimized, exhibiting natural hierarchical anatomies and morphologies with multi-scale dimensions and components [8]. In recent years, the preparation of nanomaterials by the plant template method has attracted widespread attention, as a material of choice can be deposited onto the templates to form a nanomaterial with the shape of the template. To date, various natural materials have been employed as templates to synthesize various biomorphic oxide materials, including wood rattan [9], cotton fibers [10], bamboo [11], eggshells [12,13], cellulose [14,15], and peanut shells [16,17], among others. In recent years, especially in the application of adsorption, studies on the preparation of metal oxides by biological template methods have also been of great interest. For example, Yu et al. synthesized ZnAl-LDH/$ZnCo_2O_4$ through a hydrothermal process with a pollen template, and the resulting material showed extraordinary adsorption ability in Congo red solution [18]. Using flax, gauze, and sisal as templates, Liu et al. successfully synthesized a ZnAl-CLDH/$FeWO_4$ material which exhibited a high adsorption activity for the removal of dyes from water [19].

In this work, Mg-Al metal oxide composites were synthesized via a filter paper template under hydrothermal conditions to obtain hierarchically porous materials. The adsorption performance of the new adsorbents was investigated by looking at the removal efficiency of SL in oilfield wastewater.

2. Experimental

2.1. *Materials*

In this work, magnesium nitrate hexahydrate ($Mg(NO_3)_2{\cdot}6H_2O$) was bought from West Asia Chemical Technology Co. Ltd. Shandong, China. Urea ($CO(NH_2)_2$) and aluminum nitrate nonahydrate ($Al(NO_3)_3{\cdot}9H_2O$) were obtained from Damao Chemical Reagent Co., Ltd. Tianjin, China. Filter paper was obtained from the local market. Sulfonated lignite was purchased from Tarim, China. Ethanol was purchased from Fuyu Fine Chemical Co., Ltd. Tianjin, China. In addition, deionized water was used for all synthesis processes and for washing the precipitate.

2.2. *Preparation of Mg-Al Metal Oxides Materials*

In our synthesis procedure (Figure 1), Mg-Al metal oxides were synthesized by a facile hydrothermal route. First, the filter paper was rinsed with deionized water and ethyl alcohol three times and then dried at 80 °C for 12 h for later use. Next, 4.10 g of $Mg(NO_3)_2{\cdot}6H_2O$, 3.01 g of $Al(NO_3)_3{\cdot}9H_2O$, and 4.36 g of urea were dispersed in 80 mL of deionized water at room temperature with vigorous stirring to form a mixed solution. After that, 1 g of the filter paper was added to the mixed solution under ultrasound for 30 min. The mixed solution was heated at 140 °C and transferred into a 500 mL autoclave pressure vessel for varying lengths of time. The as-prepared products were taken out of the vessel after cooling to room temperature, washed with distilled water 3 times, and dried in the drying furnace at 80 °C for 12 h. Eventually, the sample was calcined under 500 °C,

600 °C, and 700 °C in an air atmosphere for 3 h (ramping rate was 10/min). Besides that, a blank sample without a template using the same experimental process was prepared.

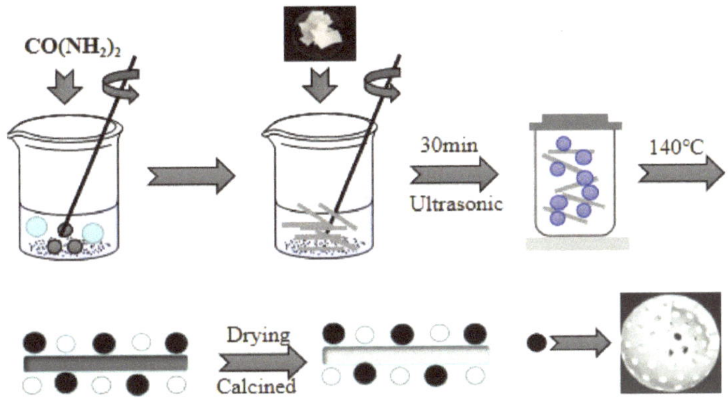

Figure 1. Preparation flow chart of Mg-Al metal oxide.

2.3. Characterization of Materials

Using a scanning electron microscope with a 20.0 kV accelerating voltage (SEM, JSM-6390A, JEOL, Tokyo, Japan) the morphology of the prepared material was elucidated. X-ray diffraction analysis (XRD), (JDX-3530, JEOL, Tokyo, Japan) using an X-ray copper tube was used for an investigation of the structures of the prepared material. Ka radiation release at accelerations to 40 mA and 40 kV were arranged at 10–90° 2θ degree with a scan speed of 2/min. Specific surface area (SSA) and pore structure measurements were carried out by liquid N_2 (−196 °C) using a BET analyzer, Micromeritics ASAP 2010 (Norcross, GA, USA), at a temperature of 77 K. SSA calculation was based on the Brunauer–Emmett–Teller (BET) method. The Barrett–Joyner–Halenda (BJH) method was used for calculating the pore size distribution. FTIR spectra were collected using a Nicolet 5700 Fourier transform infrared (Thermo Electron Co., West Palm Beach, FL, USA) and 1 mg of the sample was homogenized with 200 mg of KBr to form powder pellets. A total of 64 scans with a resolution of 4 cm^{-1} were used for FTIR measurements.

2.4. Adsorption Experiment

This work used sulfonated lignite (SL) as a pollutant to study the adsorption capacity of the obtained materials. A certain amount of adsorbent was added to the 50 mL SL solution with an initial concentration of 200 $mg \cdot L^{-1}$. The mixture was stirred for 12 h to ensure the adsorption equilibrium at room temperature (303 K). Then, the solutions were filtered and the residual concentrations of SL were measured by a visible UV-vis spectrophotometer Avantes (Apeldoorn, The Netherlands) with a maximum wavelength of 300 nm and also ultraviolet radiation. The effects of different parameters—such as the amount of adsorbent and pH value—on the adsorption process in the solution were studied. With the addition of 0.1 $mol \cdot L^{-1}$ HCl or 0.1 $mol \cdot L^{-1}$ NaOH, the pH value of the solution was altered. To improve the accuracy of the adsorption findings to obtain average values, the experiments were repeated at least three times. Based on the formula given below, the amount of SL adsorbed was calculated:

$$q_t = \frac{(C_0 - C_t)V}{M} \tag{1}$$

$$q_e = \frac{(C_0 - C_e)V}{M} \tag{2}$$

where C_e (mg·L^{-1}) and C_0 are the equilibrium and initial SL concentration, M is the mass of the adsorbent (g), and V is the volume of the solution (mL).

3. Results and Discussion

3.1. Structural Characterization

3.1.1. SEM Analysis

Figure 2 shows the SEM micrographs of the as-prepared samples. Figure 2a,b are SEM images of the filter paper under different magnifications. It can be concluded that the microstructure of the filter paper is a three-dimensional network structure composed of many micron-sized nanotubes. Figure 2c,d show the morphology of Mg-Al LDHs derived from the filter paper template. In the hydrothermal processes, Mg^{2+} and Al^{3+} cations were dispersed on the filter paper surfaces; subsequently, the increase in temperature caused the decomposition of urea to produce OH^- and $CO_3{}^{2-}$, which led to the deposition of magnesium nitrate and aluminum nitrate, forming Mg-Al LDH on the filter paper templates.

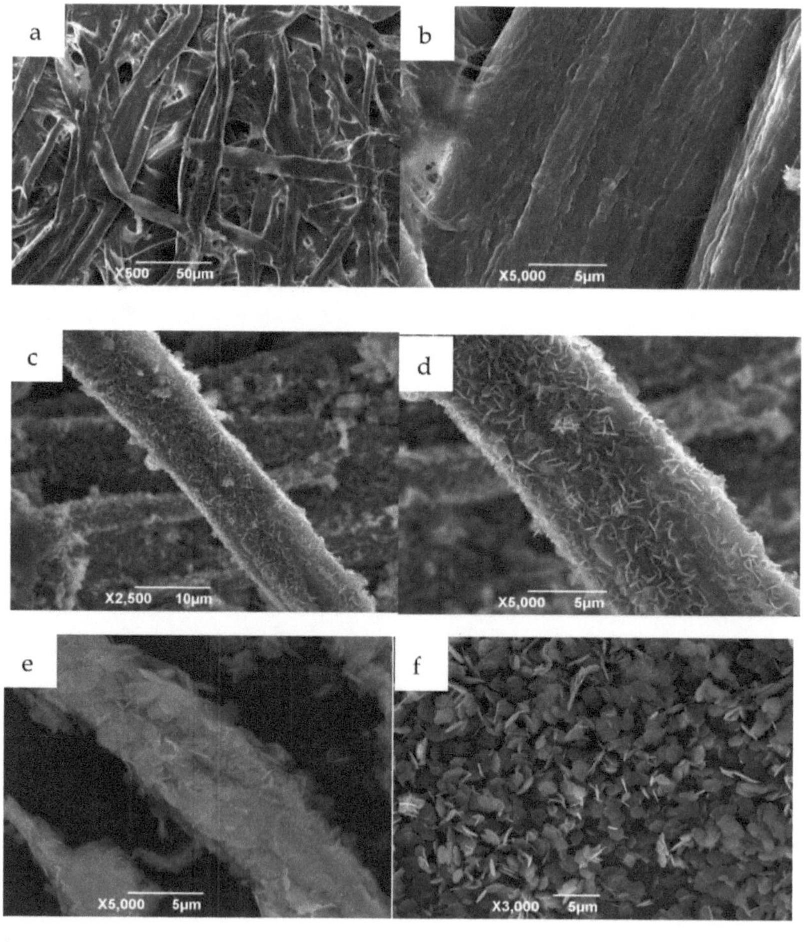

Figure 2. SEM images of filter paper (**a**,**b**); filter paper-templated Mg-Al LDH (**c**,**d**); filter paper-templated MgAl metal oxides (600 °C, 3 h) (**e**); Mg-Al metal oxides crystallized without a template (**f**).

Furthermore, it was found that the layer double hydroxide (LDH) nanoplatelets vertically aligned on the surface of the filter paper fibers in an orderly fashion, indicating that the mechanism of the crystal growth process is governed by self-assembly and in situ crystallization [20]. Moreover, the filter paper fiber shape is well-maintained, even upon increasing the calcination temperature to 600 °C, as shown in Figure 2e. These results demonstrate that the evolutionarily optimized microstructure of filter paper was inherited by the template-assisted synthesized Mg-Al metal oxides after the complete removal of the biotemplate. Figure 2f presents an image of Mg-Al metal oxides derived from Mg-Al LDH without the template. It is clearly visible from the figure that the Mg-Al metal oxides exhibited the flat, hexagonal, sheet-like morphology typical of such oxides.

3.1.2. Specific Surface Area and Porosity Analysis

Calculating the specific surface area (SSA) using a BET analysis based on nitrogen adsorption–desorption isotherms is the most common way to identify the porosity of various materials [21]. In Figure 3, the N_2 adsorption–desorption isotherms and pore size distribution curves of different samples can be seen. It was observed that the N_2 adsorption–desorption isotherm of the Mg-Al metal oxides synthesized with the template exhibits a type IV isotherm with an H3-type hysteresis loop ($P/P0 > 0.4$), indicating that slit-like pores are formed by the accumulation of flake particles [22]. A sharp increase is seen at high pressure ($P/P0 = 0.9$–1.0). This higher adsorption at high relative pressure indicates that macropores are also present [23]. On the curve of the biomorphic Mg-Al metal oxides, it can be seen that the distribution range of the filter paper-templated particles is 1.78–120 nm. As illustrated in Table 1, the introduction of paper fiber templates causes a distinct increase in BET surface area, from 117.04 to 178.84 m^2/g, and in pore volume from 0.19 to 0.32 cm^3/g, which provides access to more adsorption sites by facilitating the diffusion of sulfonated lignite.

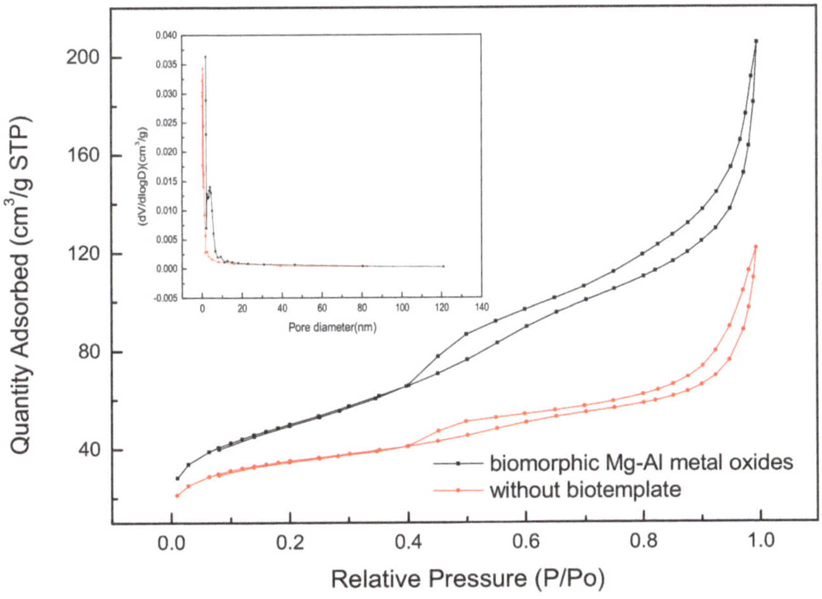

Figure 3. Nitrogen adsorption-desorption isotherms and corresponding BJH pore size distribution curves of the samples.

Table 1. BET specific surface area (SSA), pore volume, and pore diameter of different samples.

Sample	BET Surface Area (m^2/g)	Pore Volume (cm^3/g)	Pore Diameter (nm)
Filter paper-templated Mg-Al metal oxides	178.84	0.3177	7.10
Without biotemplate	117.04	0.1879	6.42

3.1.3. XRD Analysis

Figures 4a and 5 show the XRD pattern of Mg-Al metal oxides and Mg-Al LDH. The main diffraction peaks of MgAl-LDH (JPCDS, no.350964) can be assigned to the (003), (006), (012), (015), (018), (110), (113), and (116) diffraction planes. These findings propose that the synthesized LDH crystallized with well-ordered, layered structures. After calcination at 500–700 °C, diffraction peaks can be observed at 2θ values of about 36.9°, 42.9°, and 62.3°, which are respectively assigned to the (111), (200), and (220) reflections of MgO (JCPDS, no.450946) [22]. The characteristic diffraction peaks of MgAl-LDH disappeared, indicating that the layered structure of MgAl-LDH was converted and destroyed into aluminum and magnesium mixed oxides. A shortage of peaks belonging to aluminum oxide means that the Al^{3+} ions are well-dispersed in the MgO. The XRD pattern of Mg-Al metal oxides after the adsorption of sulfonated lignite is also shown in Figure 4b. The characteristic diffraction peaks of hydrotalcite are clearly visible. This indicates that the Mg-Al metal oxide has undergone the reconstruction process of hydrotalcite, which is also known as the "memory effect" of hydrotalcite.

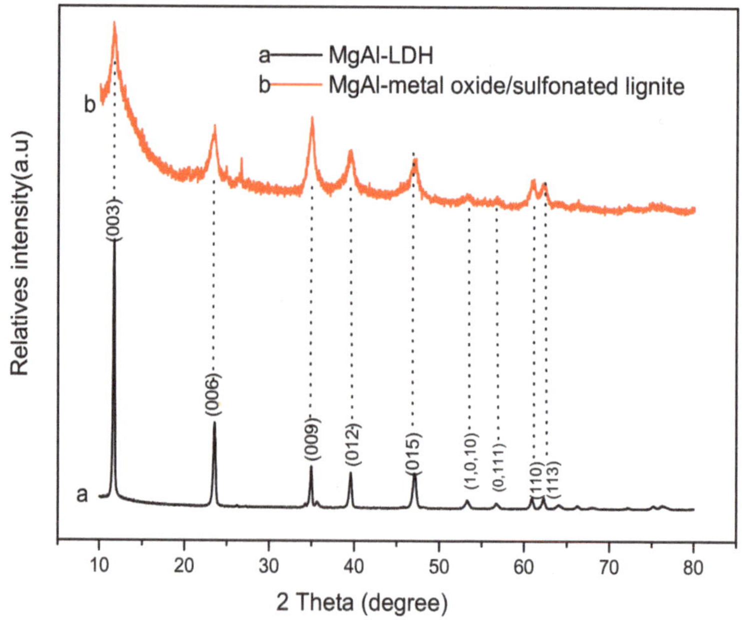

Figure 4. XRD patterns of (**a**) Mg-Al LDH and (**b**) Mg-Al metal oxides after the adsorption of sulfonated lignite.

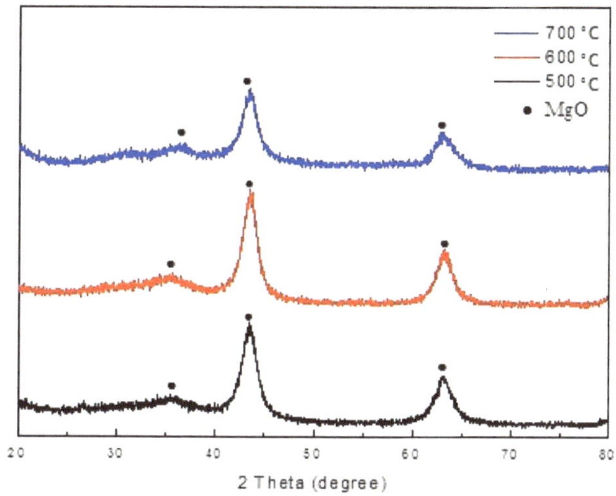

Figure 5. XRD pattern of filter paper-templated Mg-Al metal oxides.

3.1.4. FTIR Analysis

The FTIR spectra of Mg-Al metal oxides before and after the adsorption of sulfonated lignite are presented in Figure 6. A broad and strong complex band at 3445 cm^{-1} found for all samples is related to the stretching vibration of –OH groups, indicating from hydrogen-bonded, interlayered H_2O molecules, or metal –OH groups. The absorption band at 1600 cm^{-1} is assigned to the –OH flexural groups of the interlayer crystal water, and the stretching vibration caused by the small peak carbonate is visible at 1361 cm^{-1}. However, from the FTIR spectra of the Mg-Al metal oxides after adsorption, it can be seen that the crystal water and hydroxyl groups at 3445 cm^{-1} and 1600 cm^{-1} are significantly weakened, and an O=S=O stretching vibration band appeared at 1035 cm^{-1}. This finding indicates that the Mg-Al metal oxides successfully adsorbed the sulfonated lignite molecules in the solution. A series of absorption bands recorded in the 770–860 cm^{-1} region were ascribed to M-O (M = Mg, Al) vibrations.

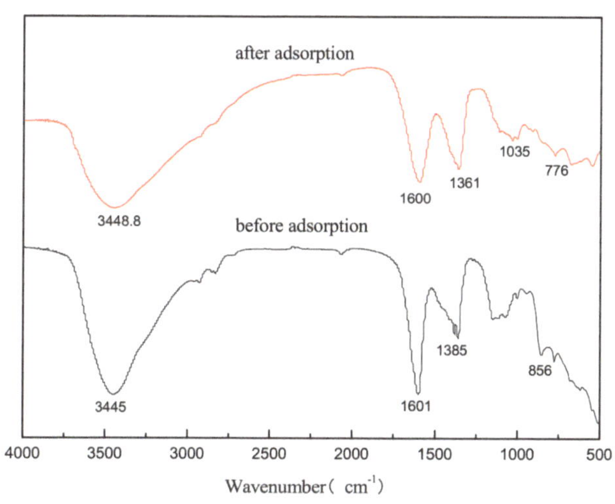

Figure 6. FTIR spectra of Mg-Al metal oxides before and after adsorption.

3.2. *Adsorption Performance*

3.2.1. Comparison of Adsorbents

Sulfonated lignite was chosen as the model pollutant to evaluate the adsorption capability of Mg-Al metal oxides. Figure 7 shows the adsorption capacities of the Mg-Al metal oxides derived from Mg-Al LDH, prepared with and without a filter paper template. The results are that a sulfonated lignite adsorption capacity of 664.83 mg/g and 807.06 mg/g was achieved for initial sulfonated lignite concentrations of 200 and 250 mg/L, respectively, by 0.3 g/L of Mg-Al oxides derived from hierarchical hydrotalcite at an initial pH of 7 and a temperature of 25 °C. Moreover, the adsorption capacities of Mg-Al metal oxides prepared without the use of a filter paper template were also investigated and found to have a lower sulfonated lignite adsorption capacity (525.43 mg/g and 665.34 mg/g, respectively). This difference can be attributed to the vertical matrix growth of MgAl-LDH nanosheets, which avoids the multilayer accumulation of nanosheets, gives the material a rich pore structure and a larger specific surface area, and provides more available active adsorption sites.

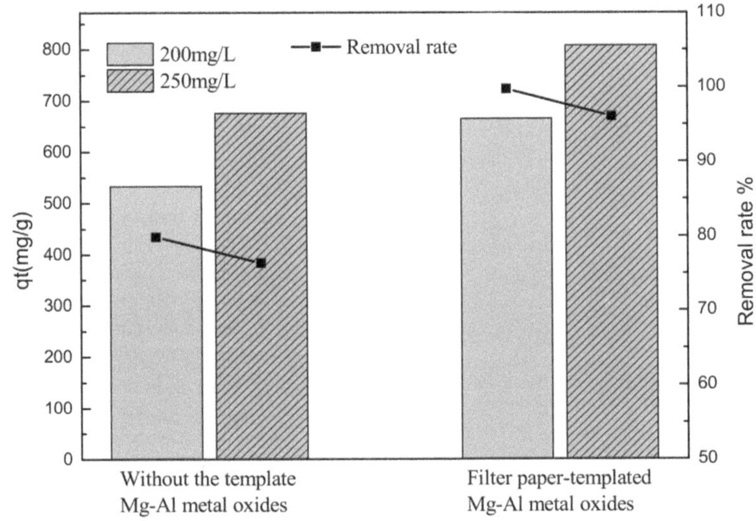

Figure 7. Comparison of the degree of adsorption of sulfonated lignite by filter paper-templated Mg-Al metal oxides and by untemplated Mg-Al metal oxides (initial sulfonated lignite concentration: 200 mg/L and 250 mg/L; adsorbent dosage: 0.015 g; initial pH: 7; temperature set to 25 °C; stirring time: 12 h).

3.2.2. Effect of Precursor Concentration, Reaction Time, Calcination Temperatures, Adsorbent Dosage, and pH

The effects of precursor concentration on the adsorption of sulfonated lignite onto filter paper-templated Mg-Al metal oxides are shown in Figure 8a. The experimental results indicate that the filter paper-templated Mg-Al metal oxides, prepared from a 0.2 mol/L precursor, exhibit the highest adsorption capacity, wherein the removal rate of SL reached 97.73%. By continuing to increase the concentration of the precursor solution, the removal rate of sulfonated lignite will decrease, due to the increase in the solution viscosity as the result of incomplete mixing with the template. This causes blocked channels in the high-temperature calcination process and decreases the surface area available for adsorption, which then leads to a decrease in removal efficiency.

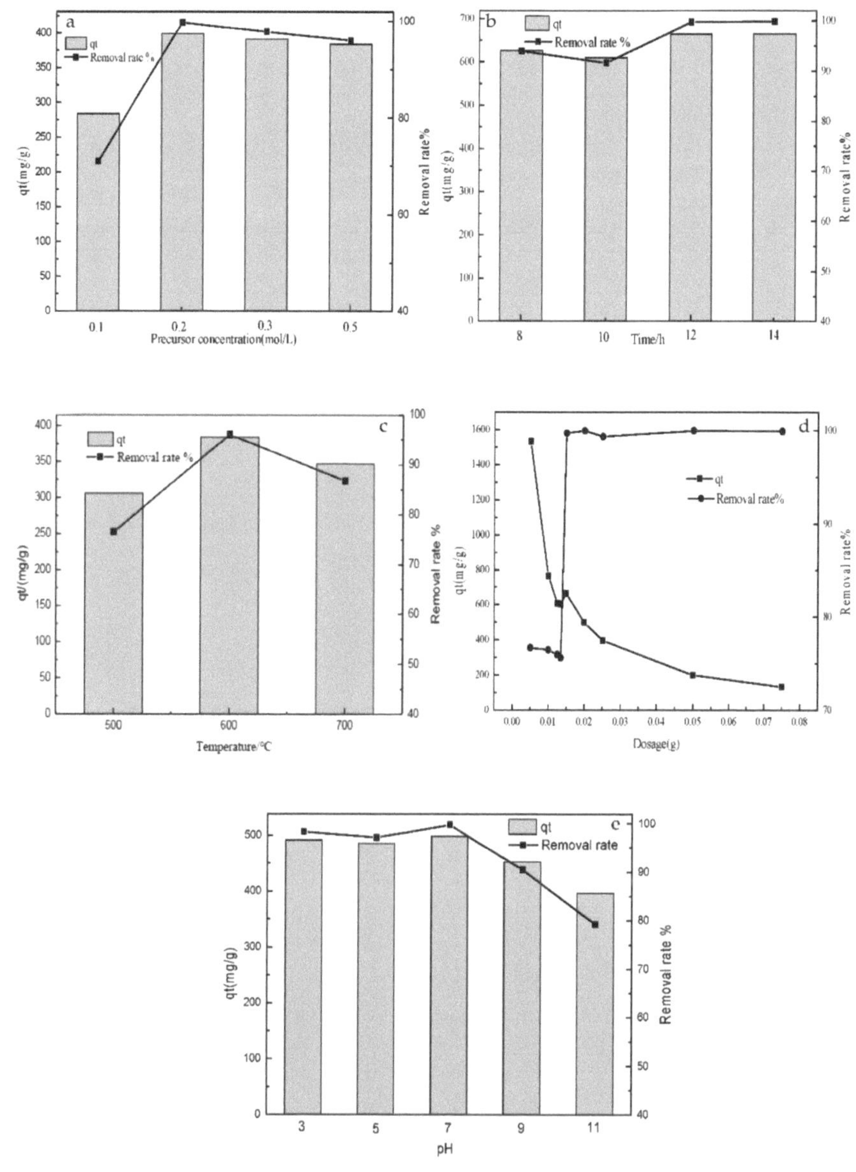

Figure 8. Effects of precursor concentration (mol/L) (**a**), reaction time (h) (**b**), calcination temperatures (°C) (**c**), adsorbent dose (g) (**d**), and pH (**e**) on the adsorption of sulfonated lignite by filter paper-templated Mg-Al metal oxides.

The effect of reaction time on the adsorption of sulfonated lignite by filter paper-templated Mg-Al metal oxides is shown in Figure 8b. The removal percentage of sulfonated lignite increased from 93.94 to 99.73% when the reaction time was prolonged from 8 to 12 h. However, upon further increasing the reaction time, the adsorption capacity of sulfonated lignite remained unchanged, indicating that the adsorption had reached equilibrium.

The adsorption capacities of filter paper-templated Mg-Al metal oxides derived at different calcination temperatures (500, 600, and 700 °C) were compared under the same adsorption conditions and are shown in Figure 8c. It can be seen that when the roasting temperature is increased from 500 °C to 600 °C, the removal rate of sulfonated lignite is increased from 76.53% to 96.06%, while a decrease is seen upon further increasing the calcination temperature to 700 °C, due to the high calcination temperature causing a collapse of the pore structure, which is not conducive to improving the degree of activation of the adsorbent.

The dose of adsorbent is an important parameter that affects the efficiency of the removal process. In this study, the effect of the adsorbent dosage on sulfonated lignite adsorption efficiency was studied by varying the adsorbent dosage from 0.1 to 1.5 g/L at an initial sulfonated lignite concentration of 200 mg/L; the results are shown in Figure 8d. The removal rate of sulfonated lignite increases from 76.69 to 99.37% when the adsorbent dosage increases from 0.1 to 0.3 g/L, which could be attributed to the large number of active sites on the surface of the adsorbent at the initial stage of adsorption, and the sulfonated lignite molecules quickly occupy these active sites [23]. However, beyond the latter dosage, decreased performance was found, and the removal efficiency remained constant even up to an adsorbent dosage of 1.5 g/L, thus indicating that a higher adsorbent dose results in a lower uptake capacity. This is presumably due to the increase in the amount of adsorbent leading to the aggregation of solid particles, the overlapping of active sites, and the reduced adsorption capacity of the adsorbent.

Solution pH plays an important role in the adsorption process due to the extent of ionization or speciation of the adsorbate and its influence on the surface charge of the adsorbent [24]. To evaluate the effect of pH on sulfonated lignite removal efficiency, experiments were conducted at pH values of 3, 5, 7, 9, and 11 using filter paper-templated Mg-Al metal oxides, and the results are summarized in Figure 8e. The maximum adsorption was 498.63 mg/g, with a removal efficiency of 99.73% at a pH value of 7 in the presence of an initial sulfonated lignite concentration of 200 mg/L. The adsorbed amount and the removal rate of sulfonated lignite remained at a high level (485 mg/g and 97% or more) at a lower pH. We attribute this to the protonation of Mg-Al metal oxides at neutral or acidic pH, and thus the formation of positively charged M-OH_2^+ [25], which in turn causes electrostatic attraction between the positively charged metal oxide surface and the negatively charged adsorbent molecule [26]. In contrast, in the alkaline range (pH = 9, 11), sulfonated lignite adsorption decreases with higher pH. In addition to the lower degree of protonation on the surface, this may also be due to the competition of OH^- with sulfonated lignite ions [27]. In addition, the increase in the number of –OH groups reduces the number of positive charges on the surface of the adsorbent, thereby reducing its adsorption capacity for sulfonated lignite.

3.2.3. Kinetics of Adsorption

To determine and interpret the mechanisms and kinetics of the sulfonated lignite adsorption process by filter paper-templated Mg-Al metal oxides, the experimental data obtained were fitted to pseudo-first-order [28,29] and pseudo-second-order kinetic models [30,31] and intra-particle diffusion models [32] according to the linear equations of these three kinetic models, as follows:

$$\log(q_e - q_t) = \log q_e - \frac{k_1 t}{2.303} \tag{3}$$

$$\frac{t}{q_t} = \frac{1}{k_2 q^2 e} + \frac{t}{q_e} \tag{4}$$

$$q_t = k_{di}\sqrt{t} + c_i \tag{5}$$

where k_1 (min^{-1}) and k_2 (g/mg·min) are the rate constants of the pseudo-first-order and pseudo-second-order models, q_t is the amount of adsorption over time (t), and k_{di} and c_i are the intra-particle diffusion model rate parameter and the intercept of stage i, respectively.

SL adsorption onto filter paper-templated Mg-Al metal oxides is plotted against contact time in solution in Figure 9, for initial sulfonated lignite concentrations of 100, 150, and 200 mg/L. For all initial concentrations, the adsorbed amount of sulfonated lignite increased rapidly at the beginning, due to plentiful vacant sites on the Mg-Al metal oxide surface, making the concentration gradient the driving force for adsorption. For a 200 mg/L initial concentration, the adsorption capacity of the sulfonated lignite was 484.14 mg/g at 240 min, constituting 97.08% of equilibrium adsorption capacity. After that, further adsorption was minimal, as the system was approaching equilibrium. Similar behavior was seen for the other initial concentrations.

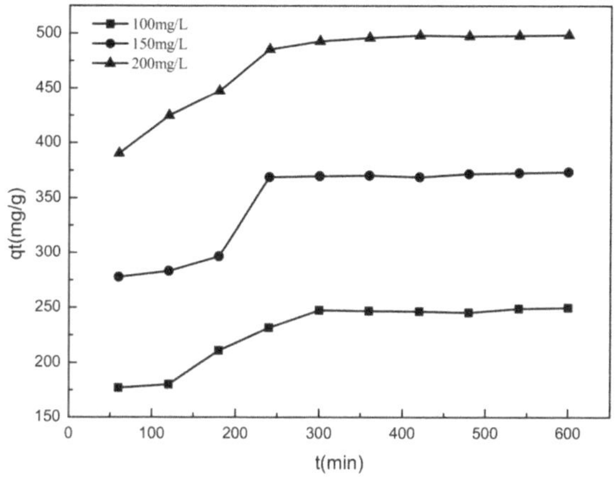

Figure 9. Effect of adsorption time on adsorption efficiency.

Figure 10 illustrates the pseudo-first-order, pseudo-second-order, and intra-particle diffusion models of sulfonated lignite adsorption by filter paper-templated Mg-Al metal oxides. From the intercept and slope of the straight line obtained, the kinetic parameters for the removal of SL were determined, as listed in Tables 2 and 3. Looking at the parameters, the pseudo-first-order model had a relatively lower correlation coefficient, and a large difference in the q_e value between the calculation (q_e, cal) and the experiment (q_e, exp) was noticed, confirming that the adsorption process did not follow the pseudo-first-order model. In contrast, the almost identical values of calculated and experimental q_e, and the high values of R^2 (>0.99), make the pseudo-second-order model a much better fit, which suggests that chemical adsorption was the main rate-limiting step in the adsorption of sulfonated lignite by the prepared Mg-Al metal oxides [33].

For the evaluation of the prediction and the mechanism of the rate-limiting steps, the intraparticle diffusion model was used. In Figure 10c, the plot exhibits a multi-straight-line nature, indicating the different stages involved in the adsorption process. The plots of q_t versus $t^{1/2}$ give a straight line and none of the lines pass through the origin, which reveals that intra-particle diffusion is not the rate-controlling step for the whole adsorption process. Table 3 shows that the k_{d1} values are larger than the k_{d2} values, which suggests that the surface diffusion step was important for the adsorption of sulfonated lignite by the biomorphic Mg-Al metal oxides [34]. Furthermore, the correlation coefficient (R^2) for the intra-particle diffusion model is lower than those obtained from the pseudo-second-order model, indicating that the adsorption kinetics are better described by the latter.

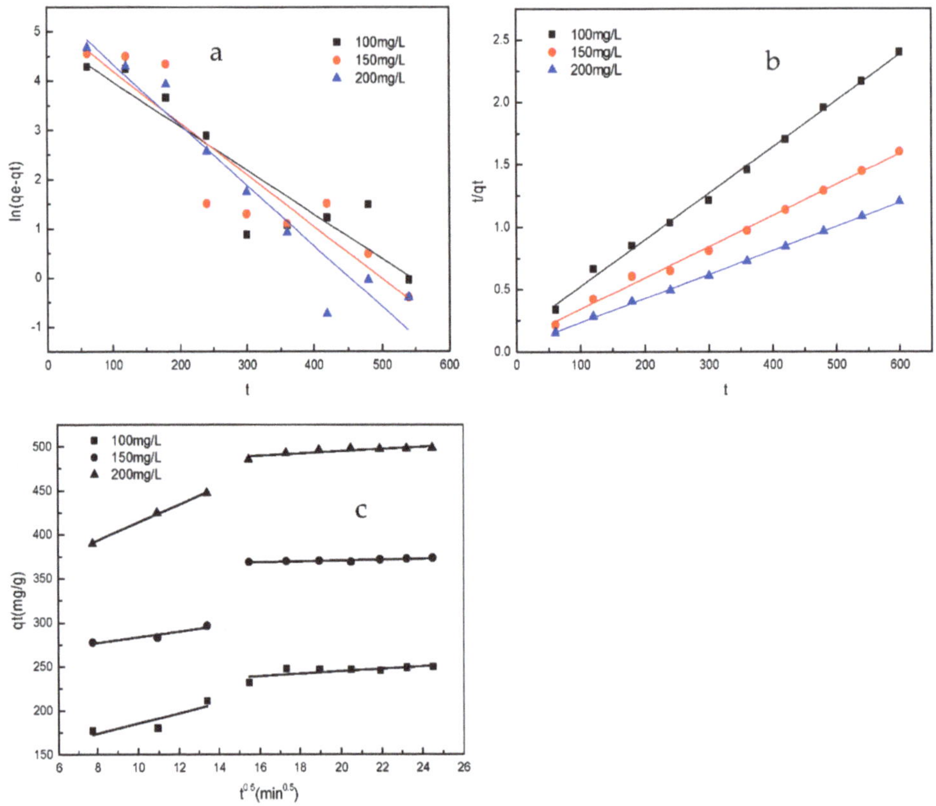

Figure 10. The fit of the pseudo-first-order (**a**), pseudo-second-order (**b**), and intra-particle diffusion (**c**) models for the adsorption of sulfonated lignite by filter paper-templated Mg-Al metal oxides.

Table 2. Adsorption kinetic parameters of the pseudo-first-order and pseudo-second-order models for the adsorption of sulfonated lignite by filter paper-templated Mg-Al metal oxides.

Concentration		Pseudo-First-Order Model			Pseudo-Second-Order Model		
C_0	qe,exp	qe,cal	$K_1 min^{-1}$	R^2	qe,cal	$K_2 min^{-1}$	R^2
100	249.88	131.63	0.0089	0.817	268.82	0.912×10^{-4}	0.997
150	373.43	193.33	0.01054	0.8325	400.00	0.678×10^{-4}	0.994
200	498.43	264.36	0.01231	0.925	520.83	0.886×10^{-4}	0.999

Table 3. Adsorption kinetic parameters of the intra-particle diffusion model for the adsorption of sulfonated lignite by filter paper-templated Mg-Al metal oxides.

Concentration	Intra-Particle Diffusion Model					
C_0	K_{d1}	K_{d2}	C_1	C_2	${R_1}^2$	${R_2}^2$
100	5.742	1.378	127.84	217.33	0.516	0.436
150	3.204	0.490	251.42	361.91	0.809	0.698
200	10.110	1.257	312.66	469.61	0.995	0.698

3.2.4. Adsorption Isotherms

To better explain the interaction between the adsorbate and the filter paper-templated Mg-Al metal oxides, the Freundlich and Langmuir adsorption isotherm models were used to fit the isotherm parameters for the adsorption of sulfonated lignite. In this work, the Langmuir (Equation (6)) and Freundlich (Equation (7)) models were used to analyze the equilibrium adsorption data. The linear expressions of these models are as follows [35,36]:

$$\frac{C_e}{q_e} = \frac{1}{q_{max}K_L} + \frac{C_e}{q_{max}} \tag{6}$$

$$\ln q_e = \ln K_F + \frac{1}{n} \ln C_e \tag{7}$$

where K_L is the Langmuir constant, K_F is the Freundlich constant related to the energy of adsorption and adsorption capacity, and q_{max} (mg·g^{-1}) is the theoretical maximum adsorption capacity.

The influence of temperature on controlling the strength of the adsorptive forces between the adsorbent and adsorbate molecules plays an important role. The equilibrium data of sulfonated lignite adsorption by Mg-Al metal oxides at temperatures of 298 and 303 K are shown in Figure 11. It can be seen that as the reaction temperature rises to 30 °C, the maximum adsorption capacity of the filter paper-templated Mg-Al metal oxides for sulfonated lignite is 1189.29 mg/g. The results show that the adsorption capacity of the sulfonated lignite is favored by higher temperatures, and thus the adsorption process is endothermic, possibly partly due to the sulfonated lignite molecules moving around more and penetrating more deeply into the surface and pores of the Mg-Al metal oxides at increasing temperatures.

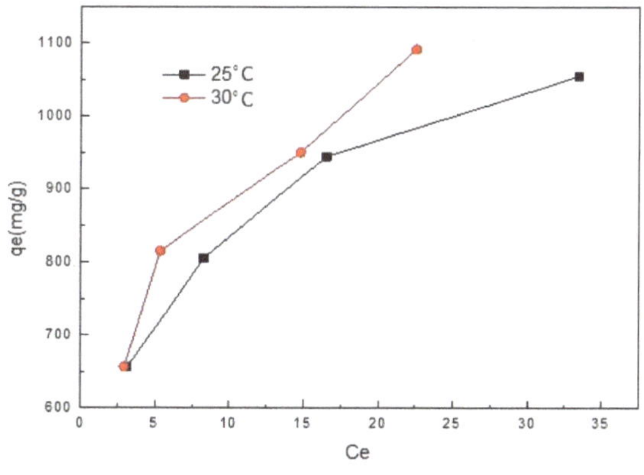

Figure 11. Equilibrium adsorption at different temperatures.

The isotherm constants of the Langmuir and Freundlich isotherms, presented in Table 4, were calculated from the slope and intercept of the linear plot obtained in Figure 12a,b respectively. The correlation coefficient values obtained for the Langmuir isotherm were all above 0.99, indicating that the Langmuir model was more qualified to explicate the adsorption isotherm of sulfonated lignite by filter paper-templated Mg-Al metal oxides, and hence that the adsorption of sulfonated lignite by metal oxides may be monolayer adsorption [37]. The maximum adsorption capacity determined from the Langmuir model was 1189.29 mg/g for a temperature of 303 K. It is known that the constant R_L in the Langmuir equation is an important index for measuring the stability of the adsorbent/adsorbate complex. $R_L = 1$ represents linear behavior; $0 < R_L < 1$, favorable to adsorption; $R_L = 0$, irreversible [38]. In this study, the separation factor R_L of the adsorption was found to be over a range of rather low values (max. 0.0149), and thus, the Langmuir isothermal model can better describe the adsorption process of sulfonated lignite by filter paper-templated Mg-Al metal oxides.

Table 4. Adsorption isotherm parameters of Langmuir and Freundlich models for the adsorption of sulfonated lignite by filter paper-templated Mg-Al metal oxides.

Temperature	Langmuir Isotherm Model				Freundlich Isotherm Model		
T (°C)	q_{max} (mg/g)	K_L (L/mg)	R_L	R^2	K_F (mg/g)(L/mg)$^{1/n}$	$1/n$	R^2
25	1139.56	0.34	0.0083–0.0149	0.997	523	0.2039	0.992
30	1189.29	0.38	0.0075–0.0129	0.99	527.63	0.2301	0.953

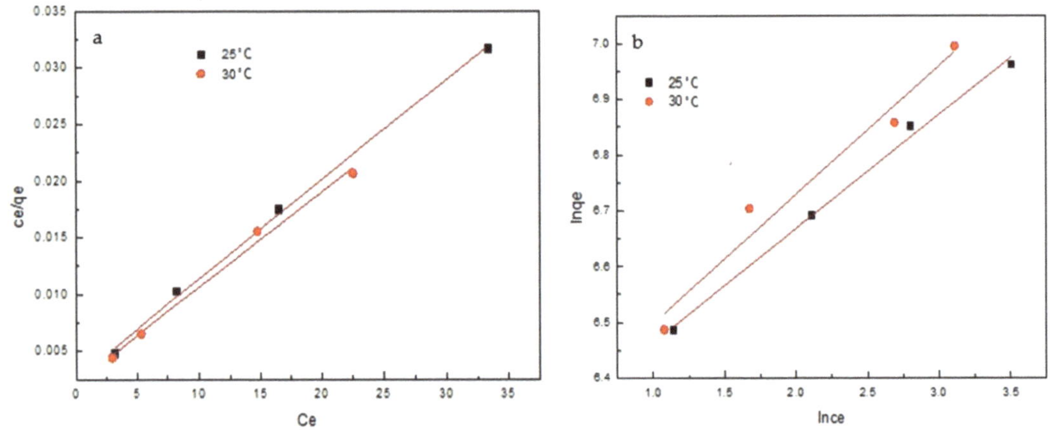

Figure 12. The Langmuir (**a**) and Freundlich (**b**) models for the adsorption of sulfonated lignite by filter paper-templated Mg-Al metal oxides at 25 and 30 °C.

3.2.5. Adsorption Thermodynamics

Thermodynamic studies can be used to determine the energy changes in the adsorption process, and are highly significant for practical applications of an adsorption process. The thermodynamic parameters Gibbs free energy (ΔG, kJ·mol^{-1}), enthalpy (ΔH, kJ·mol^{-1}), and entropy (ΔS, kJ·mol^{-1}·K^{-1}) were calculated using the following equations [39]:

$$\Delta G = -RT \ln K_d \tag{8}$$

$$\ln K_d = \frac{\Delta S}{R} - \frac{\Delta H}{RT} \tag{9}$$

$$K_d = \frac{q_e}{C_e} \tag{10}$$

where ΔG represents the standard Gibbs free energy change, R (8.314 J/mol k) is the universal gas constant, T is the absolute temperature (K), and K_d is the distribution coefficient. ΔH and ΔS can be calculated from the slope and intercept of the Van 't Hoff plots of lnK_d versus 1/T (Figure 13).

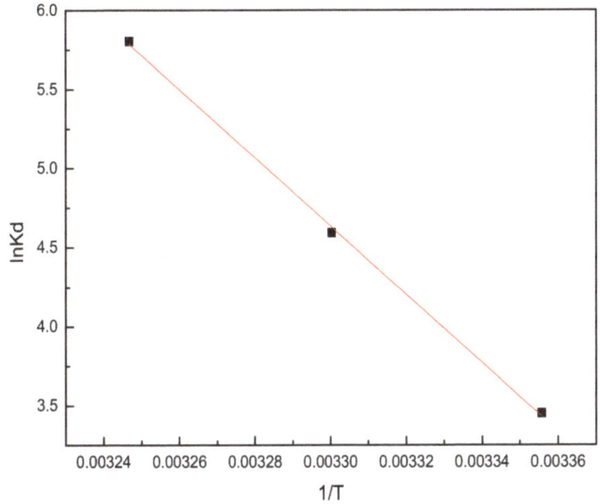

Figure 13. Effect of temperature on the adsorption of sulfonated lignite by filter paper-templated Mg-Al metal oxides (initial sulfonated lignite concentration: 350 mg/L; adsorbent dosage: 0.015; initial pH: 7; stirring time: 12 h).

The results obtained are summarized in Table 5. It can be seen that the ΔG values are negative, and increased in their absolute values with temperature, becoming increasingly negative with the increase in temperature (−8.51, −11.66, and −14.81 kJ/mol at 298, 303, and 308 K, respectively), indicating that high temperature facilitated the adsorption of sulfonated lignite by filter paper-templated metal oxides. Moreover, the positive value of ΔH indicates that the adsorption of sulfonated lignite is an endothermic process, which is consistent with the phenomenon of increased adsorption capacities with increasing temperature, as commented upon earlier in connection with Figure 11. The positive values of entropy indicate increasing freedom of movement of lignite at the solid/solution interface upon adsorption [40].

Table 5. Thermodynamic parameters for the adsorption of sulfonated lignite onto filter paper-templated Mg-Al metal oxides.

Temperature (K)	ΔG (KJ·mol^{-1})	ΔH (KJ·mol^{-1})	ΔS (J·mol^{-1}·K^{-1})
298	−8.51		
303	−11.66	179.43	630.67
308	−14.81		

3.2.6. Desorption and Regeneration

The stabilization and reutilization of adsorbents are some of the most important factors in the practical application of adsorbing materials, and they play an important role in the treatment of oilfield wastewater. Experiments were performed to investigate the activity on reuse, in which the adsorbent was regenerated by adding it to 1 mol/L, 50 mL of sodium carbonate solution under stirring at room temperature for 12 h. Then, the adsorbent was washed to neutral by deionized water, dried in a drying oven at 80 °C, calcined at 500 °C for 4h, and collected for reuse. It can be seen from Figure 14 that after three such cycles, the adsorption efficiency of Mg-Al metal oxides remained at 95.89%, indicating its good regeneration ability.

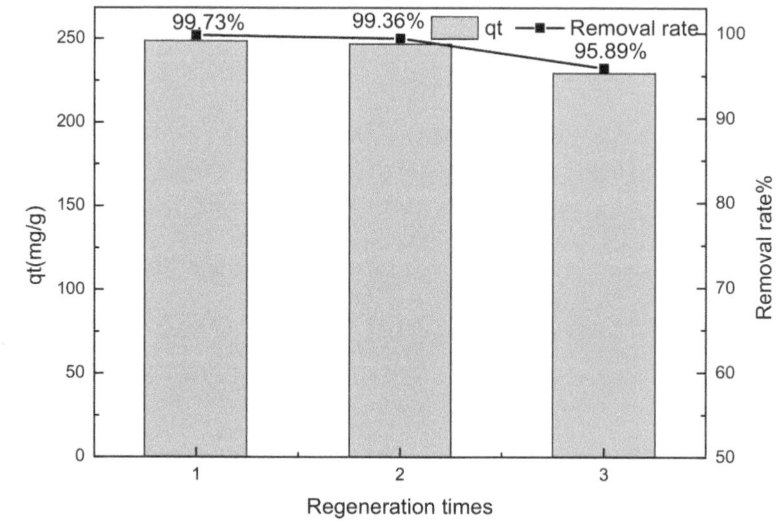

Figure 14. Influence of regeneration cycle on the adsorption of sulfonated lignite by filter paper-templated Mg-Al metal oxides (initial sulfonated lignite concentration: 200 mg/L; adsorbent dosage: 0.015; initial pH: 7; temperature set to 25 °C; stirring time: 12 h).

3.3. Formation Mechanism

The FTIR spectrum of sulfonated lignite adsorbed by Mg-Al metal oxides was investigated in order to propose a possible mechanism for the adsorption process. On comparing the FTIR spectra of Mg-Al metal oxides before and after sulfonated lignite adsorption, it is seen that the absorption band of Mg-Al metal oxides at 3445 cm^{-1}, corresponding to the O-H stretching vibrations of hydrogen-bonded hydroxyl, is slightly shifted to 3448 cm^{-1}, indicating that hydroxyl groups of Mg-Al metal oxides play an important role in the adsorption process, which we ascribe to the formation of hydrogen bonds by their interaction with the hydroxyl of sulfonated lignite. The peak shift from 1380 cm^{-1} to 1361 cm^{-1} could be ascribed to the intercalation of $-SO_3^-$ of sulfonated lignite. Moreover, after adsorption, the characteristic band of $-SO_3^-$ near 1035 cm^{-1} was very clear, which was clear evidence of the adsorption of sulfonated lignite by Mg-Al metal oxides.

Furthermore, calcined MgAl-LDHs have the ability to reconstruct their original layered structure after adsorption ("memory effect"); the XRD pattern of Mg-Al metal oxides/sulfonated lignite (Figure 4b) also exhibited the characteristic diffractions of hydrotalcite, indicating that the MgAl-LDHs were successfully reconstructed after the adsorption of sulfonated lignite. Moreover, the interlayer spacing of Mg-Al metal oxide after the adsorption of sulfonated lignite (0.75 nm) is larger than the interlayer spacing of MgAl-LDH (0.75 nm), indicating that layer expansion takes place upon adsorption. Therefore, in the reconstitution process, $-SO_3^-$ is intercalated into the layered structure, and CO_3^{2-} ions are released simultaneously.

In conclusion, it can be speculated that the adsorption mechanism of Mg-Al metal oxides for sulfonated lignite is mainly involved in two processes of electrostatic attraction (Figure 15a) and ion exchange (Figure 15b).

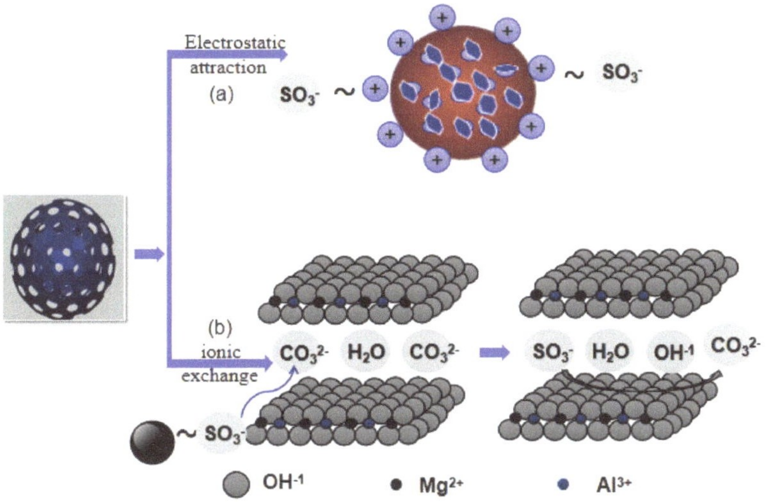

Figure 15. A possible mechanism for the adsorption of sulfonated lignite by biomorphic filter paper-templated Mg-Al metal oxides; (**a**) electrostatic attraction; (**b**) ionic exchange.

4. Conclusions

In summary, a straightforward method for successfully synthesizing a series of Mg-Al metal oxides by using filter paper as a biotemplate has been developed. XRD, FTIR, SEM, and N_2 adsorption–desorption analyses were applied to characterize and investigate the physicochemical properties of the synthesized samples. The obtained metal oxides all replicated the original macroarchitecture of the filter paper and had a hierarchically porous structure whose specific surface area was 178.84 mg/g. In addition, upon dosing 50 mL of 200 mg/L sulfonated lignite solution with 0.3 g/L of the adsorbent at a temperature of 298 K, the templated Mg-Al metal oxides showed improved adsorption performance, with 99.37% adsorption efficiency for sulfonated lignite. The reaction results were in accordance with the pseudo-second-order kinetic model and the Langmuir isotherm model. The thermodynamic parameters of ΔG and ΔH indicate that the adsorption of sulfonated lignite by Mg-Al metal oxides is a spontaneous endothermic process, and $\Delta S > 0$, i.e., there is an increase in entropy during adsorption. This study provides a feasible and simple approach for the preparation of a high-performance material for oilfield wastewater treatment in a cost-effective way.

Author Contributions: Z.L.: data curation, investigation. J.Z.: validation, conceptualization, supervision. C.Q.: data curation, investigation, formal analysis. Y.T.: investigation, methodology, data curation, writing—original draft, project administration, funding acquisition. M.S.: investigation, validation, supervision, writing—review and editing. All authors have read and agreed to the published version of the manuscript.

Funding: This research was funded by National Science Foundation of China (21763030), Open Fund of Shaanxi Key Laboratory of Carbon Dioxide Storage and Enhanced Oil Recovery (YJSYZX20SKF0008) and Slovak Research and Development Agency (APVV-18-0075 and APVV-19-0490).

Institutional Review Board Statement: Not applicable.

Informed Consent Statement: Not applicable.

Data Availability Statement: The data presented in this study are available on request from the corresponding authors.

Acknowledgments: We thank the support of Youth Innovation Team of Shaanxi University, the Center of Advanced Analysis and Testing at Xi'an Shiyou University. Michal Slaný acknowledges the financial support of this research by the Slovak Research and Development Agency (APVV-18-0075 and APVV-19-0490).

Conflicts of Interest: The authors declare no conflict of interest.

References

1. Zhang, R.; Wang, B.; Ma, H. Studies on Chromium (VI) adsorption on sulfonated lignite. *Desalination* **2010**, *255*, 61–66. [CrossRef]
2. Can, O. COD removal from fruit-juice production wastewater by electrooxidation electrocoagulation and electro-Fenton processes. *Desalin. Water. Treat.* **2014**, *52*, 65–73. [CrossRef]
3. Xia, T.; Li, Q.; Liu, X.; Meng, A.J.; Cao, X. Morphology-Controllable Synthesis and Characterization of Single-Crystal Molybdenum Trioxide. *J. Phys. Chem. B* **2006**, *110*, 2006–2012. [CrossRef]
4. Sun, M.-H.; Huang, S.-Z.; Chen, L.-H.; Li, Y.; Yang, X.-Y.; Yuan, Z.-Y.; Su, B.-L. ChemInform Abstract: Applications of Hierarchically Structured Porous Materials from Energy Storage and Conversion, Catalysis, Photocatalysis, Adsorption, Separation, and Sensing to Biomedicine. *ChemInform Soc. Rev.* **2016**, *47*, 3479–3563. [CrossRef]
5. Liu, Y.; Wang, G.; Yang, W.; Yang, J.; Li, J. Biotemplated synthesis of hierarchically porous ZnAl-CLDH/FeWO4 for effective removal of dyes from water. *Water Air. Soil.* **2019**, *230*, 1–13. [CrossRef]
6. Wong, W.T.; Chen, B.H.; Maggay, I.V.B.; Liu, C.; Duh, J.G.; Liu, W.R. Inside cover: Synthesis and electrochemical properties of hierarchically porous Zn(Co1−x Mnx)2O4 anodes for Li-ion batteries. *Energy Technol.* **2017**, *9*, 1516. [CrossRef]
7. Li, H.; Li, J.; Yang, P.; Ng, D.H.; Cui, X.; Ji, F. Three-dimensional biogenic C-doped Bi2MoO6/In2O3-ZnO Z-scheme heterojunctions derived from a layered precursor. *J. Environ. Sci.* **2018**, *79*, 54–66. [CrossRef]
8. Fang, Y.; Chen, V.W.; Cai, Y.; Berrigan, J.D.; Marder, S.R.; Perry, J.W.; Sandhage, K.H. Berrigan, Biologically Enabled Syntheses of freestanding metallic structures pos-sessing subwavelength pore arrays for extraordinary (surface plasmon-mediated) infrared transmission. *Adv. Funct. Mater.* **2012**, *22*, 2550–2559. [CrossRef]
9. Matovic, B.; Babic, B.; Egelja, A.; Radosavljevic-Mihajlovic, A.; Logar, V.; Saponjic, A.; Boskovic, S. Preparation of Porous Silica Ceramics Using the Wood Template. *Mater. Manuf. Process.* **2009**, *24*, 1109–1113. [CrossRef]
10. Gao, L.; Li, Q.; Hu, X.; Wang, X.; Song, H.; Yan, L.; Xiao, H. One-pot synthesis of biomorphic Mg-Al mixed metal oxides with enhanced methyl orange adsorption properties. *Appl. Clay Sci.* **2016**, *126*, 299–305. [CrossRef]
11. Pandoli, O.G.; Neto, R.J.; Oliveira, N.R.; Fingolo, A.C.; Corrêa, C.C.; Ghavami, K.; Santhiago, M. Ultra-highly conductive hollow channels guided by a bamboo bio-template for electric and electrochemical devices. *J. Mater. Chem. A* **2020**, *8*, 4030–4039. [CrossRef]
12. Wen-kun, Z.; Xue-gang, L.; Chi, Z.; Tao, D.; Jian, Z. Regulation of Microstructure of Calcium Carbonate Crystals by Egg White Protein. *Chem. Res. Chin. Univ.* **2021**, *2*, 180–185.
13. Sun, R.-Q.; Sun, L.-B.; Chun, Y.; Xu, Q.-H.; Wu, H. Synthesizing nanocrystal-assembled mesoporous magnesium oxide using cotton fibres as exotemplate. *Microporous Mesoporous Mater.* **2008**, *111*, 314–322. [CrossRef]
14. Liu, X.; Gu, Y.; Huang, J. Hierarchical, Titania-Coated, Carbon Nanofibrous Material Derived from a Natural Cellulosic Substance. *Chem. A Eur. J.* **2010**, *16*, 7730–7740. [CrossRef] [PubMed]
15. Parfen'Eva, L.S.; Orlova, T.S.; Smirnov, B.I.; Smirnov, I.A.; Misiorek, H.; Mucha, J.; Jezowski, A.; Rodríguez, R.C.; Ramirez-Rico, J. Thermal conductivity of high-porosity heavily doped biomorphic silicon carbide prepared from sapele wood biocarbon. *Phys. Solid State* **2012**, *54*, 1732–1739. [CrossRef]
16. Dias, J.M.; Alvim-Ferraz, M.C.; Almeida, M.F.; Rivera-Utrilla, J.; Sanchez-Polo, M. Waste materials for activated carbon preparation and its use in aqueous-phase treatment: A review. *J. Environ. Manag.* **2008**, *85*, 833–846. [CrossRef]
17. Xu, Z.P.; Zhang, J.; Adebajo, M.; Zhang, H.; Zhou, C. Catalytic applications of layered double hydroxides and derivatives. *Appl. Clay Sci.* **2011**, *53*, 139–150. [CrossRef]
18. Yu, J.; Lu, L.; Li, J.; Song, P. Biotemplated hierarchical porous-structure of ZnAl-LDH/ZnCo2O4 composites with enhanced adsorption and photocatalytic performance. *RSC Adv.* **2018**, *6*, 12797–12808. [CrossRef]
19. Yu, J.; Lu, L.; Li, J.; Song, P. Magnetic calcinated cobalt ferrite/magnesium aluminum hydrotalcite composite for enhanced adsorption of methyl orange-sciencedirect. *J. Alloy. Compd.* **2016**, *688*, 101–112.
20. Kim, K.C.; Yoon, T.-U.; Bae, Y.-S. Applicability of using CO_2 adsorption isotherms to determine BET surface areas of microporous materials. *Microporous Mesoporous Mater.* **2016**, *224*, 294–301. [CrossRef]
21. Kumar, K.Y.; Muralidhara, H.; Nayaka, Y.A.; Balasubramanyam, J.; Hanumanthappa, H. Low-cost synthesis of metal oxide nanoparticles and their application in adsorption of commercial dye and heavy metal ion in aqueous solution. *Powder Technol.* **2013**, *246*, 125–136. [CrossRef]
22. Tian, P.; Han, X.-Y.; Ning, G.-L.; Fang, H.-X.; Ye, J.-W.; Gong, W.-T.; Lin, Y. Synthesis of Porous Hierarchical MgO and Its Superb Adsorption Properties. *ACS Appl. Mater. Interfaces* **2013**, *5*, 12411–12418. [CrossRef] [PubMed]
23. Calatayud, M.; Markovits, A.; Ménétrey, M.; Mguig, B.; Minot, C. Adsorption on perfect and reduced surfaces of metal oxides. *Catal. Today* **2003**, *85*, 125–143. [CrossRef]
24. Yuan, X.; Wang, Y.; Wang, J.; Zhou, C.; Tang, Q.; Rao, X. Calcined graphene/MgAl-layered double hydroxides for enhanced Cr(VI) removal. *Chem. Eng. J.* **2013**, *221*, 204–213. [CrossRef]

25. Duan, N.; Li, Q.; Liu, J.; Xiao, H. Enhanced adsorption performance of CuO-Al_2O_3 composite derived from cotton template. *Can. J. Chem. Eng.* **2015**, *93*, 2015–2023. [CrossRef]
26. Yuan, D.; Zhou, L.; Fu, D. Adsorption of methyl orange from aqueous solutions by calcined ZnMgAl hydrotalcite. *Appl. Phys. A* **2017**, *123*, 146. [CrossRef]
27. Bing, X.; Li, J.; Liu, J.; Cui, X.; Ji, F. Biomimetic synthesis of Bi2O3/Bi2WO6/MgAl-CLDH hybrids from lotus pollen and their enhanced adsorption and photocatalysis performance. *J. Photoch. Photobio. A* **2018**, *364*, 449–460. [CrossRef]
28. Li, H.-Q.; Huang, G.-H.; An, C.-J.; Zhang, W.-X. Kinetic and equilibrium studies on the adsorption of calcium lignosulfonate from aqueous solution by coal fly ash. *Chem. Eng. J.* **2012**, *200-202*, 275–282. [CrossRef]
29. Ho, Y. The kinetics of sorption of divalent metal ions onto sphagnum moss peat. *Water Res.* **2000**, *34*, 735–742. [CrossRef]
30. Lorenc-Grabowska, E.; Gryglewicz, G. Adsorption of lignite-derived humic acids on coal-based mesoporous activated carbons. *J. Colloid Interface Sci.* **2005**, *284*, 416–423. [CrossRef]
31. Guo, Y.; Zhu, Z.; Qiu, Y.; Zhao, J. Enhanced adsorption of acid brown 14 dye on calcined Mg/Fe layered double hydroxide with memory effect. *Chem. Eng. J.* **2013**, *219*, 69–77. [CrossRef]
32. Gupta, V.K.; Pathania, D.; Sharma, S.; Singh, P. Preparation of bio-based porous carbon by microwave assisted phosphoric acid activation and its use for adsorption of Cr(VI). *J. Colloid Interface Sci.* **2013**, *401*, 125–132. [CrossRef] [PubMed]
33. Tanzifi, M.; Hosseini, S.H.; Kiadehi, A.D.; Olazar, M.; Karimipour, K.; Rezaiemehr, R.; Ali, D.I. Artificial neural network optimization for methyl orange adsorption onto polyaniline nano-adsorbent: Kinetic, isotherm and thermodynamic studies. *J. Mol. Liq.* **2017**, *244*, 189–200. [CrossRef]
34. Lv, L.; He, J.; Wei, M.; Evans, D.G.; Duan, X. Uptake of chloride ion from aqueous solution by calcined layered double hy-droxides: Equilibrium and kinetic studies. *Water Res.* **2006**, *40*, 735–743. [CrossRef] [PubMed]
35. Irving, L. Adsorption of gases on glass, mica and platinum. *J. Am. Chem. Soc.* **1918**, *40*, 361–1403.
36. Freundlich, H.M.F. Over the adsorption in solution. *J. Phys. Chem.* **1906**, *57*, 1100–1107.
37. Liu, Z.; Chen, G.; Xu, L.; Hu, F.; Duan, X. Removal of Cr(VI) from wastewater by a novel adsorbent of magnetic goethite: Ad-sorption performance and adsorbent characterisation. *Chemistry. Select.* **2019**, *47*, 13817–13827. [CrossRef]
38. Ma, Y.-Z.; Zheng, D.-F.; Mo, Z.-Y.; Dong, R.-J.; Qiu, X.-Q. Magnetic lignin-based carbon nanoparticles and the adsorption for removal of methyl orange. *Colloids Surf. A: Physicochem. Eng. Asp.* **2018**, *559*, 226–234. [CrossRef]
39. Yamaguchi, N.U.; Bergamasco, R.; Hamoudi, S. Magnetic $MnFe_2O_4$ –graphene hybrid composite for efficient removal of glyphosate from water. *Chem. Eng. J.* **2016**, *295*, 391–402. [CrossRef]
40. Tan, X.; Liu, S.; Liu, Y.; Gu, Y.; Zeng, G.; Cai, X.; Chen, B. One-pot synthesis of carbon supported calcined-Mg/Al layered double hydroxides for antibiotic removal by slow pyrolysis of biomass waste. *Sci. Rep.* **2016**, *6*, 1–12. [CrossRef]

MDPI AG
Grosspeteranlage 5
4052 Basel
Switzerland
Tel.: +41 61 683 77 34

Materials Editorial Office
E-mail: materials@mdpi.com
www.mdpi.com/journal/materials

www.ingramcontent.com/pod-product-compliance
Lightning Source LLC
Chambersburg PA
CBHW042012260926
51850CB00020B/3270